Theories of Nonlinear Systems for Power Converters

Bertrand Rivet

Theories of Nonlinear Systems for Power Converters

Foreword by Benoît Peron

Springer

Bertrand Rivet
Saint-Cyr-sur-Loire, France

ISBN 978-3-032-25061-2 ISBN 978-3-032-25062-9 (eBook)
https://doi.org/10.1007/978-3-032-25062-9

This Springer imprint is published by the registered company Springer Nature Switzerland AG
The registered company address is: Gewerbestrasse 11, 6330 Cham, Switzerland

Foreword

Nowadays, all the electronic devices used by everyone such as mobile phones, computers, TVs, battery chargers, etc., embed several power converters to supply the intelligent devices like microprocessors, and electronics interface systems. In the future, the power electronics devices will be used more and more, in particular to replace the traditional gasoline powered vehicles by electric cars. Because of this trend, intense research and huge development efforts have been made by the electronics companies around the world, in many areas such as the semiconductor technologies, silicon designs, magnetic components, electro-optical material, packaging material, cooling systems. All these power systems, which are more complex than the current ones, should be driven, according to control laws in line with the needs of the application. Specific control methods should be developed to ensure the stability of systems that are usually nonlinear, exhibiting a variety of complex behaviours, such as unexpected changes of operating regime and chaotic behaviours.

The theoretical study of the nonlinear system in the field of power converters began in 1970, when the power electronics appeared on the market due to the introduction of cheap power semiconductor. Mr. R. D. Middlebrook in 1975 was considered as the pioneer of the modelling of power converters. The Averaged method associated with the linearization technique around the stable point was developed, and today, it is still used by both engineers and modern software simulators. Nevertheless, this averaged method cannot explain all phenomena such as the chaotic behaviours, and instabilities as for example the thermal runaway of a nonlinear thermal system. Sometimes, in some power electronics systems, it is necessary to optimise, that is to say, to maximise or to minimise the power losses, the energy transferred, the charge time of a capacitor, etc. In the same way, the usual methods are not applicable for those kind problems.

Power electronics engineers are always dealing with these topics in the course of developing power electronics systems. The engineer's job is to make the circuits work in the expected operating area. The averaged model is used by the engineer to guarantee the stability in this condition. Nevertheless, in the undesirables operating areas such as the soft-start or the transient from the heavy load variation, empiric ways adjusting circuit components and/or software parameters after several tests are used but they are not sufficient to cover all the operating cases. Thus, to respect both a robust design and the reliability requirement, the traditional empirical methods cannot be used and it is necessary for engineers to understand thoroughly the behaviour of the systems being designed under all possible operating conditions. The physical understanding of the phenomena in the power electronics discipline becomes mandatory to solve the behaviour of the nonlinear devices.

This book has been written to describe and to provide some theoretical tools to the engineers, graduate students and researchers to understand unwanted behaviours in the power electronics. It emphasises some control methods to stabilise the systems using Lyapunov control, sliding control, robust control, optimal control, chaotic control, direct nonlinear control etc. In my professional career, I have used some of these tools to solve non-intuitive problems to control some nonlinear systems. For example, in the high voltage insulated nonlinear fed-current power converter with several outputs, the Direct nonlinear control was used. In another example, Lyapunov control, and sliding mode control was set to drive several bidirectional fed-current power converters working together. Of course, it is possible to provide additional examples such as the management of the thermal runaway, minimization of power losses during the charge and the discharge of the storage capacitors used in the railway systems.

In the first part, the author gives an overview of the basics of the nonlinear systems outlining some important rules such as the defining of the attraction domain and the equilibrium point. These definitions help to understand, the mathematical point of view of the nonlinear systems, leading to Lyapunov's and Routh's criteria. Furthermore, in this part are reviewed the main principles of the optimal control illustrated with numerous examples. The author continues by describing a great number of robust controls such as Lyapunov, Backstepping, Sliding and Predictive controls. The last section of the first part is dedicated to the control techniques used to stabilise a chaotic system. The second part begins by describing the Averaged, the Hybrid and the discrete models, which are the three possible ways of simulating the dynamics of a power converter working in continuous and discontinuous modes. In addition, a harmonics model is described using a resonant power converter. Several examples, using the Control methods shown in the first chapter, are applied to power converter models. A step-by-step procedure using the Averaged models and the linear compensation correctors are proposed and applied to solve some common issues like the design control of a power converter having a Right Half Pole Zero (RHPZ). This latter causes some instability, especially, in the power converters requiring a closed-loop at high frequency (200 kHz) or when it has a high-power capability (100 kW/400 V). In addition, part two emphasises the instability of nonlinear systems in power converters such as the thermal runaway issues of nonlinear thermal systems. Examples showing the interest of the optimal control in the power converter field are given, e.g. the way to optimise the charge of the storage elements such as the batteries or super capacitors. The very interesting direct control, based on the averaged method is introduced to show that is possible to keep the equilibrium point whatever the biasing of the power converter even during the transient modes. Buck and sepic power converters are used as examples. Robust controls such as Lyapunov, Backstepping, Inverse Optimal and Sliding controls are used in the regulation of a buck power converter to show the manner to regulate the system, even if some parameters are unknown. Several interesting controls (Linear, Predictive, Ott Grebogi and Yporke (OGY)) presented in the first part but not used in the field of Power conversion are used to stabilise a buck converter with a chaotic behaviour.

This book presents a great number of methods to regulate the complex behaviours of the power electronic systems used today. It will help engineers, researchers, and students to understand and to imagine other methods to control power converters or to optimise some operating behaviours. Thanks to numerical circuits like the Digital Signal Processing (DSP), it should be easier to implement the best control methods described in this book, in any given operating situation.

We can thank the author for proposing a summary of the control methods often met in other domains that can be applied to the power electronics. This book opens up the way to a wide range of investigations, an interesting follow-up would be to compare the different control techniques and to find the most appropriate one for any given specification.

Bordeaux, France Benoît Peron

Benoît Peron joined the semiconductor company STMicroelectronics in 1995 as Application Engineer and Strategic Marketing Engineer. He participated in the development of the discrete power components based on Silicon and Silicon Carbide dedicated to power conversion applications. He also contributed to power integration development, mixing together several Pulse Width Modulation controllers in the same tiny QFN package, for the major players in LCD screen market. The aim was to develop new topologies in order to find the smallest solution with the best cost effectiveness.

Between 2011 and 2014, Benoît joined Faiveley Transport Company as Principal Engineer in charge of the development of high-power conversion products dedicated to railway applications. He introduced and developed the first high voltage insulated, multiple output current-fed power converter. He also developed a high power interleaved soft switching sepic converter. In addition, he introduced the 'Direct Control' method using the Digital Signal Processing to control several complex power converters like the sepic.

In 2015, he joined the Adetel Group as Architect and Expert for the power application. He is currently developing a compact, soft switching, bidirectional current-fed power converter based on the hybrid technology mixing the Si IGBT and SiC Mosfet. The first application of this technology is for heavy autonomous vehicles like trains without catenary. Moreover, he also introduced a high voltage, insulated, soft switching, bidirectional current-fed power converter (400kW) to recover the kinetic energy on the heavy rolling stock systems.

Benoît holds an engineering degree in Electronics from the "Conservatoire National des Arts et Métiers". He has published numerous articles and conference paper and has applied for 25 patents.

Competing Interests The author has no competing interests to declare that are relevant to the content of this manuscript.

About This Book

In a power converter, there is a great number of non-linear systems. Better knowledge of the theories of non-linear systems will contribute to optimize the design of such converters. This book consists of two parts composed of several chapters.

In Part I, the author reviews a great number of theoretical tools useful for the study of non-linear systems. After providing some definitions and generalities in Chap. 1, he examines the methods for determining the stability limits of a non-linear system. He devotes Chap. 3 to hybrid and chaotic systems and in Chap. 4, he provides examples to illustrate the concepts presented in the first three chapters. The remaining chapters of Part I are devoted to linear controls, optimal control, the control of chaotic systems and robust controls respectively.

In Part II, the concepts previously described are applied to the field of power converters. The author starts by giving dynamic models of switching power converters then, he proposes a step-by-step procedure to design a power converter controller. In the following two chapters, he analyzes common issues of control and gives examples of stability study in the field of power converters. In the last chapters, the different methods of control described in the Chap. 1 are used to realize different controls of power converters.

In this book, many examples are given to illustrate each theoretical concepts. The examples chosen are as simple as possible so as to facilitate the assimilation of the different concepts by the reader. The large number of theoretical tools presented in this book can be used to solve basic problems such as how to determine the stability areas of a thermal system or how to dimension corrector circuits. However, the author hopes that this book will be also a source of inspiration for young (and less young) engineers to help them to explore new ways of optimization in the field of power converters.

About the Author

Bertrand Rivet obtained his engineering degree from the “Conservatoire National des Arts et Métiers”. In 1989, he joined STMicroelectronics in Tours after working for smaller companies for two years. For thirty years, he led an application engineer team responsible for the development of new power components. In the last 35 years, he has participated in numerous development programmes and their implementations in industrial designs He is the author and co-author of numerous publications and papers. He filed approximately 40 patents, some of which were commercial successes such as the “Tandem diode” and a soft start circuit used in vacuum cleaners.

Introduction

A power converter comprises a large number of nonlinear systems. A good understanding of the theories of nonlinear systems will help optimise the converters design. During my career, I lead an applications engineer team and I tried to teach both theoretical and practical approach. It is very surprising to note that very often, the young engineers who spent many years learning to master theoretical tools seem to forget all they have learned when they start their professional careers. We are the heirs of all these great mathematicians, physicists and other distinguished scholars, and it is a shame not to use their legacy.

The objective of this book is to provide a synthesis of useful theoretical tools to study and optimise the designs of power converters. In Part I, having recalled the definitions and generalities of nonlinear systems, we will give theorems to define the areas of stability of a nonlinear system. These chapters are crucial for a better understanding, particularly with regard to the thermal instability issues frequently encountered in power conversion. One section is devoted to the hybrid and chaotic system. In Chap. 4, examples will be given to illustrate the concepts presented in the first three chapters. In the last four chapters of the Part I will be treated: linear control, optimal control, control of chaotic system and robust control. In this Part I, the author has voluntarily chosen illustrative examples outside the field of power converters to show that these tools are generic and can be used in a variety of areas.

In Part II, the concepts previously described will be applied to the power converters domain. We start by given dynamic models of power converters then we will propose a step-by-step controller design procedure. In the followings chapters, will be analysed common problematics of control and examples of stability study. In the last chapters, the different methods of control described in Part I will be used to implement different regulation techniques of power converters. In this book, numerous examples are given to illustrate each theoretical concept. These examples are chosen as simple as possible to facilitate the assimilation of the different notions by the reader. The large number of theoretical tools presented in this book can be used to solve basic problems, such as determining stability areas of a thermal system or sizing corrector circuits. However, the author hope that this book will be also a source of inspiration for young (and less young) engineers in order to explore new ways of optimization in the field of power converters. Today the control modes are increasingly digital which offers the possibility to embark innovative algorithms. The author has provided several avenues for further investigation, such as optimal control to minimize dissipated energy during start-up phases, the use of small control signals to suppress chaotic behaviour or the very interesting nonlinear direct control presented in Chap. 14.

I would like to warmly thank my former colleague and friend, Benoît Peron, for the time he spent diligently rereading this book and for its numerous and helpful suggestions. I also extend my thanks to Yohan Potaux for his contribution to making the mathematical formalism more rigorous.

Contents

Abbreviations

$\vec{p}$	Adjoint vector
T_{amb}	Ambient temperature
v_{AR}	Artificial ramp generator
$G_{BO}(f)$	Bode plan gain of H_{BO}
$Arg(H_{BO}(f))$	Bode plan phase of H_{BO}
$\det(sI - J) = 0$	Characteristic equation
$H_{BC}(s)$	Closed-loop transfer function
t_{on}	Conduction time of the switch
$\vec{X}$	Continuous state vector
$\vec{E}_r$	Control error vector
C	Controllability matrix
$\vec{U}$	Control vector
V_{ctr}	Control voltage
C_r	Convergent region
$C_r(s)$	Corrector transfer function
$J\left(\vec{U}\right)$	Cost functional
f_c	Crossover frequency
$\vec{X}_k$	Discrete state vector
$D(s)$	Disturbance transfer function
d	Duty cycle of the switch
$\vec{X}_e$	Equilibrium state vector
$\vec{U}_{eq}$	Equivalent control vector
$\widehat{\varphi}$	Estimated value of θ
$\tilde{\theta}$	Estimation error of θ
F	Floquet matrix
g_m	Gain margin
H	Hamiltonian
$\vec{X}_0$	Initial condition state vector
V_{IN}	Input voltage
J	Jacobian matrix

T_J	Junction temperature
L	Lagragian
s	Laplace variable
R_L	Load resistor
λ_i	Lyapunov exponent
$V(X_1, \ldots X_n)$	Lyapunov function
V_{mains}	Mains voltage
f_{mains}	Mains frequency
M	Monodromy matrix
V_A	Magnitude of v_{AR}
V_M	Magnitude of V_{mains}
V_P	Magnitude of V_{sth}
u	Number of control parameters
m	Number of equilibrium points
n	Number of state parameters
p	Number of system parameters
O	Observability matrix
$H_{BO}(s)$	Open-loop transfer function
V_{OUT}	Output voltage
$\vec{Y}$	Output vector
V_{OUT}	Output voltage
φ_m	Phase margin
$\vec{X}_{ref}$	Reference vector
V_{REF}	Reference voltage
R_A	Region of attraction
f_{RHPZ}	Right-half-plane zero frequency
s_i	Root of $\det(sI - J)$
z_i	Root of $\det(zI - J)$
$sgn(X)$	Sign of X
$\vec{\sigma}\left(\vec{X}\right) = \vec{0}$	Sliding surface
$\vec{x}$	Small variation of the state vector around $\vec{X}_e$
V_{sth}	Sawtooth generator
$\overrightarrow{\Delta x_0}$	Small perturbation of the state vector around $\vec{X}_e$
F_s	Switching frequency
$\vec{\varphi}\left(\vec{X}, \vec{p}\right)$	Switching function
T_s	Switching period
S	System of n state parameters
S_e	System of n state parameters at the equilibrium
$\vec{P}$	System parameter vector
$H_P(s)$	System transfer function
V_{OUTtar}	Targeted output voltage
$Cth_{(x-y)}$	Thermal capacitor between x and y
$Rth_{(x-y)}$	Thermal resistor between x and y

t	Time variable
t_H	Time horizon
$\vec{W}$	Uncertainty vector
$\vec{F}$	Vector field describing the dynamics of the system
z	z variable

Part I
Theories of Nonlinear Systems

Chapter 1
Some Definitions and Properties of Nonlinear Systems

In this chapter, we discuss some definitions and properties of nonlinear systems (Strogatz, 2015).

1.1 Classification

The classification of nonlinear systems proposed in this book is not intended to cover all nonlinear systems. We only deal with the nonlinear systems useful for studying dynamic behaviour of power converters. Figure 1.1 summarises this classification. In this chapter, we consider the continuous and discrete nonlinear systems defined by a set of differential equations or recurrence relations given in Para. 1.2. The definition of hybrid systems will be given in Para. 3.1. Each of these systems can generate chaotic behaviours. Chaotic systems will be studied in Para. 3.2.

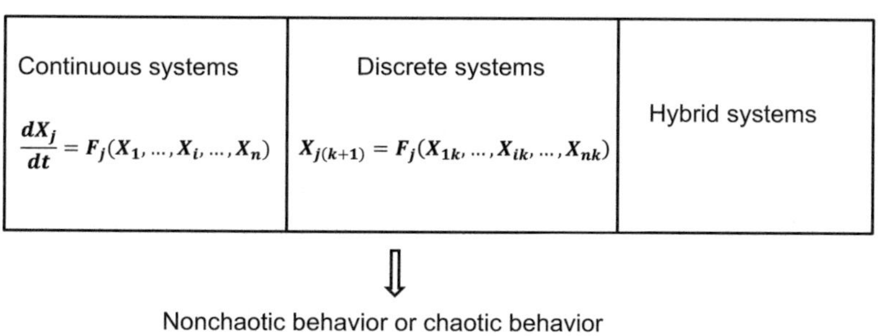

Fig. 1.1 Classification of nonlinear systems

B. Rivet, *Theories of Nonlinear Systems for Power Converters*,
https://doi.org/10.1007/978-3-032-25062-9_1

1.2 Description of Nonlinear Systems

In this chapter, we will address both continuous and discrete nonlinear systems. Consider the closed system S in Fig. 1.2 of order n defined by n state parameters, a set of n equations (differential or recurrence) with n initial conditions, p system parameters (considered as constant in a first step) and m equilibrium points. If the system is controllable, we will also consider u control parameters. The n state parameters can be represented by a real continuous state vector:

$$\vec{X}(t) = (X_1(t), \ldots, X_i(t), \ldots, X_n(t)) \tag{1.1}$$

or by a discrete state vector:

$$\vec{X}_k = (X_{1k}, \ldots, X_{ik}, \ldots, X_{nk}). \tag{1.2}$$

A state vector is the minimal set of parameters that accurately describes the system. So, if the state parameters are known, the system will be fully known. For an electrical system, the state parameters are typically voltages across capacitors and currents through inductors. The number of capacitors and inductors determines the order n of an electrical system. For a mechanical system, the state parameters will be, for example, positions and speeds.

For an electrical system, examples of system parameters are resistors, capacitors and current and voltage generators. For a mechanical system, system parameters are typically masses, friction coefficients and potential fields. The p system parameters and the u control parameters can also be represented by:

$$\begin{aligned} &\vec{P}(t) = (P_1(t), \ldots, P_j(t), \ldots, P_p(t)) \quad \text{and} \\ &\vec{U}(t) = (U_1(t), \ldots, U_i(t), \ldots, U_u(t)) \end{aligned} \tag{1.3}$$

for a continuous system and by:

$$\vec{P}_k = \left(P_{1k}, \ldots, P_{jk}, \ldots, P_{pk}\right) \quad \text{and}$$

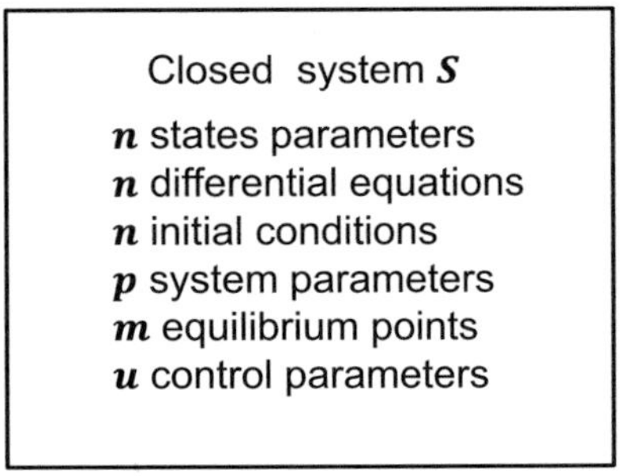

Fig. 1.2 Parameters of a nonlinear closed system of order n

$$\vec{U}_k = (U_{1k}, \ldots, U_{ik}, \ldots, U_{uk}) \tag{1.4}$$

for a discrete system. In this book, we will consider continuous systems whose dynamics is given by the following system S :

$$S : \begin{cases} \frac{dX_1}{dt} = F_1\big(X_1, \ldots, X_i, \ldots, X_n, P_1, \ldots, P_j, \ldots, P_p, U_1, \ldots, U_l, \ldots, U_u\big) \\ \vdots \\ \frac{dX_j}{dt} = F_j\big(X_1, \ldots, X_i, \ldots, X_n, P_1, \ldots, P_j, \ldots, P_p, U_1, \ldots, U_l, \ldots, U_u\big) \\ \vdots \\ \frac{dX_n}{dt} = F_n\big(X_1, \ldots, X_i, \ldots, X_n, P_1, \ldots, P_j, \ldots, P_p, U_1, \ldots, U_l, \ldots, U_u\big) \end{cases} . \tag{1.5}$$

For a discrete system, the system S becomes:

$$S : \begin{cases} X_{1(k+1)} = F_1\big(X_{1k}, \ldots, X_{ik}, \ldots, X_{nk}, P_{1k}, \ldots, P_{jk}, \ldots, P_{pk}, U_{1k}, \ldots, U_{lk}, \ldots, U_{uk}\big) \\ \vdots \\ X_{j(k+1)} = F_j\big(X_{1k}, \ldots, X_{ik}, \ldots, X_{nk}, P_{1k}, \ldots, P_{jk}, \ldots, P_{pk}, U_{1k}, \ldots, U_{lk}, \ldots, U_{uk}\big) \\ \vdots \\ X_{n(k+1)} = F_n\big(X_{1k}, \ldots, X_{ik}, \ldots, X_{nk}, P_{1k}, \ldots, P_{jk}, \ldots, P_{pk}, U_{1k}, \ldots, U_{lk}, \ldots, U_{uk}\big) \end{cases} . \tag{1.6}$$

The F_j's are nonlinear functions of the state, system and control variables from $\mathbb{R}^{n+p+u}$ to $\mathbb{R}$. This can be expressed more simply by:

$$\vec{F} : \mathbb{R}^{n+p+u} \rightarrow \mathbb{R}^n$$

$$\left(\vec{X}(t), \vec{P}(t), \vec{U}(t)\right) \rightarrow \vec{F}\left(\vec{X}(t), \vec{P}(t), \vec{U}(t)\right) = \frac{d\vec{X}}{dt} \tag{1.7}$$

for a continuous system and

$$\vec{F} : \mathbb{R}^{n+p+u} \rightarrow \mathbb{R}^n$$

$$\left(\vec{X}_k, \vec{P}_k, \vec{U}_k\right) \rightarrow \vec{F}\left(\vec{X}_k, \vec{P}_k, \vec{U}_k\right) = \overrightarrow{X_{k+1}} \tag{1.8}$$

for a discrete system. In this book, these functions do not depend explicitly on time. In this case, the system is called autonomous. If all the functions F_j are linear, the system S is also linear. These functions are found by applying basic physical laws such as Ohm's law, Kirchoff's law or Newton's law. We will define

$$\vec{X}_0 = (X_{10}, \ldots, X_{i0}, \ldots, X_{n0}) \tag{1.9}$$

as the initial condition of $\vec{X}(t)$ and at $t = 0$ and $k = 0$. We can also define a y dimensional output vector $\vec{Y}$ as a nonlinear function of the state parameters:

$$\vec{Y} = \vec{G}\left(\vec{X}\right). \tag{1.10}$$

1.3 Transformation of a Nonautonomous System into an Autonomous One

In a nonautonomous system, the functions F_1 to F_n depend on time. It is possible to transform a nonautonomous system into an autonomous system by replacing t (or k) with the following $(n+1)$th state parameter:

$$t = X_{n+1}(t) \quad \text{or} \quad k = X_{(n+1)k}. \tag{1.11}$$

The system becomes a system of order $n+1$ with the added differential equation:

$$\frac{\mathrm{d}X_{n+1}}{\mathrm{d}t} = F_{n+1} = 1 \quad \text{or} \quad X_{(n+1)(k+1)} = F_{n+1} = X_{(n+1)k} + 1 \tag{1.12}$$

and the initial condition $X_{(n+1)0} = 0$.

1.4 Temporal Representation and Representation in the State Space

The behaviour of a system can be represented in two different ways. The first one consists of representing the variation of each state parameter with respect to time t (or k):

$$X_i = X_i(t) \quad \text{or} \quad X_i = X_{ik}. \tag{1.13}$$

This is the temporal representation. The second representation does not depend on time. A trajectory is drawn in the n dimensional space of state parameters. This representation, as we shall see later, is very interesting to study, for instance, the stability area, even if it is difficult to visualise it for $n > 3$. In this representation the p system parameters are kept constant. Figure 1.3 shows an example of these two representations for a second-order system (both continuous and discrete). In this example, a constant period T_S separates two successive indices k.

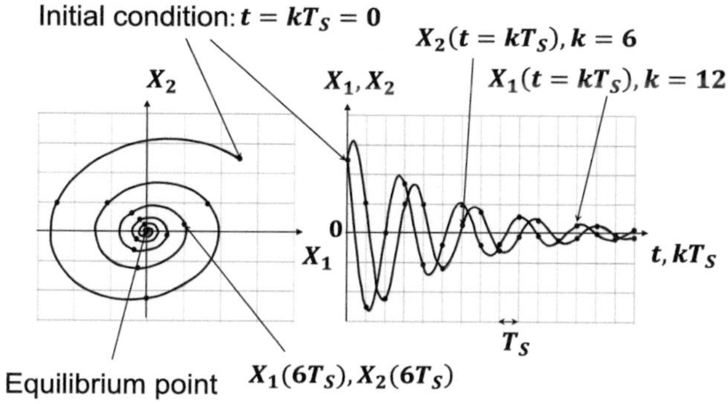

Fig. 1.3 Two possible representations of the behaviour of a system

1.5 Equilibrium Points

A system S has m equilibrium points. Each of them can be separately stable or unstable. If $m = 0$, the system is unstable. When the system reaches an equilibrium point, each state parameter becomes independent of time $\left(\frac{dX_j}{dt} = 0\right)$ or from k $\left(X_{j(k+1)} = X_{jk}\right)$. Thus the m equilibrium points are found by solving the continuous system S_e below (the system and control parameters being fixed):

$$S_e : \begin{cases} 0 = F_1(X_1, \ldots, X_i, \ldots, X_n) \\ \vdots \\ 0 = F_j(X_1, \ldots, X_i, \ldots, X_n) \\ \vdots \\ 0 = F_n(X_1, \ldots, X_i, \ldots, X_n) \end{cases} . \tag{1.14}$$

For a discrete system we get:

$$S_e : \begin{cases} X_{1(k+1)} = X_{1k} = F_1(X_{1k}, \ldots, X_{ik}, \ldots, X_{nk}) \\ \vdots \\ X_{j(k+1)} = X_{ik} = F_j(X_{1k}, \ldots, X_{ik}, \ldots, X_{nk}) \\ \vdots \\ X_{n(k+1)} = X_{nk} = F_n(X_{1k}, \ldots, X_{ik}, \ldots, X_{nk}) \end{cases} . \tag{1.15}$$

For $h \in (1, \ldots, m)$, the corresponding equilibrium point is:

$$\vec{X}_{he} = (X_{1he}, \ldots, X_{ihe}, \ldots, X_{nhe}). \tag{1.16}$$

1.6 Types of Stabilities and Attractors

Having calculated the coordinates of the equilibrium points by solving the system S_e, we have to determine the type of local stability around each of them. The basic idea to study the local stability consists of generating at $t = 0$ ($k = 0$) a small perturbation of the state parameters near the considered equilibrium point and observing the behaviour of the system.

Figure 1.4 illustrates the different possibilities in the space of state parameters. Each $\vec{X}_{ol}$ represents the initial conditions corresponding to small perturbations around the equilibrium point $\vec{X}_e$. The first stability to consider is the Lyapunov stability as represented by Fig. 1.4c. Lyapunov stability means that the solutions starting 'close enough' to the equilibrium remain 'close enough' forever. Note that this must be true for any perturbation. In mathematical terms, it can be expressed as follows: the equilibrium point $\vec{X}_{he}$ is stable in the sense of Lyapunov if for any $\rho > 0$ there is $r > 0$ such that if

$$\|(X_{10}, \ldots, X_{i0}, \ldots, X_{n0}) - (X_{1e}, \ldots, X_{ie}, \ldots, X_{ne})\| < r \tag{1.17}$$

then

$$\forall t > 0\ \|(X_1(t), \ldots, X_i(t), \ldots, X_n(t)) - (X_{1e}, \ldots, X_{ie}, \ldots, X_{ne})\| < \rho. \tag{1.18}$$

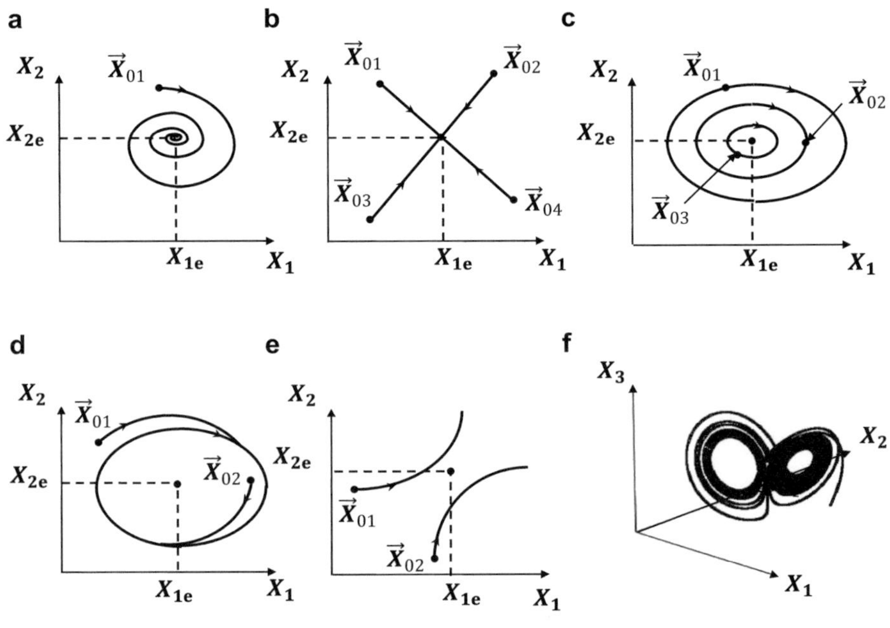

Fig. 1.4 Stability types

$\|\vec{X}\|$ being the norm of $\vec{X}$. A system can have a unique periodic orbit that does not depend on the initial condition $\vec{X}_{0i}$ as shown in Fig. 1.4d or different periodic orbits depending on the initial conditions (see Fig. 1.4c). Figure 1.4a,b is examples of asymptotic stability. This type of stability means that solutions that start close enough to the equilibrium point not only remain close enough but also eventually to converge towards this equilibrium point.

The mathematical definition is that the equilibrium point $\vec{X}_{he}$ is asymptotically stable if it is stable in the sense of Lyapunov and if there is $r > 0$ such that if

$$\|(X_{10}, \ldots, X_{i0}, \ldots, X_{n0}) - (X_{1e}, \ldots, X_{ie}, \ldots, X_{ne})\| < r \tag{1.19}$$

then

$$\lim_{t \to \infty} \left\| \vec{X}(t) - \vec{X}_{he} \right\| = 0. \tag{1.20}$$

We can also define the exponential stability as follows: the equilibrium point $\vec{X}_{he}$ is exponentially stable if it is asymptotically stable and if there are $r, k, \alpha > 0$ so that if

$$\|(X_{10}, \ldots, X_{i0}, \ldots, X_{n0}) - (X_{1e}, \ldots, X_{ie}, \ldots, X_{ne})\| < r \tag{1.21}$$

then

$$\forall t > 0, \left\| \vec{X}(t) - \vec{X}_{he} \right\| < \left\| \vec{X_0} - \vec{X}_{he} k e^{\alpha t} \right\|. \tag{1.22}$$

Obviously, if the state parameters move away from the equilibrium point (Fig. 1.4e) or reach another equilibrium point, the system will become locally unstable around this point. A system can also be stable without ever reaching an equilibrium point or an orbit, the trajectory staying inside a space area. Some stable chaotic systems like the well-known Lorenz system belong to this category (Fig. 1.4f). More information is given about chaotic systems in Para. 3.2.

An attractor is a region in the phase space where the trajectories converge regardless of their initial conditions. An attractor can be a point (Fig. 1.4a, b), a periodic curve with one frequency (Fig. 1.4c, d). They may be periodic trajectories with k frequencies. In this case, the attractor is a k torus. For a chaotic system, we talk about strange attractors (Fig. 1.4f).

1.7 Basin of Attraction of an Attractor

When the properties of the local stability of each point are well-known, the next step is to determine the area of stability for each of them. A stable area is a subspace of the space of state parameters around the considered stable point (or around other attractor) where the system remains stable for every initial condition within. The greater stable area is called the basin of attraction. If it is equal to $\mathbb{R}^n$, the system is globally stable. Obviously, from a physical point of view, such a system does not exist. A basin of attraction can be either a closed space or an open space. It can also be a solution of the system S; in this case, we talk about a limit cycle of stability. The theoretical tools used to find these areas are given in Chap. 2. These methods will be applied to analyse the nonlinear system examples treated in Para. 4.

1.8 Linearisation Around an Equilibrium Point

Around each equilibrium point $\vec{X}_{he}$ with $h \in (1, \ldots, m)$, it is possible to linearise the nonlinear system. The basic idea is to consider small variations of each state parameter x_i around the equilibrium point $\vec{X}_{he}$ which can be translated mathematically speaking by:

$$X_i(t) = X_{ihe} + x_i(t) \tag{1.23}$$

for a continuous system and:

$$X_{ik} = X_{ihe} + x_{ik} \tag{1.24}$$

for a discrete system, with $i \in (1, \ldots, n)$. It can be also useful to consider small variations of the p system parameters and the u control parameters. In the same way, we get:

$$P_j(t) = P_{je} + p_j(t) \quad \text{and} \quad U_l(t) = U_{le} + u_l(t). \tag{1.25}$$

for a continuous system and

$$P_{jk} = P_{je} + p_{ik} \quad \text{and} \quad U_{lk} = U_{le} + u_{lk} \tag{1.26}$$

for a discrete system, with $j \in (1, \ldots, p)$ and $l \in (1, \ldots, u)$. Using the Taylor series, neglecting the second order term and taking into account the system S_e, we get for a continuous S system:

$$\dot{x}_i = \sum_{j=1}^{j=n} \frac{\partial F_i}{\partial X_j} x_j + \sum_{j=1}^{j=p} \frac{\partial F_i}{\partial P_j} p_j + \sum_{l=1}^{l=u} \frac{\partial F_i}{\partial U_l} u_l \tag{1.27}$$

and for a discrete system:

$$x_{i(k+1)} = \sum_{j=1}^{j=n} \frac{\partial F_i}{\partial X_{jk}} x_{jk} + \sum_{j=1}^{j=p} \frac{\partial F_i}{\partial P_{jk}} p_{jk} + \sum_{l=1}^{l=u} \frac{\partial F_i}{\partial U_{lk}} u_{lk} \tag{1.28}$$

This can be expressed in matrix form:

$$\begin{pmatrix} \dot{x}_1(t) \\ \vdots \\ \dot{x}_i(t) \\ \vdots \\ \dot{x}_n(t) \end{pmatrix} = \begin{pmatrix} \frac{\partial F_1}{\partial X_1} & \cdots & \frac{\partial F_1}{\partial X_j} & \cdots & \frac{\partial F_1}{\partial X_n} \\ \vdots & \ddots & \vdots & \ddots & \vdots \\ \frac{\partial F_i}{\partial X_1} & \cdots & \frac{\partial F_i}{\partial X_j} & \cdots & \frac{\partial F_i}{\partial X_n} \\ \vdots & \ddots & \vdots & \ddots & \vdots \\ \frac{\partial F_n}{\partial X_1} & \cdots & \frac{\partial F_n}{\partial X_j} & \cdots & \frac{\partial F_1}{\partial X_n} \cdots \end{pmatrix} \begin{pmatrix} x_1(t) \\ \vdots \\ x_j(t) \\ \vdots \\ x_n(t) \end{pmatrix}$$
$$+ \begin{pmatrix} \frac{\partial F_1}{\partial P_1} & \cdots & \frac{\partial F_1}{\partial P_j} & \cdots & \frac{\partial F_1}{\partial P_p} \\ \vdots & \ddots & \vdots & \ddots & \vdots \\ \frac{\partial F_i}{\partial P_1} & \cdots & \frac{\partial F_i}{\partial P_j} & \cdots & \frac{\partial F_i}{\partial P_p} \\ \vdots & \ddots & \vdots & \ddots & \vdots \\ \frac{\partial F_n}{\partial P_1} & \cdots & \frac{\partial F_n}{\partial P_j} & \cdots & \frac{\partial F_n}{\partial P_p} \end{pmatrix} \begin{pmatrix} p_1(t) \\ \vdots \\ p_j(t) \\ \vdots \\ p_p(t) \end{pmatrix} + \begin{pmatrix} \frac{\partial F_1}{\partial U_1} & \cdots & \frac{\partial F_1}{\partial U_l} & \cdots & \frac{\partial F_1}{\partial U_u} \\ \vdots & \ddots & \vdots & \ddots & \vdots \\ \frac{\partial F_i}{\partial U_1} & \cdots & \frac{\partial F_i}{\partial U_l} & \cdots & \frac{\partial F_i}{\partial U_u} \\ \vdots & \ddots & \vdots & \ddots & \vdots \\ \frac{\partial F_n}{\partial U_1} & \cdots & \frac{\partial F_n}{\partial U_l} & \cdots & \frac{\partial F_1}{\partial U_u} \end{pmatrix} \begin{pmatrix} u_1(t) \\ \vdots \\ u_l(t) \\ \vdots \\ u_u(t) \end{pmatrix} \tag{1.29}$$

or more simply by:

$$\overrightarrow{\dot{x}}(t) = J\vec{x}(t) + M_p\vec{p}(t) + M_u\vec{u}(t). \tag{1.30}$$

The same matrix system for discrete systems is given by:

$$\vec{x}_{k+1} = J\vec{x}_k + M_p\vec{p}_k + M_u\vec{u}_k. \tag{1.31}$$

J is the Jacobian matrix. Each partial derivative is calculated at the following equilibrium point:

$$\left(X_{1he}, \ldots, X_{ihe}, \ldots, X_{nhe}, P_{1e}, \ldots, P_{je}, \ldots, P_{pe}, U_{1e}, \ldots, U_{le}, \ldots, U_{ue}\right). \tag{1.32}$$

It is also possible to linearise the output vector $\vec{Y}(t)$ by considering small variations $\vec{y}(t)$ around a nominal value. It is possible to express $\vec{y}(t)$ with the $y \times n$ Jacobian matrix C associated to the function G:

$$\vec{y}(t) = C\vec{x}(t). \tag{1.33}$$

Taking into account (1.28) and (1.31), a linearised system can be represented as shown in Fig. 1.5.

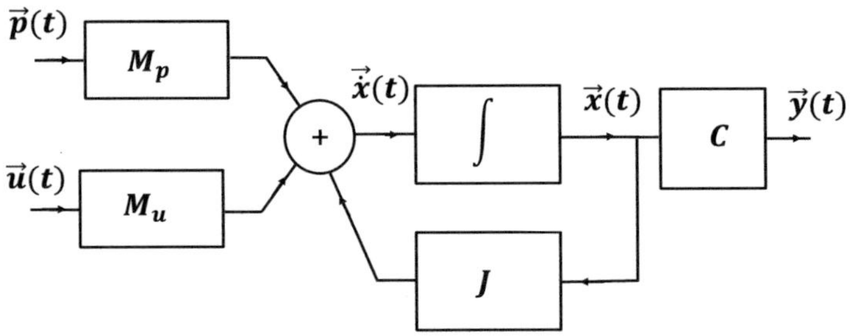

Fig. 1.5 Representation of a linearised system

1.9 Automation Representation and Transfer Functions

From the linearised system previously described, it is possible to get an automation representation, using the Laplace transform and eliminating the undesirable variables. For example, for a system with input and output signals $V_{\text{IN}}(t)$ and $V_{\text{OUT}}(t)$, the transfer function can be written in the form:

$$V_{\text{OUT}}(s) = \left(\sum_{j=1}^{j=p} H_{Pj}(s)P_j(s) + H_{V\text{IN}}(s)H_P(s)V_{\text{IN}}(s) \right) \frac{1}{1 + H_P(s)C_r(s)}. \quad (1.34)$$

s is the Laplace variable. All the transfer functions $H_{Pj}(s)$, $H_P(s)$, $C_r(s)$, $H_{V\text{IN}}(s)$ depend on the considered equilibrium point. $C_r(s)$ is the transfer function of the corrector. We will have to determine $C_r(s)$ in order to meet the specifications of the regulated system. A typical specification could minimise the static error, ensure stability and provide a good fast transient response.

From (1.32) it is possible to derive the general automation representation shown in Fig. 1.6.

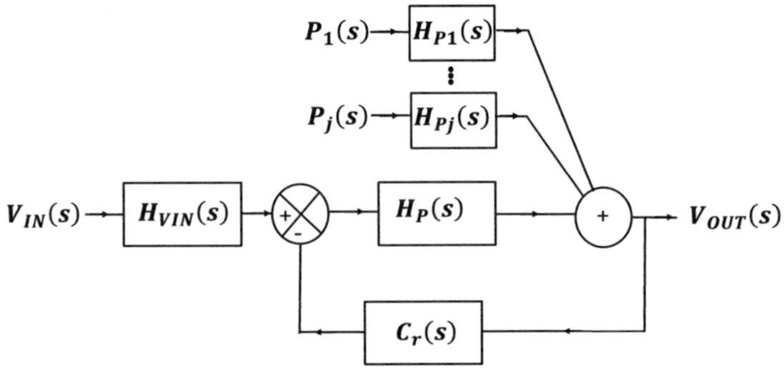

Fig. 1.6 Automation system representation

1.10 Poincaré Section and Map

The Poincaré section and map are important concepts useful in the study of nonlinear systems (Poincaré, 1892–1899). The phase trajectory is often difficult to determine, particularly if the order of the system is large. Conventionally, instead of analysing the continuous flow, we observe the dynamics induced by the trajectory on a particular section S of the phase space. The chosen section is called the Poincare section and helps visualise the issue. The concept of the Poincare section is useful to study cyclic attractors.

For example, as illustrated in Fig. 1.7, if one of the variables, say X_3, passes repeatedly through some particular level ($X_3 = h$), we can define a Poincaré map by collecting the values of the two other variables (X_1, X_2) every time X_3 passes through the selected level. Instead of directly studying the flow, we consider its intersection with the Poincare section S. The points $\vec{P}_0, \vec{P}_1, \vec{P}_2 \ldots$ form the Poincaré map. The Poincaré map reduces a continuous flow to a discrete-time mapping. We need to experiment with a few Poincaré sections before selecting the most interesting one. It is not necessarily orthogonal to a coordinate axis.

The Poincaré map is a continuous mapping T of the Poincaré section S into itself. Since the system is deterministic, $\vec{P}_0$ determines $\vec{P}_1$, $\vec{P}_1$ determines $\vec{P}_2$ etc. We get:

$$\vec{P}_{k+1} = T\left(\vec{P}_k\right) = T\left(T\left(\vec{P}_{k-1}\right)\right) = T^2\left(\vec{P}_{k-1}\right) = T^j\left(\vec{P}_{k-(j-1)}\right). \tag{1.35}$$

Obviously, the time interval between two points is not necessarily constant. Figure 1.8 shows a Poincaré map when the attractor is a periodic orbit with one frequency (Fig. 1.8a) and a two-frequency torus (Fig. 1.8b). For the first case, the Poincaré map is a fixed point:

$$\vec{P}_0 = T\left(\vec{P}_0\right) = T^2\left(\vec{P}_0\right) = \cdots = T^j\left(\vec{P}_0\right). \tag{1.36}$$

Fig. 1.7 Poincaré section

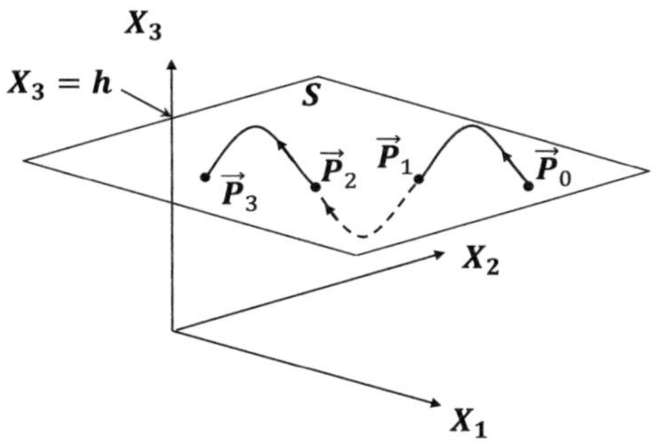

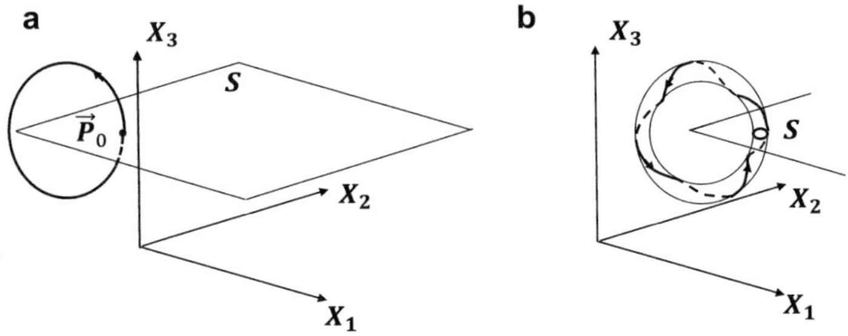

Fig. 1.8 Poincaré map of one and two-periodic limit cycles

1.11 Lyapunov Exponents and Spectrum

Lyapunov exponents and the Lyapunov spectrum (Wolf et al., 1985) are quantities that characterise the rate of separation of infinitesimally close trajectories $\overrightarrow{\delta_X}(t)$. Given a continuous dynamical system in an n-dimensional phase space, we monitor the long-term behaviour of an infinitesimal n sphere of initial conditions as illustrated in Fig. 1.9. The ith Lyapunov exponent λ_i is defined by:

$$\delta_{Xi}(t) = \delta_{Xi0} 2^{\lambda_i t}. \tag{1.37}$$

This number can be determined as follows:

$$\lambda_i = \lim_{t \to \infty} \left(\frac{1}{t} \log_2 \left(\frac{\delta_{Xi}(t)}{\delta_{Xi0}} \right) \right). \tag{1.38}$$

The sphere will become an ellipsoid due to the local deformation of the flow.

The Lyapunov spectrum is the set of the n exponents $(\lambda_1, \ldots \lambda_i, \ldots \lambda_n)$. The largest exponent is often called Lyapunov exponent. The signs of the Lyapunov exponents provide a qualitative picture of system dynamics. A positive Lyapunov exponent determines chaos in the system. For a first-order system

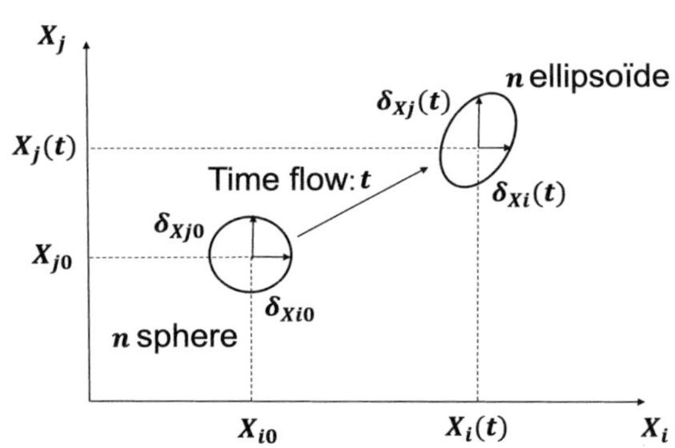

Fig. 1.9 Definition of Lyapunov exponents

($n = 1$), there is only one Lyuapunov exponent ($\lambda = \lambda_1$). If $\lambda_1 = 0$ the system is stable in the sense of Lyapunov. If $\lambda_1 < 0$ the attractor is a periodic orbit or a fixed point and if $\lambda_1 > 0$, the system is chaotic. For an order three system ($n = 3$), the Lyapunov spectrum is $(\lambda_1, \lambda_2, \lambda_3)$. If the signs are $(> 0, 0, < 0)$, $(0, 0, < 0)$, $(0, < 0, < 0)$, $(< 0, < 0, < 0)$ we get respectively a strange attractor, a two frequencies torus, a limit cycle and a fixed point.

The Lyapunov exponents can be calculated easily for a system with known differential equations; however, they cannot be determined analytically and, in most cases, numerical techniques must be used. Extracting these exponents from a time series of experimental data is a complex problem and requires care. These numerical methods will not be described in this book. Examples of the Lyapunov spectrum for conventional chaotic systems are given in Chap. 3.

1.12 Time Horizon

When a system has a positive λ_i, there is a time horizon t_H beyond which prediction breaks down. Suppose, we measure the initial condition of the corresponding state parameter very accurately with an error δ_{Xi0}. After time t the error becomes:

$$\delta_{Xi}(t) = \delta_{Xi0} 2^{\lambda_i t}. \tag{1.39}$$

If t_{ol} is the tolerance, the prediction becomes incorrect when

$$\delta_{Xi}(t) > t_{ol} \tag{1.40}$$

and this occurs after the time horizon:

$$t_H = \frac{1}{\lambda_i} \log_2\left(\frac{t_{ol}}{\delta_{Xi0}}\right). \tag{1.41}$$

1.13 Methodology of Study of a Nonlinear System

To summarise this section, we can propose the following steps to study a nonlinear system:

- Define the system S by using basic physical laws (Ohm, Newton…). Identify the n state, the p system and the u control parameters.
- Calculate with the system S_e, the m equilibrium points.
- For each equilibrium point, define the corresponding linearised system.
- With the theoretical tools given in Chap. 2, study the nature of equilibrium points (stable in the sense of Lyapunov, stable orbit, asymptotically stable, unstable…) and the nature of the attractor (fixed point, periodic orbit, strange attractor…).

- For each stable point, define the basin of attraction or, at minimum, an area of stability (see Chap. 2).
- For a controlled system, define the control laws using one of the different strategies of control (Linear, Optimal, OGY, Predictive…).

Chapter 2
Stability of Nonlinear Systems

In this chapter, we will review some of the basic theoretical tools available to study the stability of nonlinear systems (Ogata, 2010).

2.1 Stability Around an Equilibrium Point

We will begin by considering the various tools at our disposal to study the stability of linear systems and nonlinear systems around an equilibrium point.

2.1.1 Stability Criteria

Consider the linearised system (continuous and discrete) defined by the Jacobian matrix J. At $t = 0$ ($k = 0$), a small perturbation of the state parameters $\overrightarrow{\Delta x_0}$ around the considered equilibrium point $\vec{X}_{he}$ is applied. Referring to Eqs. (1.27) and (1.28) and using the Laplace and z transforms of the small variations $\vec{x}(t)$ and $\vec{x}_k$ we get:

$$\vec{X}(s)s - \overrightarrow{\Delta x_0} = J\vec{X}(s), \quad \vec{X}(z)z - \overrightarrow{\Delta x_0}z = J\vec{X}(z). \tag{2.1}$$

The resolution of these systems gives:

$$\forall \in (1, \ldots n) \;\; X_j(s) = \sum_{i=1}^{n} \frac{A_{ji}}{s - s_i}, \quad X_j(z) = \sum_{i=1}^{n} \frac{zA_{ji}}{z - z_i}. \tag{2.2}$$

The complex numbers s_i and z_i are respectively the roots of $\det(sI - J)$ and $\det(zI - J)$. The A_{ji} are real coefficients. Returning to the time domain, we find:

B. Rivet, *Theories of Nonlinear Systems for Power Converters*,
https://doi.org/10.1007/978-3-032-25062-9_2

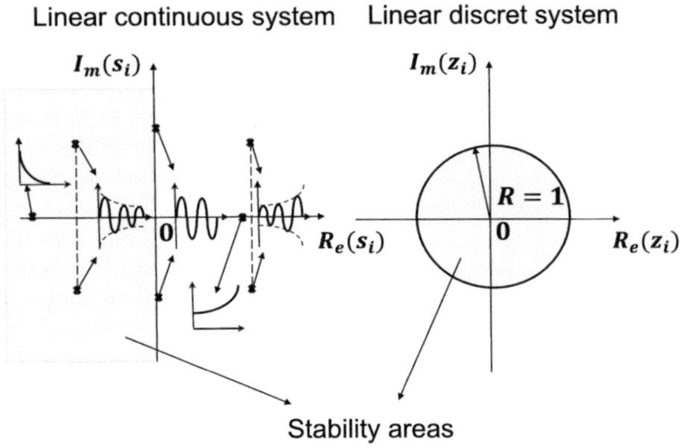

Fig. 2.1 Stability regions of continuous and discrete systems

$$\forall j \in (1, \ldots n) \quad x_j(t) = \sum_{i=1}^{n} A_{ji} e^{\alpha_i t}, \quad x_{jk} = \sum_{i=1}^{n} A_{ji} z_i^k. \tag{2.3}$$

α_i is the real part of s_i. $\det(sI - J) = 0$ is called the characteristic equation. From Eq. (2.3), we can deduce the following well-known properties.

- For a linear continuous system, if all the real parts of the roots α_i of $\det(sI - J)$ are strictly negative and for a discrete system, if all the z_i are inside the circle of centre 0 and radius $R = 1$, the system is locally asymptotically stable.
- If at least one real part α_i is positive or if at least one complex number z_i is outside the stability circle, the system is locally unstable.
- In other cases, we cannot draw any conclusions concerning the local stability of the nonlinear system around the considered equilibrium point.

Figure 2.1 shows the stability regions in the s_i and z_i planes for linear continuous and discrete systems.
The same results can be obtained using transfer functions. Figure 2.2 shows the basic representation of a continuous closed system where $H_{BO}(s)$ is the open-loop transfer function.

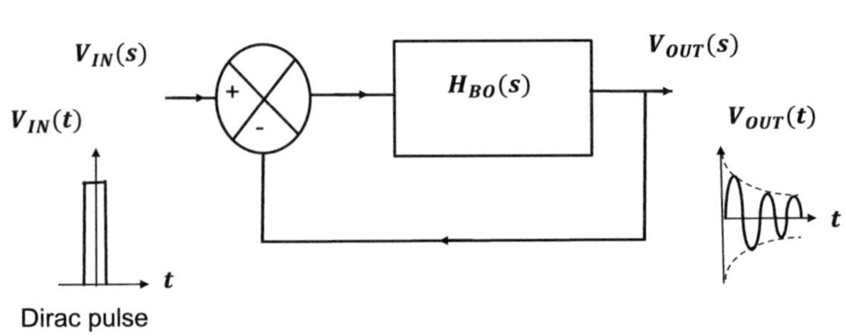

Fig. 2.2 Basic representation of a closed-loop system

The closed-loop transfer function $H_{BC}(s)$ can be easily deduced:

$$H_{BC}(s) = \frac{V_{\text{OUT}}(s)}{V_{\text{IN}}(s)} = \frac{H_{BO}(s)}{1 + H_{BO}(s)}. \tag{2.4}$$

By using the method given in Para. 1.9 to switch from the linearised system to an automation representation, it can shown that the roots of $\det(sI - J)$ correspond to the poles of the closed-loop transfer function$(H_{BC}(s))$. Suppose that a Dirac pulse, whose Laplace transform is 1, is applied at $t = 0$ to the system represented in Fig. 2.2. The output will be of the form:

$$V_{\text{OUT}}(s) = H_{BC}(s)V_{\text{IN}}(s) = \frac{H_{BO}(s)V_{\text{IN}}(s)}{1 + H_{BO}(s)} = \sum_{i=1}^{n} \frac{B_i}{s - s_i}. \tag{2.5}$$

It follows that the same properties regarding local stability can be inferred. Figure 2.1 shows the different responses for a Dirac input in relation to the positioning of s_i in the complex plane. The same reasoning can be applied to a discrete system. In the next section, we will focus on the stability criteria of a linear system.

2.1.2 Routh Stability Criterion

For a system of the first or the second order, the roots of $\det(sI - J)$ can be easily calculated. For higher orders systems, it is often not possible to find these roots. The well-known Routh stability criterion (Routh, 1877) provides a method to determine the number of roots within the unstable right half-plane and the number of roots in the stable left half-plane. It may also allow us to find the number of roots on the imaginary axis. This method does not locate the roots. The characteristic equation being tested is of the form:

$$\det(sI - J) = a_n s^n + a_{n-1} s^{n-1} + \cdots + a_1 s^1 + a_0 = 0. \tag{2.6}$$

The first necessary condition to apply this criterion is that all the coefficients a_i are positive. The Routh Array is constructed as follows:

$$\begin{array}{lllllll}
 & & \text{Col 1} & \text{Col 2} & \text{Col 3} & \text{Col 4} & \cdots \\
\text{Row 0} & s^n & a_n & a_{n-2} & a_{n-4} & a_{n-6} & \cdots \\
\text{Row 1} & s^{n-1} & a_{n-1} & a_{n-3} & a_{n-5} & a_{n-7} & \cdots \\
\text{Row 2} & s^{n-2} & b_1 & b_2 & b_3 & b_4 & \cdots \\
\text{Row 3} & s^{n-3} & c_1 & c_2 & \cdots & \cdots & \cdots \\
. & \cdots & \cdots & \cdots & \cdots & \cdots & \cdots \\
\text{Row } n-1 & s^1 & y_1 & y_2 & & & \\
\text{Row } n & s^0 & z_1. & & & &
\end{array} \tag{2.7}$$

The coefficients are:

$$\begin{aligned} b_1 &= (a_{n-1}a_{n-2} - a_n a_{n-3})/a_{n-1} \\ b_2 &= (a_{n-1}a_{n-4} - a_n a_{n-5})/a_{n-1} \\ c_1 &= (b_1 a_{n-3} - a_{n-1}b_2)/b_1 \\ c_2 &= (b_1 a_{n-5} - a_{n-1}b_3)/b_1 \ldots \text{etc.} \end{aligned} \tag{2.8}$$

The coefficient will be set to 0 if the row is too short to complete the calculation. The last row will have just one element. All the roots of the characteristic equation have negative real parts only if all elements in column 1 have the same sign. The number of sign changes in column 1 is equal to the number of roots with positive real parts.

Example Consider the following characteristic equation:

$$\det(sI - J) = s^3 + 5s^2 + 4s + 2 = 0. \tag{2.9}$$

The corresponding Routh array is constructed as follows:

$$\begin{array}{lccc} & \text{col 1} & \text{col 2} & \text{col 3} \\ \text{Row 0 } s^3 & 1 & 4 & 0 \\ \text{Row 1 } s^2 & 5 & 2 & 0 \\ \text{Row 2 } s^1 & \frac{18}{5} & 0 & \\ \text{Row 3 } s^0 & 2 & & \end{array} \tag{2.10}$$

There is no change of sign in column 1, so the root of $\det(sI - J)$ are all real and negative and the system is stable.

Case of a Zero in the First Column

If a zero appears in the first column, the previous method cannot be used because the divisor would be zero. A convenient technique consists in replacing the zero with a small positive number γ and proceeding with the operation (taking into account the limit as $\gamma \to 0$).

Example Study the stability of the system having the characteristic equation below.

$$\det(sI - J) = s^5 + 2s^4 + 3s^3 + 6s^2 + 4s + 2 = 0. \tag{2.11}$$

The corresponding Routh array becomes:

$$\begin{array}{lcccc} & \text{Col 1} & \text{Col 2} & \text{Col 3} & \text{Col 4} \\ \text{Row 0 } s^5 & 1 & 3 & 4 & 0 \\ \text{Row 1 } s^4 & 2 & 6 & 2 & 0 \\ \text{Row 2 } s^3 & \gamma & 3 & 0 & 0 \\ \text{Row 3 } s^2 & -6/\gamma & 2 & 0 & \\ \text{Row 4 } s^1 & 3 & 0 & & \\ \text{Row 5 } s^0 & 2 & 0 & & \end{array} \tag{2.12}$$

There are two sign changes in the first column and therefore two roots with positive real parts, so the system is unstable.

2.1.3 *Bilinear Transform of a Digital System*

One common method of analysing the stability of a digital system in the z domain is to use the bilinear transformation to convert the characteristic polynomial:

$$K(z) = k_n z^n + k_{n-1} z^{n-1} + \cdots + k_1 z^1 + k_0 z^0 = 0 \tag{2.13}$$

from the z domain to the s domain. The bilinear transformation is given by:

$$z = \frac{1+s}{1-s}. \tag{2.14}$$

We obtain $H(s) = h_n s^n + h_{n-1} s^{n-1} + \ldots + h_1 s^1 + h_0$ and then we apply the Routh criterion. Obviously, $K(z)$ is stable if and only if $H(s)$ is stable.

2.1.4 *Jury Stability Criterion*

Jury's criterion (Jury, 1961) is similar to the Routh's criterion in the z domain. Again, we consider the characteristic polynomial K(z) of the discrete system, k_n must be positive (if k_n is negative, multiply $K(z)$ by -1). The system is considered stable only if it passes all the following tests, ensuring that no root lies outside the unit circle.

- $K(1) > 0$
- $(-1)^n K(-1) > 0$
- $k_n > |k_0|$

$$\begin{aligned} |b_0| &> |b_{n-1}| \\ |c_0| &> |c_{n-2}| \\ |d_0| &> |d_{n-3}| \end{aligned} \tag{2.15}$$

and so on, up the last row of the array. The numbers $b_0, c_0, d_0, b_{n-1}, c_{n-2}, d_{n-3}$ are given by the Jury Array which is constructed as follows:

Row	z^0	z^1	z^2	$\ldots$	z^{n-k}	$\ldots$	z^{n-1}	z^n
1	k_0	k_1	k_2	$\ldots$	k_{n-k}	$\ldots$	k_{n-1}	k_n
2	k_n	k_{n-1}	k_{n-2}	$\ldots$	k_k	$\ldots$	k_1	k_0

(continued)

(continued)

3 4	b_0 b_{n-1}	b_1 b_{n-2}	b_2 b_{n-3}	$\dots$ $\dots$	b_{n-k} b_k	$\dots$ $\dots$	b_{n-1} b_0	
5 6	c_0 c_{n-2}	c_1 b_{n-3}	c_2 c_{n-4}	$\dots$ $\dots$	$\dots$ $\dots$	$\dots$ $\dots$	c_{n-2} c_0	
$\vdots$	$\vdots$	$\vdots$	$\vdots$	$\dots$	$\dots$			
$2n-5$ $2n-4$	s_0 s_3	s_1 s_2	s_2 s_1	s_3 s_0				
$2n-3$	r_0	r_1	r_2					

The coefficients are calculated using the following formulas:

$$
\begin{aligned}
b_i &= \begin{vmatrix} k_0 & k_{n-i} \\ k_n & k_i \end{vmatrix} \quad i = 0, 1, \dots, n-1, \\
c_i &= \begin{vmatrix} b_0 & b_{n-i-1} \\ b_{n-1} & b_i \end{vmatrix} \quad i = 0, 1, \dots, n-1, \\
&\vdots \\
r_0 &= \begin{vmatrix} s_0 & s_3 \\ s_3 & s_0 \end{vmatrix}, \quad r_1 = \begin{vmatrix} s_0 & s_2 \\ s_3 & s_1 \end{vmatrix}, \quad r_2 = \begin{vmatrix} s_0 & s_1 \\ s_3 & s_2 \end{vmatrix}.
\end{aligned}
\tag{2.16}
$$

As an illustration, test the stability of the polynomial:

$$K(z) = z^5 - 0.1z^4 - 0.75z^3 - 0.023z^2 + 0.125z + 0.012. \tag{2.17}$$

The corresponding Jury array is:

Row	z^0	z^1	z^2	z^3	z^4	z^5
1 2	0.012 1	0.125 – 0.1	– 0.023 – 0.75	– 0.75 – 0.023	– 0.1 0.125	1 0.012
3 4	– 0.9999 – 0.1262	0.1015 0.014	0.7497 0.7497	0.014 0.1015	– 0.1262 – 0.9999	
5 6	0.9838 – 0.0012	– 0.0997 – 0.655	– 0.655 – 0.0997	– 0.0012 0.9838		
7	0.9678	– 0.0989	– 0.6444			

The results of the tests are:

$$
\begin{aligned}
&K(1) = 0.264 > 0,\ (-1)^n K(-1) = (-1)^5(-0.486) > 0,\ 1 > |0.012| \\
&|-0.9999| > |-0.1262|,\ |0.9838| > |-0.0012|,\ |0.9678| > |-0.6444|.
\end{aligned}
\tag{2.18}
$$

The system has passed all tests so, it can be considered stable. This result is confirmed by the factorization of $K(z)$:

$$K(z) = (z - 0.5)(z + 0.5)(z - 0.8)(z + 0.6)(z + 0.1), \tag{2.19}$$

showing that all the roots are inside the unit circle.

2.1.5 *Generalised Nyquist Criterion with Proof*

In 1932, applying Cauchy's residue theorem, Nyquist established a fundamental result that enables the determination of the stability of a closed-loop system by analysing the zeros and poles of the open-loop transfer function (Khalil, 2002; Nyquist, 1932; Strogatz, 2015). A proof of this criterion is provided here to help the reader better understanding it.

Proof of the Generalised Nyquist Criterion

First, let's examine Cauchy's residue Theorem. Consider a closed and oriented contour C in the complex plane s and an analytic function $H(s)$ of the form:

$$H(s) = K\frac{\prod_{i=1}^{i=z}(s - z_i)}{s^l \prod_{i=1}^{i=p}(s - p_i)}. \tag{2.20}$$

We assume no poles of $H(s)$ lie on C. The positive orientation of the contour is defined as counterclockwise. Thanks to Cauchy's residue Theorem, we know that, when considering the image of the contour C under the function $H(s)$, the number of counterclockwise encirclements of the origin $(0, 0)$ is equal to $Z - P$, where Z and P are respectively the number of zeros and poles inside C. Take the following example as an illustration:

$$H(s) = \frac{(s - z_1)(s - z_2)}{s(s - p_1)} \tag{2.21}$$

with $z_1 = 1.5 + 1.5i, z_2 = -0.5i, p_1 = 0.5 + 0.5i, p_2 = 0$.
C is a circle of radius $R = 1$ and centre $(0, 0)$ defined in the s plane by:

$$(R\cos(\theta), R\sin(\theta)) \quad \text{with } \theta \in [0, 2\pi]. \tag{2.22}$$

Figure 2.3 shows that there are one zero and two poles inside C (none on the contour), so $Z = 1$ and $P = 2$. Consequently, the number of counterclockwise encirclements of the origin is $1 - 2 = -1$. The points a, b, c on the circle C moving in the positive direction correspond to the points A, B, C in the $H(s)$ plane moving in the opposite (negative) direction. Figure 2.3 confirms that the number of encirclements is equal to -1.

The stability of the closed-loop system represented in Fig. 2.2 is determined by the zeros of $1 + H_{BO}(s)$, while the stability of the open-loop system is given by the poles of $H_{BO}(s)$. It is possible that the closed-loop system may be stable even if the

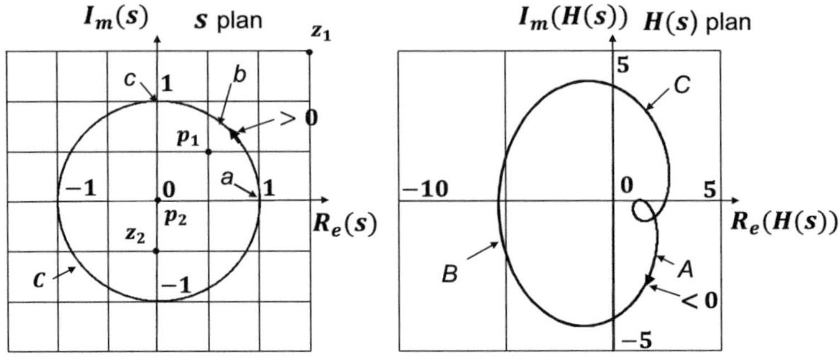

Fig. 2.3 Cauchy's residue theorem

open-loop system is unstable. Given $H_{BO}(s)$, we are usually faced with the problem of finding the roots of $1 + H_{BO}$. The Nyquist criterion allows us to determine without factoring $1 + H_{BO}(s)$ whether or not the closed-loop system is stable and, if it is unstable, the number of unstable poles (i.e. the number of zeros of $1 + H_{BO}(s)$ in the right half plane). To delimit the unstable right half plane, we construct the Nyquist contour in the s plane as shown in Fig. 2.4.

While constructing the contour C, we should bear in mind that the different portions of the Nyquist contour take up the following values. On the imaginary axis:

$$s = iw, \tag{2.23}$$

on a small indentation around a pole on the positive imaginary axis:

Fig. 2.4 Nyquist contour

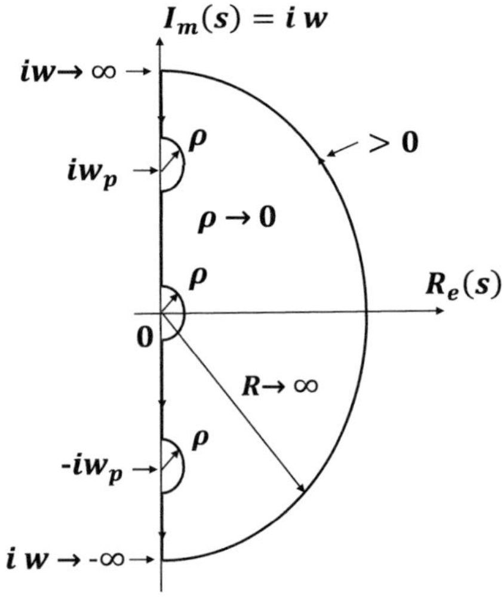

$$s = iw_p + \rho e^{i\theta} \quad \text{with} \quad -\frac{\pi}{2} \leq \theta \leq \frac{\pi}{2}, \tag{2.24}$$

on the large semi-circle:

$$s = Re^{i\theta} \quad \text{with} \quad -\frac{\pi}{2} \leq \theta \leq \frac{\pi}{2}, \tag{2.25}$$

on a small indentation around a pole on the negative imaginary axis:

$$s = -iw_p + \rho e^{i\theta} \quad \text{with} \quad -\frac{\pi}{2} \leq \theta \leq \frac{\pi}{2}, \tag{2.26}$$

on the indentation around the origin:

$$s = \rho e^{i\theta} \quad \text{with} \quad -\frac{\pi}{2} \leq \theta \leq \frac{\pi}{2}. \tag{2.27}$$

This contour and $1 + H_{BO}(s)$ correspond, respectively, to the contour C and the function $H(s)$ in the residue theorem. The small half-circles are drawn around each pole of $1 + H_{BO}(s)$ on the imaginary axis to satisfy the hypothesis of the theorem (no pole of $H(s)$ on the C contour). When the radius of the small circles ρ tend to zero and the larger radius R tend to ∞, the Nyquist path delimits the instability area (right half plane). Note that the poles of $1 + H_{BO}(s)$ are the poles of $H_{BO}(s)$. Thanks to Cauchy's residue theorem, we can deduce the generalised Nyquist criterion.

Generalised Nyquist Criterion

The number of unstable poles (Z) of the closed-loop system (i.e. number of zeros of $1 + H_{BO}(s)$) is equal of the number of encirclements of the origin (N) (in the image of the Nyquist path under the transfer function $1 + H_{BO}(s)$ plus the number of unstable poles of $H_{BO}(s)(P)$), that is $Z = N + P$.

Examples of the Use of the Generalised Nyquist Criterion

As a first example, consider a simple linear system having the open-loop transfer function:

$$H_{BO}(s) = \frac{1}{1+s}. \tag{2.28}$$

There are no poles on the Nyquist path. The image of this path under the transfer function$1+H_{BO}(s)$ is shown in Fig. 2.5. The points a, b, c on the Nyquist path moving in the positive direction correspond to the points A, B, C moving also in the positive direction in the $1 + H_{BO}(s)$ plane. We can see that the number of encirclements of the origin is zero. The number of unstable poles of $H_{BO}(s)$ being equal to zero, we can conclude there is no unstable pole in the corresponding closed-loop system. In this example, it is easy to calculate the zero of $1 + H_{BO}(s)$. We get only one zero equal to -2, confirming the previous result.

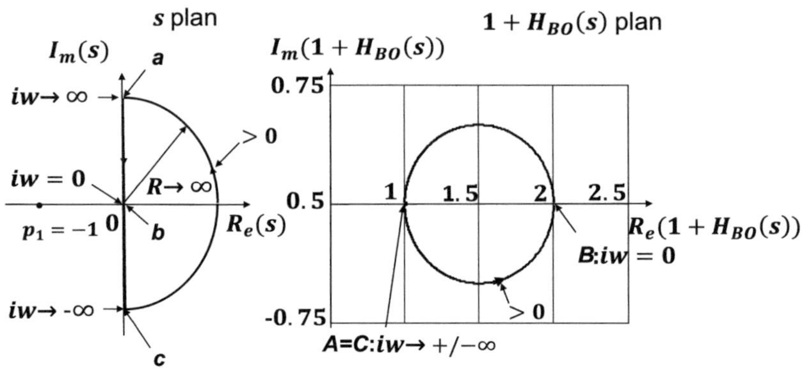

Fig. 2.5 Nyquist contour for the transfer function: $1 + \frac{1}{1+s}$

The second example is defined by the following open-loop transfer function:

$$H_{B0}(s) = \frac{1}{s^3(1+s)} \tag{2.29}$$

It is more complex because there is a third-order pole ($s = 0$) on the Nyquist path. We need to consider the indentation around the origin:

$$s = \rho e^{i\theta} \quad \text{with} \quad -\frac{\pi}{2} \leq \theta \leq \frac{\pi}{2} \tag{2.30}$$

as represented in Fig. 2.6.

This figure, in which $\rho = 0.1$ and $R = 10$, shows there are two positive encirclements of the origin in the image of the Nyquist path under the transfer function. The points a, b, c, d, e correspond to the points A, B, C, D, E. There is no pole of $H_{B0}(s)$ inside the region enclosed by the Nyquist path. So, the number of unstable poles of the closed-loop system is equal to two. Calculation of these poles gives:

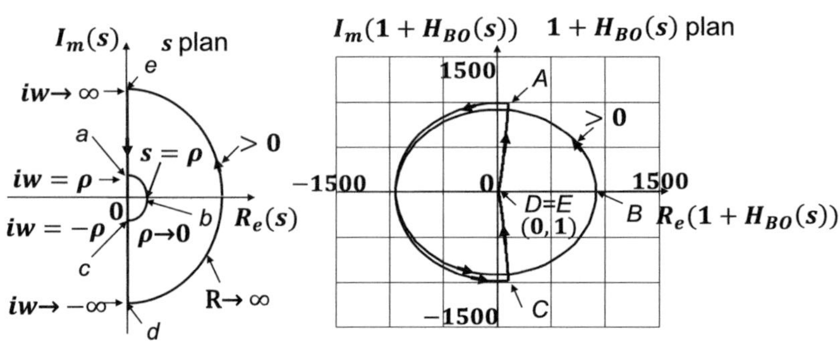

Fig. 2.6 Nyquist contour for the transfer function: $1 + \frac{1}{s^3(1+s)}$

$$p_1 \simeq 0.233 - 0.793i, \quad p_2 \simeq 0.233 + 0.793i, \quad p_3 \simeq -1.466. \tag{2.31}$$

confirming the presence of the two unstable poles.

2.1.6 Simplified Nyquist Criterion

The above criterion can be simplified by plotting only the function $H_{BO}(s)$ instead of $1 + H_{BO}(s)$ and count the encirclements of the Nyquist plot of $H_{BO}(s)$ around the point $(-1, 0i)$. That way, the Nyquist criterion takes the following form: the number of unstable closed-loop poles is equal to the number of unstable open-loop poles plus the number of encirclements of the point $(-1, 0i)$ of the Nyquist plot of $H_{BO}(s)$. If $H_{BO}(s)$ has no poles in the right half plane and on the imaginary axis, then the closed-loop system is stable if and only if the closed contour given by $H_{BO}(iw)$, with w moving from $-\infty$ to $+\infty$ has no net encirclement of the critical point $(-1, 0i)$. There is a symmetry of the mapping $H_{BO}(iw)$ with respect to real axis when w is transformed to $-w$ (see Fig. 2.6). So, we can consider only the positive frequencies (or pulsations) when applying the Nyquist criteria. An alternative to explicitly computing the Nyquist plot is to determine the frequency response using the Bode plot which gives the Nyquist curve for $s = iw = 2\pi fi$ and $f \geq 0$. Bode plots gain:

$$G_{BO}(f) = 20 \log|H_{BO}(2\pi fi)| \tag{2.32}$$

and phase:

$$\mathrm{Arg}(H_{BO}(2\pi fi)) \tag{2.33}$$

are the most widely used means because they are log–log plots, displaying a wide range of frequencies on the horizontal axis and a wide range of gains on the vertical axis, all within a viewable format. If $H_{BO}(s)$ has poles on the imaginary axis or on the right half plane, it is better to use the generalised Nyquist criterion.

2.1.7 Stability Margins

In practice, we also need to know the margins of stability of the system under consideration and its robustness to perturbations. Two of the most common stability margin criteria, inspired by Nyquist's stability criterion, are the gain margin and phase margin. The gain margin g_m of a sytem is defined as the smallest amplification we have to apply to the open-loop before the closed-loop system becomes unstable. The gain margin g_m is defined at the phase crossover frequency f_{pc}, where the phase of $H_{BO}(2\pi fi)$ is equal to -180°. We get:

$$\left|H_{BO}\left(i2\pi f_{pc}\right)\right| g_m = 1 \tag{2.34}$$

and in decibel:

$$20\log(g_m) = -20\log\left|H_{BO}\left(i2\pi f_{pc}\right)\right|. \tag{2.35}$$

Similarly, the phase margin is the amount of additional phase shift required to reach the stability limit. The phase margin φ_m is defined at the gain crossover frequency f_{gc} where the module of $H_{BO}\left(i2\pi f_{gc}\right)$ is equal to 1. The phase margin is given by:

$$\varphi_m = \pi + \operatorname{Arg}\left(H_{BO}\left(i2\pi f_{pc}\right)\right). \tag{2.36}$$

These two margins are illustrated in Fig. 2.7 on the Nyquist and the Bode plots. The gain and phase margins are classical robustness measures that have been used for a long time in control system design. However, these margins can be acceptable but the system may still not be robust. Consider the following open-loop transfer function:

$$H_{B0}(s) = \frac{0.38\left(s^2 + 0.1s + 0.55\right)}{s(1+s)\left(s^2 + 0.06s + 0.5\right)}. \tag{2.37}$$

Figure 2.8 shows the Nyquist and bode plot of this transfer function. The phase margin ($\varphi_m = 70°$) and the gain margin ($g_m > 1000$) are good but the Nyquist curve is very close to the critical point $(-1, 0)$. That is the reason why it is also interesting to consider the stability margin d_m, which is the shortest distance from the Nyquist curve to the critical point (see Fig. 2.8). In this example, the stability margin is $d_m = 0.27$, which is low. The step response of the corresponding closed system represented in Fig. 2.9 is highly oscillatory. In critical cases, it is best to use

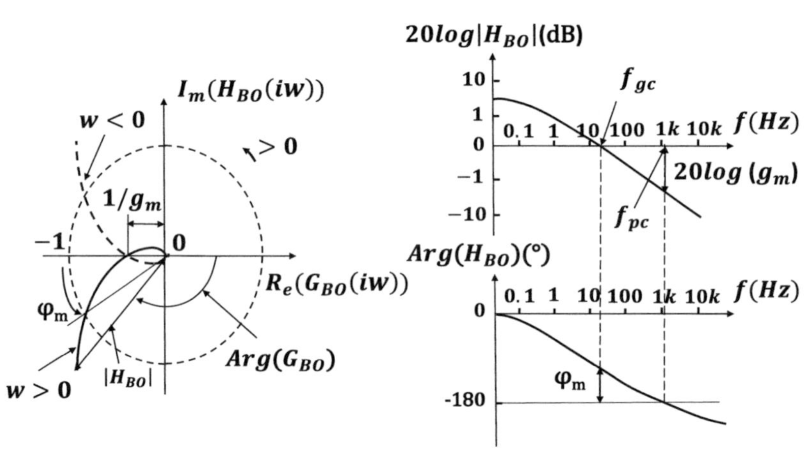

Fig. 2.7 Gain and phase margins on Nyquist and Bode plots

the Nyquist plot to check stability since it provides more complete information than the Bode plot. Acceptable values of the margins are:

$$\varphi_m \in \left[30°, 70°\right], g_m \in [2, 10], \quad \text{and} \quad d_m \in [0.5, 0.8]. \tag{2.38}$$

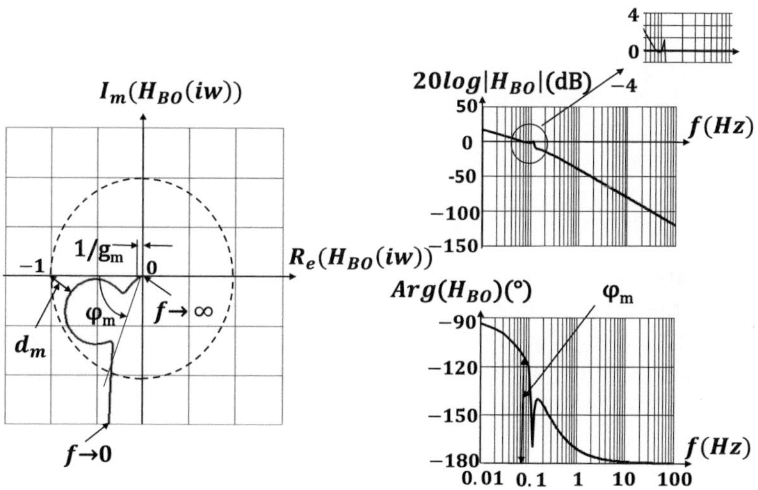

Fig. 2.8 System with good gain and phase margin but poor stability margin

Fig. 2.9 Step response of a system with a poor stability margin

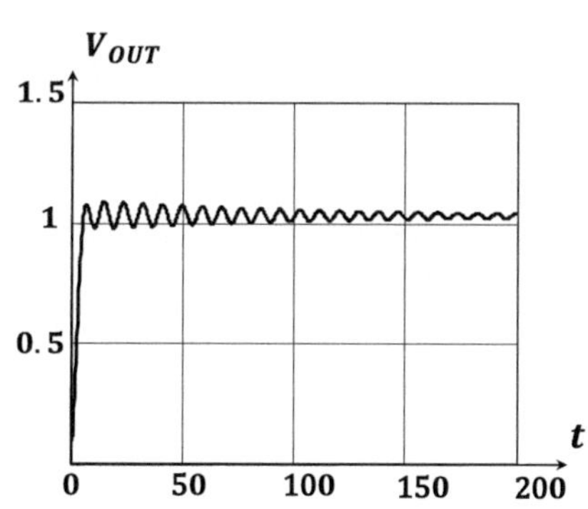

2.2 Stability of a Periodic Orbit

2.2.1 Stability Definition

Consider the dynamics $\frac{d\vec{X}}{dt} = \vec{F}\left(\vec{X}\right)$ of a system that has a periodic solution $\vec{X}_P(t)$. As explained in Chap. 1, the Poincaré map is a fixed point $\vec{P}_0$ in the section S (see Fig. 1.8a). For an orbit $\vec{\varphi}\left(\vec{X}_0, t\right)$ starting at $\vec{X}_0$ close to $\vec{P}_0$ in the section S, the phase flow will return to S according to the Poincaré map P defined by:

$$\begin{aligned} &P : S \to S \\ &\vec{X}_0 \to P\left(\vec{X}_0\right) = \vec{\varphi}\left(\vec{X}_0, T\right). \end{aligned} \tag{2.39}$$

T is the period of the orbit. In this paragraph, we will consider only the case of a constant period T. Note that $\vec{P}_0$ being a fixed point of the map P, the study of a periodic orbit $\vec{X}_P(t)$ is reduced to the study of the stability of the fixed point $\vec{P}_0$. We can use the same stability definitions given in Chap. 1 for a fixed point. The periodic solution $\vec{X}_P(t)$ is stable in the sense of Lyapunov if for any $\rho > 0$ there is $r > 0$ such that if

$$\left\| \overrightarrow{OX}_0 - \overrightarrow{OP}_0 \right\| < r \tag{2.40}$$

then

$$\forall n \in \mathbb{N} \quad \left\| P^n\left(\vec{X}_0\right) - \overrightarrow{OP}_0 \right\| < \rho. \tag{2.41}$$

The periodic solution $\vec{X}_P(t)$ is asymptotically stable if it is stable in the sense of Lyapunov and there is $r > 0$ so that if

$$\left\| \overrightarrow{OX}_0 - \overrightarrow{OP}_0 \right\| < r \tag{2.42}$$

then

$$\lim_{n \to \infty} P^n\left(\vec{X}_0\right) = \vec{P}_0. \tag{2.43}$$

2.2.2 Stability Criterion Defined by a Floquet Matrix

Let the Jacobian matrix of the map P at the fixed point $\vec{P}_0$ be $\frac{\partial P}{\partial \vec{X}_0}\left(\vec{P}_0\right)$. This linearised Poincaré map is called the Floquet matrix:

$$\frac{\partial P}{\partial \vec{X}_0}\left(\vec{P}_0\right). \tag{2.44}$$

It can be used to study the stability of $\vec{P}_0$ and consequently the stability of its corresponding periodic orbit (Panardo, 2022). We get:

$$P\left(\vec{X}_0\right) \simeq \vec{P}_0 + \frac{\partial P}{\partial \vec{X}_0}\left(\vec{P}_0\right)\left(\vec{X}_0 - \vec{P}_0\right). \tag{2.45}$$

Let $\boldsymbol{\lambda}_1, \ldots, \boldsymbol{\lambda}_{n-1}$ be the eigenvalues of the Floquet matrix. If the moduli of all eigen values are ≤ 1, then $\vec{P}_0$ is stable. If the modulus of at least one of the eigenvalues is larger than 1, then $\vec{P}_0$ is unstable.

2.2.3 Stability Criterion Defined by a Monodromy Matrix

Another way to study the stability of a periodic orbit is to use the Monodromy matrix (Asahara & Kousaka, 2021).

Consider a point $\vec{X}_0$ in the n-dimensional phase space. A trajectory that starts from $\vec{X}_0$ with a small perturbation $\overrightarrow{\delta X}_0$ will be displaced after one period T by:

$$\overrightarrow{\delta X}(T) = \vec{\varphi}\left(\vec{X}_0 + \overrightarrow{\delta X}_0, T\right) - \vec{\varphi}\left(\vec{X}_0, T\right). \tag{2.46}$$

In the first order, this displacement is:

$$\overrightarrow{\delta X}(T) = \frac{\partial \vec{\varphi}\left(\vec{X}_0, T\right)}{\partial \vec{X}_0} \overrightarrow{\delta X}_0. \tag{2.47}$$

The Monodromy matrix is given by:

$$M = \frac{\partial \vec{\varphi}\left(\vec{X}_0, T\right)}{\partial \vec{X}_0}. \tag{2.48}$$

The eigenvalues are called characteristic multipliers. One eigenvalue is always equal to 1, the moduli of the $n - 1$ other eigenvalues determine the stability of the orbit. Determining the Monodromy or the Floquet Matrix is not always an easy task and in practice it is usually necessary to resort to numerical methods.

2.2.4 *Stability of an* LC *Circuit*

The concepts of Floquet and Monodromy matrices are difficult to conceptualise. To make it easier to understand we will consider the simple circuit of an inductor L and a capacitor C connected in parallel (see Fig. 2.10). The state vector is $\vec{X} = \begin{pmatrix} I_L \\ V_C \end{pmatrix}$ with the initial conditions $\vec{X}_0 = \begin{pmatrix} I_{L0} \\ V_{C0} \end{pmatrix}$. The dynamics of this system is given by:

$$\begin{cases} \frac{\mathrm{d}I_L}{\mathrm{d}t} = \frac{V_C}{L} \\ \frac{\mathrm{d}V_C}{\mathrm{d}t} = -\frac{I_L}{C} \end{cases} \tag{2.49}$$

Obviously, the corresponding attractor is a periodic orbit with the constant period:

$$T = 2\pi\sqrt{LC}. \tag{2.50}$$

We get:

$$\overrightarrow{\varphi\left(\vec{X}_0, t\right)} = \begin{pmatrix} I_L(t) = I_{L0}\cos\left(\frac{t}{\sqrt{LC}}\right) + \sqrt{\frac{\mathrm{C}}{\mathrm{L}}}V_{C0}\sin\left(\frac{t}{\sqrt{LC}}\right) \\ V_C(t) = -\sqrt{\frac{\mathrm{L}}{\mathrm{C}}}I_{L0}\sin\left(\frac{t}{\sqrt{LC}}\right) + V_{C0}\cos\left(\frac{t}{\sqrt{LC}}\right) \end{pmatrix}, \tag{2.51}$$

so

$$\overrightarrow{\varphi\left(\vec{X}_0, T\right)} = \begin{pmatrix} I_L(T) \\ V_C(T) \end{pmatrix} = \begin{pmatrix} I_{L0} \\ V_{C0} \end{pmatrix}. \tag{2.52}$$

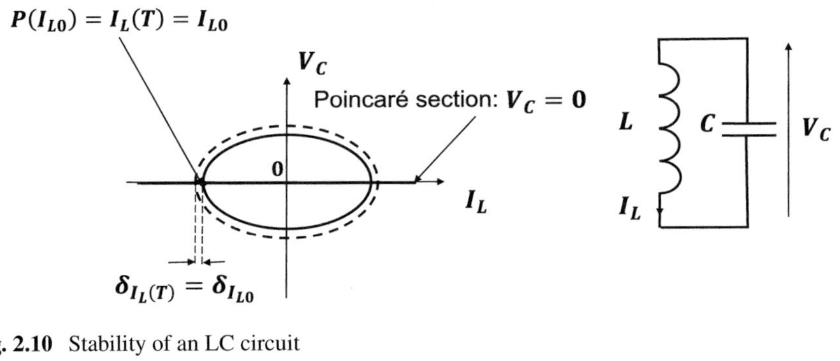

Fig. 2.10 Stability of an LC circuit

The Monodromy matrix is:

$$M = \begin{pmatrix} \frac{\partial I_L(T)}{\partial I_{L0}} & \frac{\partial I_L(T)}{\partial V_{C0}} \\ \frac{\partial V_C(T)}{\partial I_{L0}} & \frac{\partial V_C(T)}{\partial V_{C0}} \end{pmatrix} = \begin{pmatrix} 1 & 0 \\ 0 & 1 \end{pmatrix}. \tag{2.53}$$

The two eigenvalues of the Monodromy matrix being equal to 1, we get

$$\forall n \in \mathbb{N} \quad \vec{X}(nT) = \vec{X}_0, \tag{2.54}$$

so, the system is stable. Consider now the Poincaré section given by $V_C = 0$ and the initial point $P_0 = I_{L0}$ in this section (see Fig. 2.10). The corresponding Poincaré map is given by

$$\begin{aligned} &P : \{V_C = 0\} \to \{V_C = 0\} \\ &I_{L0} \to P(I_{L0}) = I_L(T) = I_{L0}. \end{aligned} \tag{2.55}$$

In this example, the Floquet matrix becomes the following coefficient:

$$F = \frac{\partial P}{\partial I_{L0}}(I_{L0}) = 1. \tag{2.56}$$

The single eigenvalue of the Floquet matrix is equal to 1, confirming that the system is stable.

2.3 Determination of a Stability Region and a Basin of Attraction

A stability region is an area in which the system, starting from initial conditions inside it, converges towards the equilibrium point (or towards another attractor). The larger one is the basin of attraction. A stability limit cycle is a closed region in the space of the state parameters solution of $\frac{d\vec{X}}{dt} = \vec{F}\left(\vec{X}\right)$. If this cycle exists, it delimits the basin of attraction. Obviously, a stability region is always inside the basin of attraction. Very often a system has no limit stability cycle but if it is locally stable, we can always determine a stability region and its basin of attraction.

2.3.1 Determination of a Stability Area by Simulation

A way to find a stability area for a continuous system can be achieved by simulating the circuit represented in Fig. 2.11.

Here, the n state parameters $(X_1, \ldots X_n)$ are the n voltages across the n capacitors $(C_1, \ldots C_n)$. The value of each capacitor is equal to $1F$. So the n current generators allow the simulation of the n differential equations of the system S

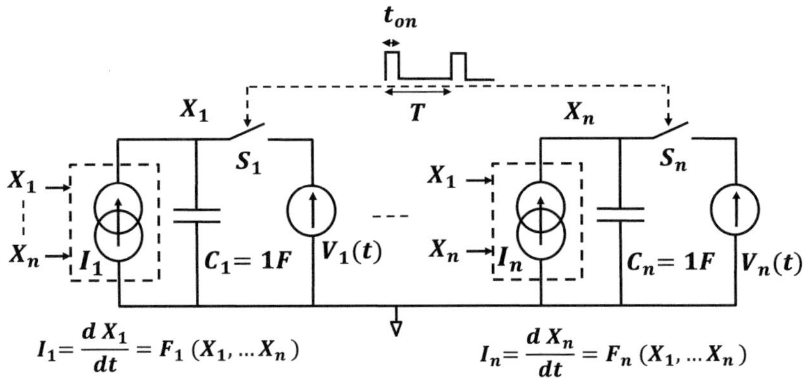

Fig. 2.11 Circuit for determining the stability region of a continuous system

$$\forall i \in (1, \ldots, n), I_i = C_i\frac{\mathrm{d}X_i}{\mathrm{d}t} = \frac{\mathrm{d}X_i}{\mathrm{d}t} = F_i(X_1, \ldots X_n). \tag{2.57}$$

At each period T, the n switches $(S_1, \ldots S_n)$ are closed during t_{on} charging each capacitor with an initial value fixed by $(V_1(t), \ldots V_n(t))$. After the pulse duration t_{on}, the n switches turn off and the system evolves, according to the dynamics given by the system S, towards the equilibrium, if the corresponding initial conditions are inside the basin of attraction. If these initial conditions are outside this basin, the simulation will diverge. T has to be long enough so that the system has the time to reach the equilibrium point or to diverge but not too long in order to reduce the simulation time. The choice of $(V_1(t), \ldots V_n(t))$ enables us to define which stability region to study. For a two-dimensional system, this area can be a rectangle or a circle. For a three-dimensional system, it can be a sphere or a cube. In Chap. 4, example 1 illustrates this type of use. The same approach can be used to find a stability region of a discrete system. Figure 2.12 shows an iterative flowchart.

With a given initial condition $(X_{10}, \ldots X_{n0})$ corresponding to the stability region to study, the discrete system evolves according to the set of n recursion series during N iterations. At the end of a cycle, the initial conditions change according to $(V_{1j}, \ldots V_{nj})$ describing the desired region. When $j = M$, the flowchart stops and allows us to determine whether this region is stable or not.

2.3.2 Determination of a Stability Area by Lyapunov Direct Theory

Here are some definitions. A continuous differentiable function:

$$V : \mathbb{R}^n \rightarrow \mathbb{R}^+ \tag{2.58}$$

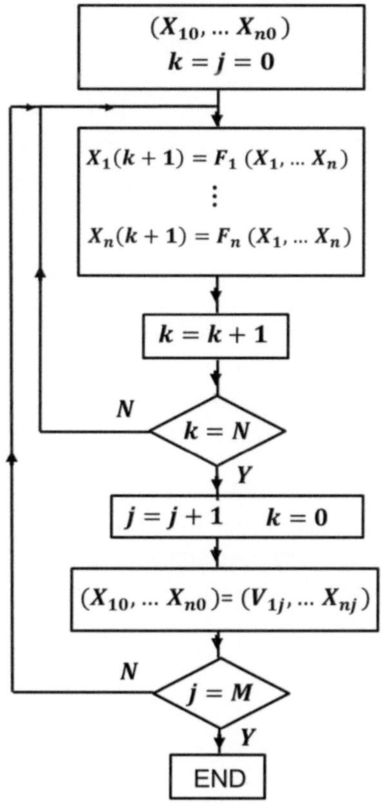

Fig. 2.12 Flowchart to determine the stability region of a discrete system

is said to be positive definite in a region G of $\mathbb{R}^n$ that contains the equilibrium point $(X_{1e}, \ldots X_{ne})$ if

$$V(X_{1e}, \ldots X_{ne}) = 0 \tag{2.59}$$

and

$$V(X_1, \ldots X_n) > 0 \quad \text{for } (X_1, \ldots X_n) \in G \backslash \{(X_{1e}, \ldots X_{ne})\}. \tag{2.60}$$

V is said to be positive semi-definite if

$$V(X_1, \ldots X_n) \geq 0 \quad \text{for } (X_1, \ldots X_n) \in G \backslash \{(X_{1e}, \ldots X_{ne})\}. \tag{2.61}$$

Conversely, if the condition Eq. (2.60) is replaced with $V(X_1, \ldots X_n) < 0$ then V is said to be negative definite. V is said to be negative semi-definite if $V(X_1, \ldots X_n) \leq 0$.

The Lyapunov direct method (Lyapunov, 1992) provides tools for studying the stability of a dynamical system around an equilibrium point. It allows the determination of a stability region without finding the trajectories (i.e. without solving the differential equation). The intuitive idea is to analyse a scalar function and its time derivative that can be understood both as a generalised energy function and the associated generalised dissipation. The process can be defined as follows: consider, as a first step, a family of closed surfaces in $\mathbb{R}^n$ defined by a function V called a Lyapunov function:

$$V(X_1, \ldots X_n) = K, \tag{2.62}$$

K being a constant. This function has to be null at the equilibrium point

$$V(X_{1e}, \ldots X_{ne}) = 0. \tag{2.63}$$

Moreover, if $K_1 < K_2$, then the corresponding surface $V(X_1, \ldots X_n) = K_1$ must be contained inside the surface defined by $V(X_1, \ldots X_n) = K_2$.
The second step consists of calculating the time derivative of the Lyapunov function using

$$\frac{\mathrm{d}V(X_1, \ldots X_n)}{\mathrm{d}t} = \frac{\partial V(X_1, \ldots X_n)}{\partial X_1}\frac{\mathrm{d}X_1}{\mathrm{d}t} + \cdots + \frac{\partial V(X_1, \ldots X_n)}{\partial X_n}\frac{\mathrm{d}X_n}{\mathrm{d}t}. \tag{2.64}$$

The last step consists of studying the sign of $V(X_1, \ldots X_n)$ and $\frac{\mathrm{d}V(X_1,\ldots X_n)}{\mathrm{d}t}$ in a region $G \subset \mathbb{R}^n$ defined by $V(X_1, \ldots X_n) \leq K$. The Lyapunov direct theory indicates that.

- If a Lyapunov function V is definite in sign in G and its derivative is definite with the opposite sign in G, then the system will be asymptotically stable in G.
- If V is definite in sign in G and its derivative is semi definite with the opposite sign in G, then the system will be stable in the sens of Lyapunov in G.
- If $V(X_1, \ldots X_n)$ is definite in sign in an area G and its derivative is definite with the same sign in G, then the system will be unstable.

These theorems are very useful to quickly determine a stability region around an equilibrium point. In Chap. 4, examples will be used to illustrate this theory.

2.3.3 *Theorem of Non-existence of Stability Limit Cycle*

This theorem allows to determine the non-existence of a stability limit cycle in a given sub space S_B of the space of state parameters (Llibre & Zhang, 2019; Naslin, 1962; Pelegrin et al., 1960). This theorem states that if the quantity

$$H(X_1, \ldots X_n) = \frac{\partial \dot{X}_1}{\partial X_1} + \cdots + \frac{\partial \dot{X}_n}{\partial X_n} = \frac{\partial F_1}{\partial X_1} + \cdots + \frac{\partial F_n}{\partial X_n} \tag{2.65}$$

has always the same sign in the sub space $S_B \subset \mathbb{R}^n$ then there is no stability limit cycle in S_B. If the sign of H changes in S_B no conclusion can be reached. Some examples of the use of this theorem are given in Chap. 4.

2.3.4 *Determination of Limit Stability Cycle with the Adjoint System*

As discuss previously a limit stability cycle is a solution for the system:

$$\frac{\mathrm{d}\vec{X}}{\mathrm{d}t} = \vec{F}\left(\vec{X}\right). \tag{2.66}$$

When such a cycle exists, a way to find it is to use the adjoint system. The adjoint system S_a is determined by replacing t with $-t$ in the system S. It allows going back in the past. Starting with a given initial condition, the adjoint system will change towards the limit stability cycle. So, the adjoint system can be expressed as:

$$S_a : \begin{cases} \frac{\mathrm{d}X_1}{\mathrm{d}t} = -F_1(X_1, \ldots, X_i, \ldots, X_n) \\ \vdots \\ \frac{\mathrm{d}X_j}{\mathrm{d}t} = -F_j(X_1, \ldots, X_i, \ldots, X_n) \\ \vdots \\ \frac{\mathrm{d}X_n}{\mathrm{d}t} = -F_n(X_1, \ldots, X_i, \ldots, X_n) \end{cases} . \tag{2.67}$$

Unfortunately, this simple method to determine the basin of attraction is not always applicable. But for each study, we can try to use it as a first step. Examples of successful uses of adjoint systems are given in Chap. 12.

Chapter 3
Hybrid and Chaotic Systems

In this chapter, we will discuss some definitions and properties of hybrid and chaotic systems (Goebel et al., 2012; van der Schaft & Schumacher, 2000).

3.1 Hybrid Systems

3.1.1 Definition

A hybrid system is characterised by interactions between continuous dynamics and discrete events (Schutter, 2005). These systems are very common in a large number of applications areas, such as robotics, air traffic control, electrical networks, power converters, and manufacturing. The generic mathematical model of a hybrid system can be represented by:

$$H = (Q, X, \text{Init}, \text{Dom}, U, F, E, G, R), \tag{3.1}$$

where

- Q is a set of n_Q discrete states.
- X is a set of n dimensional continuous state vectors $\vec{X} : \mathbb{R} \to \mathbb{R}^n$.
- Init $\in Q \times \mathbb{R}^n$ is a set of $1 + n$ initial conditions.
- $Dom = \{D_q, q \in Q\}$ is a set of domains of $\mathbb{R}^n$: $\forall q \in QD_q \subset \mathbb{R}^n$.
- $U = \left\{\overrightarrow{U_q}, q \in Q\right\}$ is a set of u dimensional continuous control vectors,

$$\forall q \in Q \quad \overrightarrow{U_q} : \mathbb{R} \to \mathbb{R}^u.$$

B. Rivet, *Theories of Nonlinear Systems for Power Converters*,
https://doi.org/10.1007/978-3-032-25062-9_3

- $F = \left\{\overrightarrow{F_q}, q \in Q\right\}$ is a set of n dimensional vector fields on $\mathbb{R}^{n+u}$ describing the dynamics of the system. If the system is in the $q \in Q$ discrete state we have: $\overrightarrow{F_q} : \mathbb{R}^{n+u} \to \mathbb{R}^n$

$$\left(\vec{X}(t), \overrightarrow{U_q}(t)\right) \to \vec{F}\left(\vec{X}(t), \overrightarrow{U_q}(t)\right) = \frac{\mathrm{d}\vec{X}}{\mathrm{d}t}$$

- $E \subset Q \times Q$ is a set of discrete transitions $e = (q, q')$.
- $G = \{G_e, e \in E\}$ is a set of guards that prescribe when a discrete state transition occurs.$\forall e = (q, q') \in E : G_e \subset D_q$.
- $R = \{R_e, e \in E\}$ is a set of reset state functions. If $e = (q, q')$, the reset function $R_e : D_q \to D_{q'}$ allows the discontinuous switching of the state parameters when the system transitions from the discrete state q to the state q'.

3.1.2 Change in Hybrid Systems

A hybrid system is characterised at each time t by the changes in the discrete and continuous state vectors. Discrete and continuous state parameters constitute the hybrid state vector $\left(q(t), \vec{X}(t)\right)$. The discrete parameter is constant between two successive switching times t_i,t_{i+1}, i.e. $q(t)$ is constant in the interval:

$$I_i = \left[t_i, t_{i+1}\right], \quad 1 \le i \le N - 1, \tag{3.2}$$

where N is the number of commutations. Execution of a hybrid system can be represented by a sequence of time intervals I_i during which the hybrid vector $\left(q, \vec{X}\right)$ changes. At $t = 0$, the initial hybrid vector is $\left(q_0, \vec{X_0}\right) \in \text{Init}$. For $t \in I_0 = [t_0, t_1]$, q_0 being constant, the hybrid system evolves according to $\overrightarrow{F_{q_0}}$. The evolution can continue until the continuous vector $\vec{X}(t)$ reaches the guard $G_e(q_0, q_1)$, or until a discrete event switch q_0 to the discrete state q_1. At $t = t_1$, the continuous state vector can change according to the reset function $R_e(q_0, q_1)$ and the whole process is repeated.

Limit cycles are common in hybrid systems. The stability of such cycles can be studied using the tools given in Chap. 2. Stability analysis examples will be considered in Chap. 4. This definition of a hybrid system is very general and covers all types of hybrid systems. In this book, we will focus on two specific classes of hybrid systems: autonomous hybrid systems and switched hybrid systems.

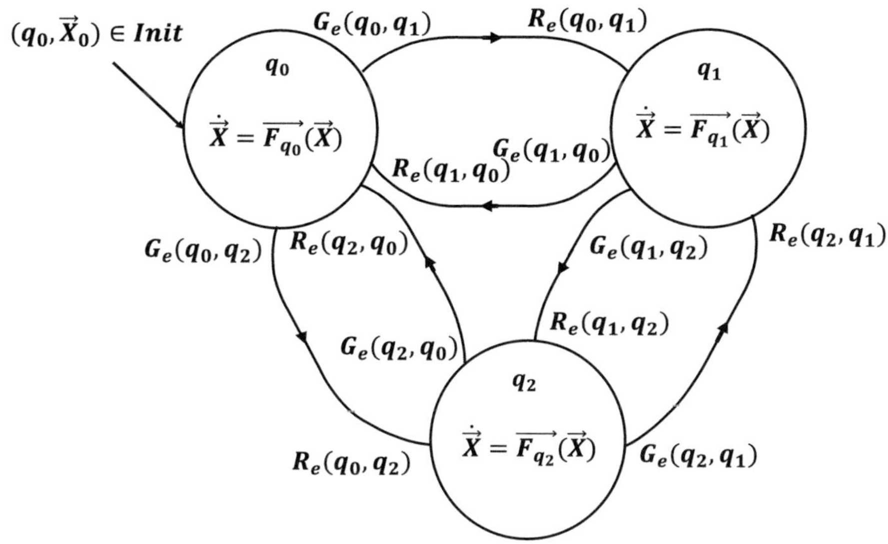

Fig. 3.1 Representation of an autonomous hybrid system

3.1.3 Autonomous Hybrid System

In an autonomous hybrid system (Obaro et al., 2018), there is no control vector.

The commutation between two discrete states is defined only by $G = \{G_e, e \in E\}$. An autonomous hybrid system can be represented by the type of diagram shown in Fig. 3.1

3.1.4 Example of a Bouncing Ball

Consider, as a first example of an autonomous hybrid system, the bouncing ball system (Sanfelice, 2011). The two continuous state parameters ($n = 2$) are X_1, the vertical position of the ball and X_2, its vertical velocity. The motion of the ball is governed by Newton's laws. The dynamics are given by:

$$\begin{cases} \frac{dX_1}{dt} = X_2 \\ \frac{dX_2}{dt} = -mg \end{cases} \tag{3.3}$$

m being the mass of the ball and g the gravitational acceleration.

Figures 3.2 and 3.3 represent, respectively, the bouncing ball system and the changes in the state parameters with respect to time. The ball bounces when $X_1 = 0$ with $X_2 < 0$. With each bounce, the ball loses a fraction of its energy. This can be described

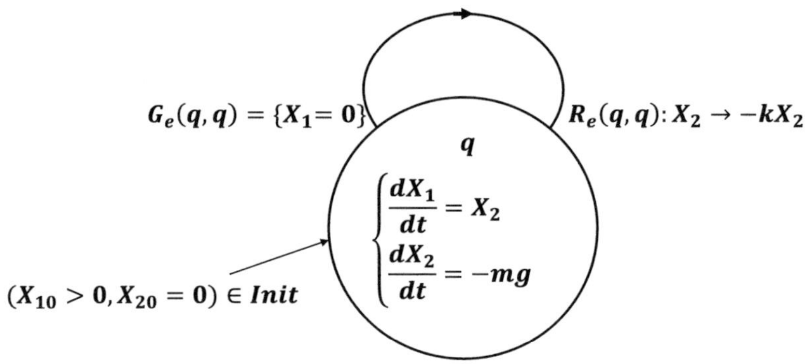

Fig. 3.2 Bouncing ball system

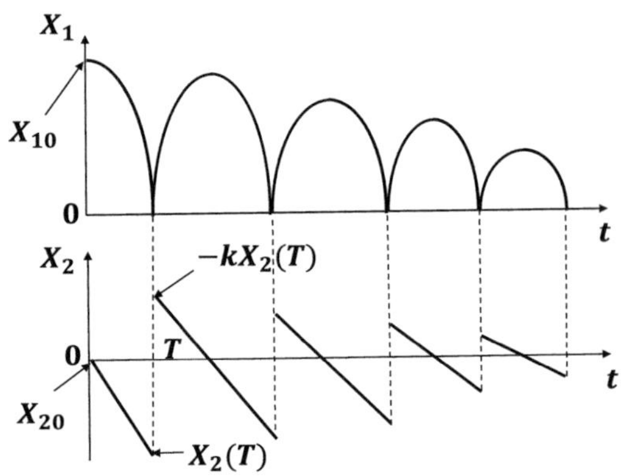

Fig. 3.3 Vertical position and velocity of the bouncing ball

by the reset function $R_e : (0, X_2) \to (0, -kX_2)$ with $k \in [0, 1]$. Here, there is one discrete state $(n_Q = 1)$ and no control parameters.

3.1.5 Example of a Thermostat

The second autonomous hybrid example is the control system of a room's temperature T_{room} (Alur et al., 1993). A thermostat is used to control T_{room}. The thermostat turns on a radiator when $T_{\text{room}} < T_{\text{roommin}}$ and turns the radiator off when $T_{\text{room}} > T_{\text{roommax}}$. Here we have two discrete parameters q_{on} and q_{off} $(n_Q = 2)$, corresponding to the radiator on or off and associated, respectively, with the continuous domains:

$$D_{q\text{off}} = [T_{\text{roommin}}, \infty[\text{and } D_{qo} =]-\infty, T_{\text{roommax}}]. \tag{3.4}$$

The only continuous state parameter is $T_{\text{room}} \cdot F$ is defined by:

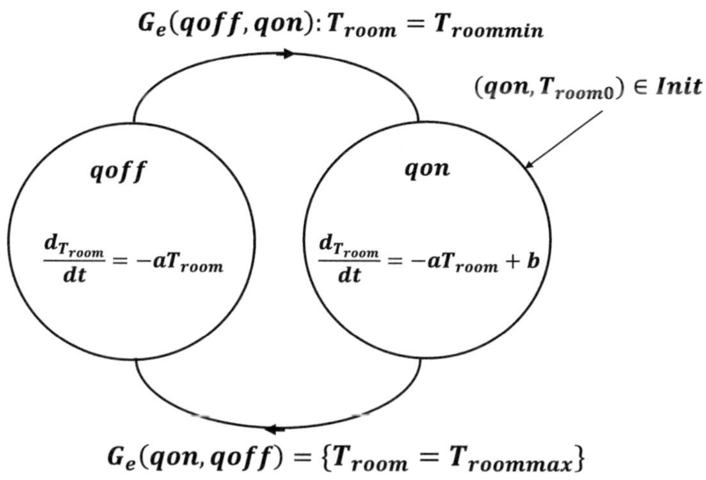

Fig. 3.4 Thermostat system

- $F_{q\text{off}} : D_{q\text{off}} \to \mathbb{R}$

$$T_{\text{room}} \to F_{q\text{off}}(T_{\text{room}}) = -aT_{\text{room}} \tag{3.5}$$

- $F_{q\text{on}} : D_{q\text{on}} \to \mathbb{R}$

$$T_{\text{room}} \to F_{q\text{on}}(T_{\text{room}}) = -aT_{\text{room}} + b \tag{3.6}$$

a, b are constant system parameters. The set of discrete transitions is

$$E = \{(q_{\text{on}}, q_{\text{off}}), (q_{\text{off}}, q_{\text{on}})\}. \tag{3.7}$$

Figure 3.4 illustrates the behaviour of this autonomous hybrid system with the guards $G_e(q_{\text{off}}, q_{\text{on}})$ and $G_e(q_{\text{on}}, q_{\text{off}})$.
In this example there is no discontinuity of the state parameters, so, the reset functions are not useful. Figure 3.4 illustrates the behaviour of this autonomous hybrid system and the Fig. 3.5 shows the evolution of T_{room}.

3.1.6 Switched Hybrid System

The switched hybrid systems considered here are defined by a n dimensional continuous state vector $\vec{X}(t)$ and a u dimensional Boolean vector $\vec{U}(t)$. The state parameter will be continuous in time (no need of reset functions). The dynamics of these systems are given by:

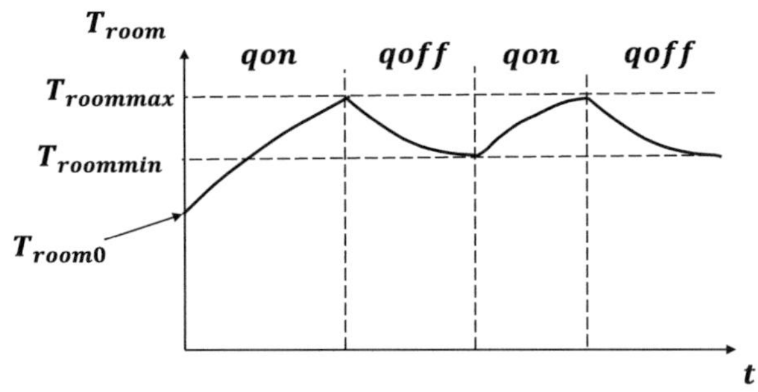

Fig. 3.5 Temporal evolution of the room temperature

$$\frac{d\vec{X}}{dt} = F\vec{X} + G\left(\vec{X}\right)\vec{U}. \tag{3.8}$$

F and G are, respectively, a $n \times n$ and a $n \times u$ matrix. The system can be written in matrix form as follows:

$$\begin{pmatrix} \dot{X}_1(t) \\ \vdots \\ \dot{X}_i(t) \\ \vdots \\ \dot{X}_n(t) \end{pmatrix} = \begin{pmatrix} a_{11} & \ldots & a_{1j} & \ldots & a_{1n} \\ \vdots & \ddots & \vdots & \ddots & \vdots \\ a_{i1} & \ldots & a_{ij} & \ldots & a_{in} \\ \vdots & \ddots & \vdots & \ddots & \vdots \\ a_{n1} & \ldots & a_{nj} & \ldots & a_{nn} \end{pmatrix} \cdot \begin{pmatrix} X_1(t) \\ \vdots \\ X_i(t \\ \vdots \\ X_n(t \end{pmatrix}$$
$$+ \begin{pmatrix} G_{11}\left(\vec{X}\right) & \ldots & G_{1j}\left(\vec{X}\right) & \ldots & G_{1u}\left(\vec{X}\right) \\ \vdots & \ddots & \vdots & \ddots & \vdots \\ G_{i1}\left(\vec{X}\right) & \ldots & G_{ij}\left(\vec{X}\right) & \cdots & G_{iu}\left(\vec{X}\right) \\ \vdots & \ddots & \vdots & \ddots & \vdots \\ G_{n1}\left(\vec{X}\right) & \cdots & G_{nj}\left(\vec{X}\right) & \ldots \vdots & G_{nu}\left(\vec{X}\right) \end{pmatrix} \begin{pmatrix} U_1(t) \\ \vdots \\ U_j(t) \\ \vdots \\ U_u(t) \end{pmatrix}. \tag{3.9}$$

The a_{ij} are real coefficients and the $G_{ij}\left(\vec{X}\right)$ are linear functions of $\vec{X}$.

3.1.7 *Example of a Two-Tank System with One Valve*

The two-tank system in Fig. 3.6 can be used to illustrate the behaviour of a switched hybrid system (Furman, 2011). The system consists of two tanks and one valve. The valve, with the single Boolean control U_1 dumps the constant inflow Q_{IN} in tank 1. The system comprises two discrete states q_1 and q_2 $\left(n_Q = 2\right)$. The two state parameters are the water levels $(X_1, X_2)(n = 2)$, with:

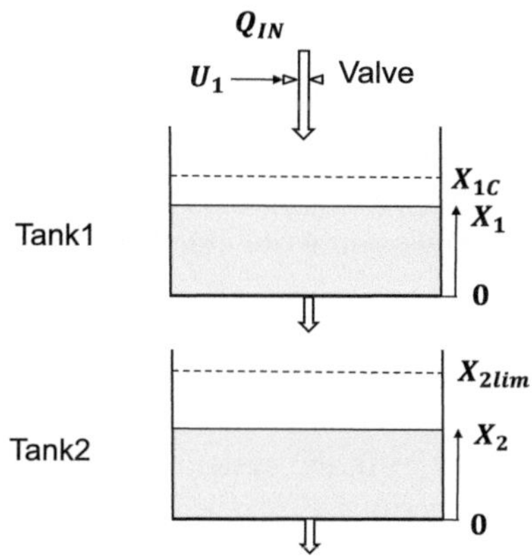

Fig. 3.6 Two-tank system with one valve

$$0 \leq X_1 \leq X_{1C}, 0 \leq X_2 \leq X_{2\text{lim}}. \tag{3.10}$$

E is defined by:

$$E = \{(q_1, q_2), (q_2, q_1)\}. \tag{3.11}$$

The Boolean control U_1 is set to 1 at the beginning of each period T, as illustrated in Fig. 3.7, and reset to 0 when X_1 reaches X_{1C}.
If $U_1 = 1$, the valve is open and the dynamics of the system are given by:

$$\begin{cases} \frac{\mathrm{d}X_1}{\mathrm{d}t} = aQ_{\text{IN}} - bX_1 \\ \frac{\mathrm{d}X_2}{\mathrm{d}t} = cX_1 - bX_2 \end{cases} \tag{3.12}$$

When $U_1 = 0$, the valve is closed and the dynamics become:

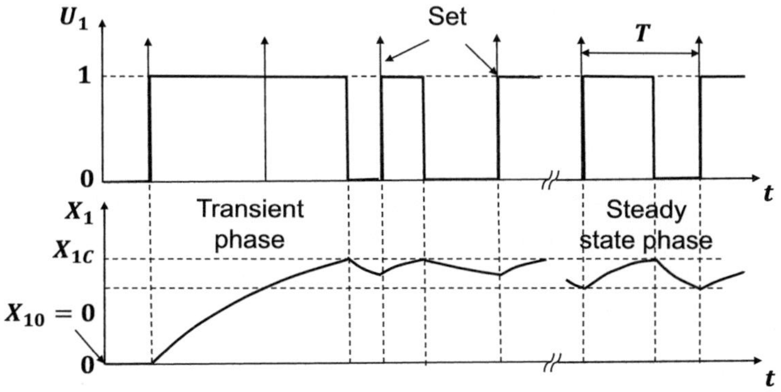

Fig. 3.7 Chronogram of X_1 and U_1

$$\begin{cases} \frac{dX_1}{dt} = -bX_1 \\ \frac{dX_2}{dt} = cX_1 - bX_2 \end{cases} \quad (3.13)$$

a, b, c are constant system parameters. In this example, the reset function is not useful because there is no discontinuity in the state parameters. Thanks to Eqs. (3.12) and (3.13) we can write more formally:

$$\begin{pmatrix} \dot{X}_1(t) \\ \dot{X}_2(t) \end{pmatrix} = \begin{pmatrix} -b & 0 \\ c & -b \end{pmatrix} \begin{pmatrix} X_1(t) \\ X_2(t) \end{pmatrix} + \begin{pmatrix} aQ_{IN} \\ o \end{pmatrix} U_1. \quad (3.14)$$

At $t = 0$, the system starts with the initial conditions $(X_{10}, X_{20}) = (0, 0)$. After some periods T, it reaches the steady state as shown in Fig. 3.7.

3.2 Chaotic Systems

3.2.1 Definition and Properties

Chaos theory, alongside relativity and quantum mechanics, is considered one of the major foundational theories of twentieth century physics. Historically, the study of chaos began in mathematics and physics. It then expanded into engineering, biology and economics. This theory analyses the behaviour of complex systems that are highly sensitive to initial conditions. This is popularly known as the "butterfly effect", the flap of a butterfly's wings in Brazil sets off a tornado in Texas (Strogatz, 2018). Although chaotic systems are deterministic, meaning that their future behaviour is perfectly determined by the initial conditions, they are unpredictable over the long term. The initial conditions can be known only to a finite degree of precision. They belong to a class of specific nonlinear systems (a linear system cannot be chaotic). As in section one, let us consider continuous and discrete systems defined by the following dynamics:

$$\frac{d\vec{X}}{dt} = \vec{F}\left(\vec{X}\right), \vec{X}_{k+1} = \vec{F}\left(\vec{X}_k\right). \quad (3.15)$$

To be chaotic, a continuous and autonomous system has to be governed by a set of at least three nonlinear differential equations. But a discrete system can be chaotic with a minimum order equal to one. A chaotic system also has an element of regularity since it is composed of a large number of pseudo-periodic orbits. A chaotic system can only be analysed by observing its responses over a large set of different initial conditions. There is currently no universally accepted mathematical definition of chaos. The behaviour of a system is referred to as chaotic if the trajectory of the state parameters has the three following properties.

- It should show high sensitivity to initial conditions (small changes can lead to radically different behaviour in the final state).
- When the system, at time t_1, reaches a given point in the state parameter space, it will not intersect this point again for any $t > t_1$.
- A trajectory will eventually reach all the points of the strange attractor. More formally, if we consider two open intervals of none zero length, a point from one interval will always reach another point in the other interval, given sufficient time.

3.2.2 *Strange Attractor*

The attractor of a chaotic system is called strange. A strange attractor is a sub-space of the state space inside which the state parameters of the chaotic system change. Figure 3.8 shows the strange attractor of the Rossler system (Ott, 2002; Rossler, 1976).

This system is defined by the set of nonlinear differential equations (3.16) where $a = 0.2, b = 0.4, c = 5.7$ are parameters of the Rossler system. In this simple example, there is only one nonlinear term X_1X_3.

$$\begin{cases} \frac{dX_1}{dt} = -(X_2 + X_3) \\ \frac{dX_2}{dt} = X_1 + aX_2 \\ \frac{dX_3}{dt} = b + X_1X_3 - cX_3 \end{cases} . \tag{3.16}$$

For some initial conditions (for example $X_1 = 0, X_2 = -4, X_3 = 0$) corresponding to Fig. 3.8, this system exhibits a strange attractor. The trajectory never converges to a stable point or a periodic orbit. Figure 3.8 shows the chaotic behaviour of this system. It should be noted that for other initial conditions, the same system can be periodic or divergent.

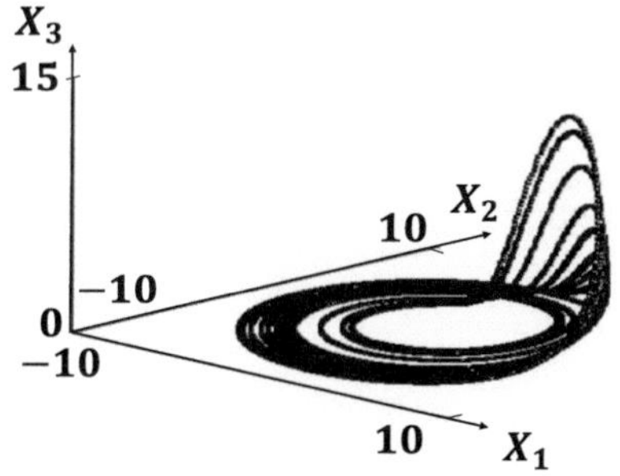

Fig. 3.8 Strange attractor of the Rossler system

3.2.3 Example of a Chaotic Poincaré Map

The concept of a Poincaré map is explained in Para. 1.11. We will now give an example of the determination of a Poincaré map for a chaotic system (Roussel, 2005). Consider the order three chaotic system defined by the following differential equations:

$$\begin{cases} \frac{dX_1}{dt} = 0.154(65 + X_3) - X_1X_2^2 - X_1 \\ \frac{dX_2}{dt} = 200(X_1X_2^2 + X_1 - X_2) \\ \frac{dX_3}{dt} = 50(X_2 - X_3) \end{cases} . \tag{3.17}$$

Figure 3.9 shows a projection of the strange attractor of this system on the plane (X_1, X_2) simulated with the initial condition (0, 12, 0).
There is a hole in the middle of this attractor. Consider a half plane defined by:

$$X_2 = 16 \tag{3.18}$$

starting inside this hole and moving towards higher values of X_1. This half plane corresponds to the Poincare section. Obviously, we decide moving to the bottom because there is a larger dispersion of X_1. in this direction. Generally, we need to experiment with a few Poincaré sections before finding the most interesting one.

A Poincaré section is not necessarily orthogonal to a coordinate axis. Very often, it is more convenient to use a section of surface that has some other orientation in phase space. The Poincaré map is represented in Fig. 3.10. It is obtained by eliminating the transient phase to be sure the state parameters have reached the strange attractor. The simulation was done for times between 100 and 150.

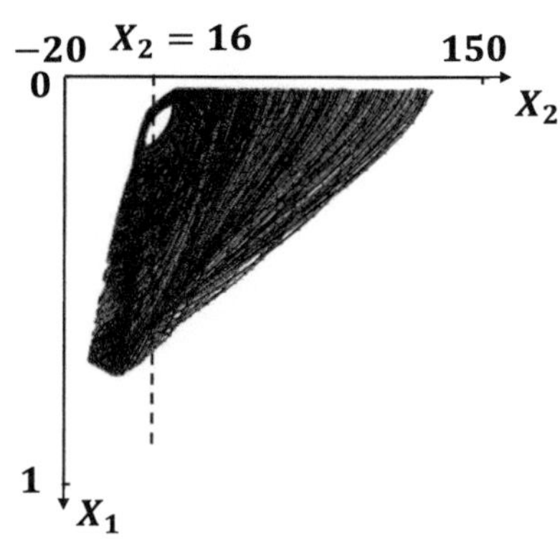

Fig. 3.9 Projection of the strange attractor onto the plane (X_1, X_2)

Fig. 3.10 Poincaré map in the plane $X_2 = 16$

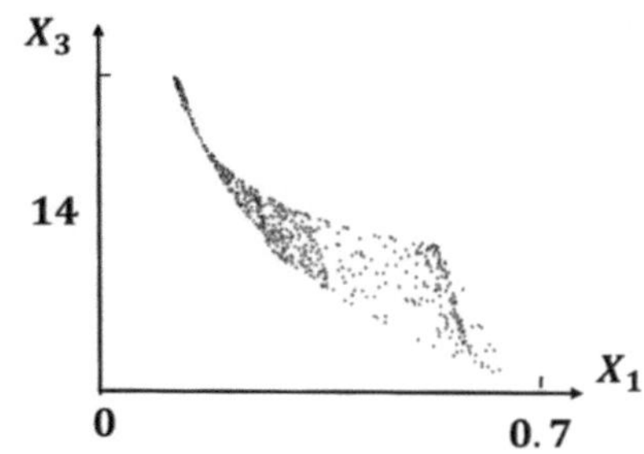

3.2.4 Bifurcation Diagram

Nonlinear systems are often subject to a phenomenon called bifurcation. The dynamics of a system can change with the change in one parameter. In a bifurcation diagram the bifurcation parameter is plotted on the abscissa and the state parameter is plotted on the ordinate.

A simple way to illustrate a bifurcation diagram is to look at a discrete system, for instance (Manneville, 2005):

$$X_{k+1} = rX_k(1 - X_k), \tag{3.19}$$

where r is a system parameter. The corresponding bifurcation diagram is represented in Fig. 3.11. This diagram is drawn for a specific initial condition. It is obtained by iterating many times at a certain r parameter value with the initial condition $X_0 = 0.5$. The first iterations are discarded to eliminate transient conditions and the last ones are plotted. For low values of r the attractor is a fixed point. The diagram displays period doubling as r increases. For higher values of r, the system becomes chaotic. The second example in this section is the well-known mathematical model for atmospheric convection developed by Edward Lorenz in 1963 (Lorenz, 1963). This system is defined by the following set of equations:

$$\begin{cases} \frac{\mathrm{d}X_1}{\mathrm{d}t} = 10(-X_1 + X_2) \\ \frac{\mathrm{d}X_2}{\mathrm{d}t} = RX_1 - X_2 - X_1X_3 \\ \frac{\mathrm{d}X_3}{\mathrm{d}t} = -\frac{8}{3}X_3 + X_1X_2 \end{cases}. \tag{3.20}$$

For $0 < R < 1$, there is only one equilibrium point(0, 0, 0). At $R = 1$ a bifurcation appears with two different equilibrium points. The bifurcation diagram shown in Fig. 3.12 is simulated with the initial condition (20, 10, 10). In this particular case, the equilibrium point changes abruptly at R approximately 14.4. For each value of the parameter R, there exists a specific basin of attraction associated with each attractor. For higher values of R, the system becomes chaotic.

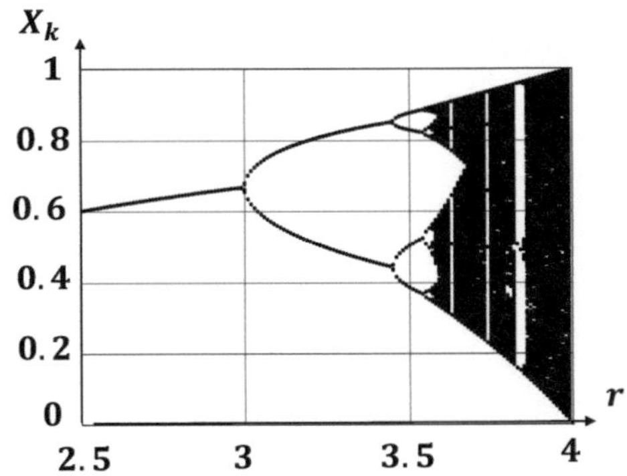

Fig. 3.11 Bifurcation diagram of $X_{k+1} = rX_k(1 - X_k)$

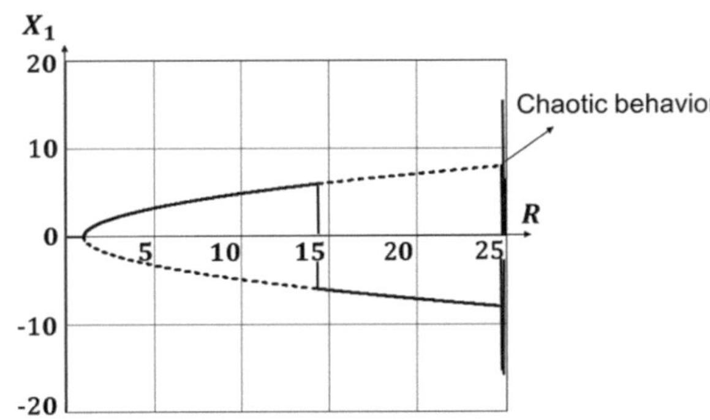

Fig. 3.12 Bifurcation diagram of Lorenz system

Figure 3.13 shows that the attractors for $R = 0.5$ and $R = 2$ are, respectively, the equilibrium points (0, 0, 0) and (1.63, 1.63, 1). For $R = 25$, we find the conventional strange attractor of the Lorenz system.

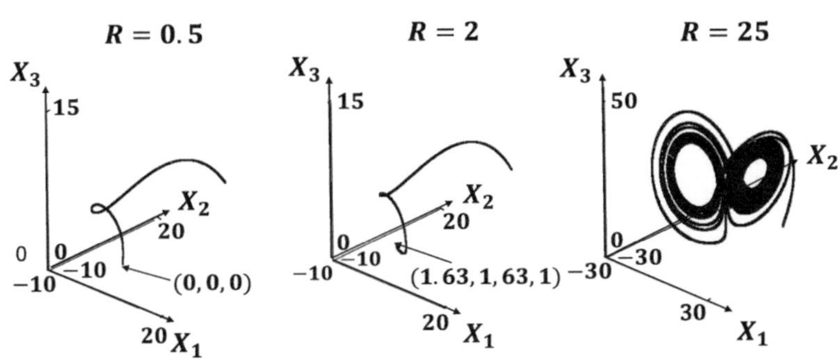

Fig. 3.13 Attractors of Lorenz system for $R = 0.5$, $R = 2$ and $R = 25$

3.2.5 Lyapunov Spectrum of Classical Chaotic Systemsa

As previously explained, a positive Lyapunov exponent determines a chaotic behaviour of the considered system. Here below are the Lyapunov spectra of Hénon, Rossler and Lorenz systems for some parameters system values (Wolf et al., 1985).

- Hénon system

$$\begin{cases} X_{1(k+1)} = 1 - aX_{1k}^2 + X_{2k} \\ X_{2(k+1)} = bX_{1k} \end{cases} \begin{cases} a = 1.4 \\ b = 0.3 \end{cases} \begin{cases} \lambda_1 = 0.603 \\ \lambda_2 = -2.34 \end{cases}, \tag{3.21}$$

- Rossler system

$$\begin{cases} \frac{dX_1}{dt} = -(X_2 + X_3) \\ \frac{dX_2}{dt} = X_1 + aX_2 \\ \frac{dX_3}{dt} = b + X_1X_3 - cX_3 \end{cases} \begin{cases} a = 0.15 \\ b = 0.2 \\ c = 10 \end{cases} \begin{cases} \lambda_1 = 0.13 \\ \lambda_2 = 0 \\ \lambda_3 = -14.1 \end{cases}, \tag{3.22}$$

- Lorenz system

$$\begin{cases} \frac{dX_1}{dt} = a(X_1 - X_2) \\ \frac{dX_2}{dt} = RX_1 - X_2 - X_1X_3 \\ \frac{dX_3}{dt} = -bX_3 + X_1X_2 \end{cases} \begin{cases} a = 16 \\ R = 45.92 \\ b = 4 \end{cases} \begin{cases} \lambda_1 = 2.16 \\ \lambda_2 = 0 \\ \lambda_3 = -32.4 \end{cases}. \tag{3.23}$$

Chapter 4
Examples of Nonlinear Systems

We will now examine some nonlinear systems to illustrate the concepts presented in the first three chapters.

4.1 Stability of a Material Point in a Potential Field

4.1.1 System Description

In this paragraph, we consider a material point M having a mass m and moving only along the X-axis. It is placed in the potential field $V(X)$, as illustrated in Fig. 4.1. $V(X)$ is defined by:

$$V(X) = X^2 - 0.005X^4. \tag{4.1}$$

This point is slowed down by a frictional force characterised by the coefficient λ. By applying Newton's law, we obtain:

$$m\frac{\mathrm{d}^2X}{\mathrm{d}t} = -\frac{\mathrm{d}V(X)}{\mathrm{d}X} - \lambda\frac{\mathrm{d}X}{\mathrm{d}t}. \tag{4.2}$$

The two state parameters ($n = 2$) are $X_1 = X$ and $X_2 = \frac{\mathrm{d}X}{\mathrm{d}t} = \dot{X}$, and the two system parameters ($p = 2$) are λ and m. In this example, we do not take the units into account and we will set $\lambda = 0.5$ and $m = 1$. The system S becomes:

$$S : \begin{cases} \frac{\mathrm{d}X}{\mathrm{d}t} = \dot{X} = F_1(X, \dot{X}) \\ \frac{\mathrm{d}\dot{X}}{\mathrm{d}t} = -2X + 0.02X^3 - 0.5\dot{X} = F_2(X, \dot{X}) \end{cases}. \tag{4.3}$$

B. Rivet, *Theories of Nonlinear Systems for Power Converters*,
https://doi.org/10.1007/978-3-032-25062-9_4

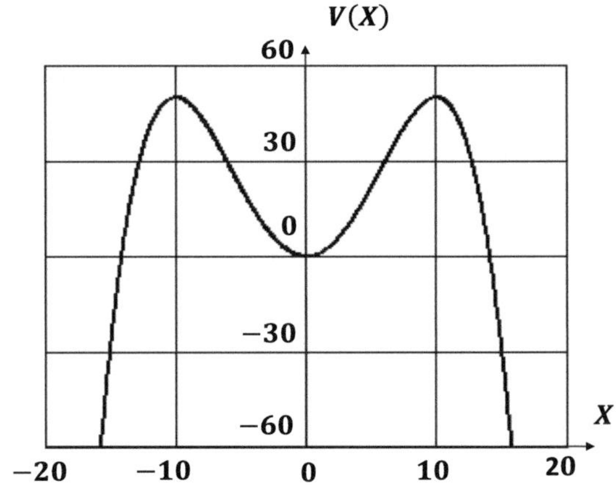

Fig. 4.1 Potential field: $V(X)$

4.1.2 Equilibrium Points

The equilibrium points are calculated by solving the associated system S_e. This mechanical system has three equilibrium points given by:

$$\left\{\left(X_{ei}, \dot{X}_{ei}\right), i \in \{1, 2, 3\}\right\} = \{(-10, 0), (0, 0), (10, 0\}. \tag{4.4}$$

4.1.3 Linearisation Around the Equilibrium Point

For each equilibrium point $\left(X_{ei}, \dot{X}_{ei}\right)$, we consider small variations $(x, \dot{x})$ around it. By replacing $X = X_{ei} + x$ in S and using the linearisation techniques described in Para. 1.8, we get the following linearised system around $\left(X_{ei}, \dot{X}_{ei}\right)$:

$$\begin{cases} \frac{\mathrm{d}x}{\mathrm{d}t} = \dot{x} \\ \frac{\mathrm{d}\dot{x}}{\mathrm{d}t} = -2x + 0.06X_{ei}^2 x - 0.5\dot{x} \end{cases}. \tag{4.5}$$

Using the Jacobian matrix J, it can be expressed by:

$$\begin{pmatrix} \dot{x} \\ \ddot{x} \end{pmatrix} = \begin{pmatrix} 0 & 1 \\ -2 + 0.06X_{ei}^2 & -0.5 \end{pmatrix} \begin{pmatrix} x \\ \dot{x} \end{pmatrix}. \tag{4.6}$$

4.1.4 Local Stability Around Each Equilibrium Point

As described in Para. 2.1.1, the first step consists of solving the characteristic equation:

$$\det(Is - J) = s^2 + 0.5s + 2 - 0.06X_{ei}^2 = 0. \tag{4.7}$$

For the equilibrium point $\left(X_{e2}, \dot{X}_{e2}\right) = (0, 0)$, the two roots are:

$$\begin{aligned} s_1 &= \frac{-0.5 + i\sqrt{7.75}}{2}, \\ s_2 &= \frac{-0.5 - i\sqrt{7.75}}{2}. \end{aligned} \tag{4.8}$$

The real parts of these two roots are negative. Therefore, the equilibrium point $\left(X_{e2}, \dot{X}_{e2}\right)$ is asymptotically stable. If, at $t = 0$, a small disturbance occurs when M is stabilised at this point, the system will return to this equilibrium point with damped oscillations given by:

$$x(t) = Ke^{-0.25t} \sin\left(\sqrt{1.939}t + \varphi\right). \tag{4.9}$$

For the equilibrium points $\left(X_{e1}, \dot{X}_{e1}\right) = (-10, 0)$ and $\left(X_{e3}, \dot{X}_{e3}\right) = (10, 0)$, the two roots are:

$$\begin{aligned} s_1 &= \frac{-0.5 + \sqrt{16.25}}{2}, \\ s_2 &= \frac{-0.5 - \sqrt{16.5}}{2}. \end{aligned} \tag{4.10}$$

one of these two roots is positive. Therefore, these two equilibrium points are locally unstable.

4.1.5 Stability Limit Cycle

In this example H is:

$$H = \frac{\partial \dot{X}}{\partial X} + \frac{\partial\left(\frac{d\dot{X}}{dt}\right)}{\partial \dot{X}} = -0.5. \tag{4.11}$$

H is always negative, so there is no limit cycle.

4.1.6 Stability Area Determined with Lyapunov Theory

In this paragraph, we will determine a stability area around the stable point (0, 0). Consider the Lyapunov function $E(X, \dot{X})$:

$$E(X, \dot{X}) = \frac{1}{2}\dot{X}^2 + V(X) = \frac{1}{2}\dot{X}^2 + X^2 - 0.005X^4. \tag{4.12}$$

E represents the total energy of the material point M. It is the sum of the kinetic and potential energies. As was said in Para. 2.3.2, it is not necessary that the Lyapunov function be an energy function. For $0 < E < 50$, $E(x, \dot{x})$ meets the following conditions to be eligible as a Lyapunov function

- $E(0, 0) = 0$.
- For $0 < E < 50$, the corresponding surface in the portrait plan $(X, \dot{X})$ defined by $E(X, \dot{X}) < E$ is a closed surface (see Fig. 4.2).
- For each $0 < E < 50$, if $E_1 < E_2$, the surface corresponding to E_1 will be contained inside the surface corresponding to E_2, as shown in Fig. 4.2.

For each $(X, \dot{X}) \neq (0, 0)$, $E(X, \dot{X}) > 0$. The next step is the calculation of $\dot{E}(X, \dot{X})$. We get:

$$\dot{E}(X, \dot{X}) = \frac{dE}{dt} = \frac{\partial E}{\partial X}\frac{dX}{dt} + \frac{\partial E}{\partial \dot{X}}\frac{d\dot{X}}{dt} = -0.5\dot{X}^2. \tag{4.13}$$

$\dot{E}(0, 0)$ being equal to zero and $\dot{E}(X, \dot{X}) < 0$ for $(X, \dot{X}) \neq (0, 0)$, we can conclude, thanks to Lyapunov's theory, that:

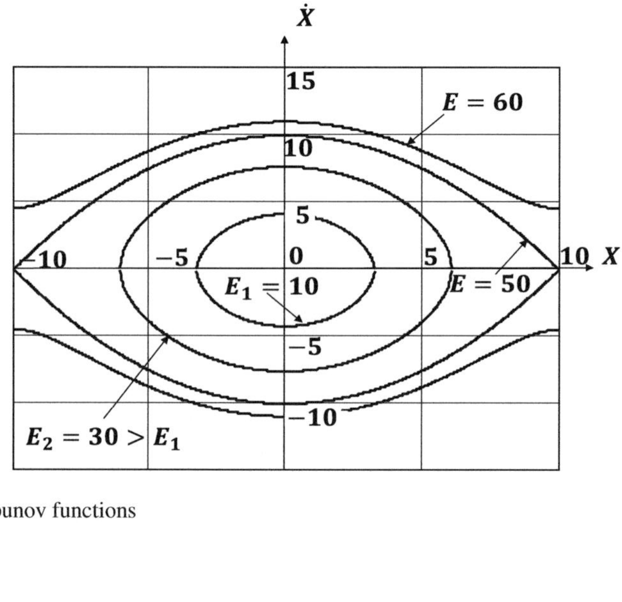

Fig. 4.2 Lyapunov functions

$$E(X,\dot{X}) = \frac{1}{2}\dot{X}^2 + X^2 - 0.005X^4 = 50 \tag{4.14}$$

delimitates a stable region. For each initial position and speed inside this area, the material point return to the equilibrium point (0, 0). This area is not the basin of attraction.

4.1.7 Greater Stability Area Determined by a Circle

In this paragraph, we will determine the largest stability region defined by a circle of radius R centred at the equilibrium point (0, 0).

Figure 4.3 shows the schematic used for the simulation. The idea of this simulation is to change the initial conditions of X and $\dot{X}$ every 200s. These initial conditions are determined by the voltage sources V_1 and V_2, which charge C_1 and C_2 when S_1 and S_2 are switched on. V_1 and V_2, are, respectively, sinusoidal and cosinusoidal voltage generators with a magnitude equal to R. The initial conditions form a circle of radius R, centred at the equilibrium point $(X_{e2}, \dot{X}_{e2}) = (0, 0)$.

After the initial conditions are stabilised,S_1 and S_2 are turned off for a period of 150 s. During this period, the system has sufficient time to either converge towards (0, 0) or diverge. The period of V_1 and V_2 (20,000 s) must be much greater than the switching period of S_1 and S_2 (200 s) in order to draw the circle using numerous points. Figure 4.4 shows the results of the simulations.

For all the initial conditions moving on the circle of radius $R = 9$, the system converges towards (0, 0). For $R = 9.1$, the system diverges after some iterations. The greater radius corresponding to the maximum stability area defined by a circle is between 9 and 9.1.

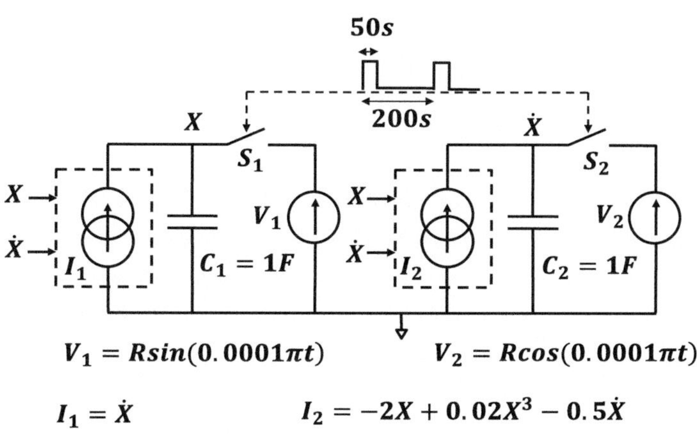

Fig. 4.3 Circuit simulation

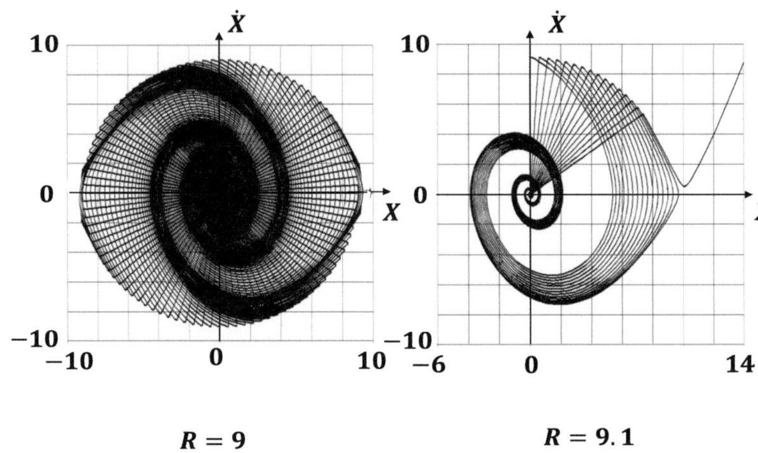

Fig. 4.4 Simulation with $R = 9$ and $R = 9.1$

4.1.8 Maximum Stability Area

The basin of attraction (maximum stability area) can be found by simulation. We can fix the initial condition of X and determine the initial condition of $\dot{X}$ that corresponds to the stability boundary. Figure 4.5, represents the basin of attraction of this system. It is not a closed area. On the same graph, the stability regions defined by the Lyapunov function $E = 50$ and by the circle of radius $R = 9$ centred at the equilibrium point $(0, 0)$, are represented. These two regions are closed. Naturally, the circle is tangential to the basin of attraction.

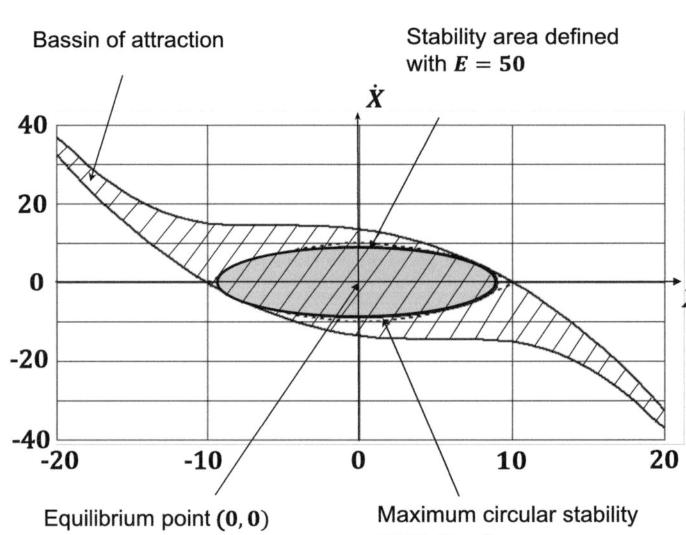

Fig. 4.5 Stability areas

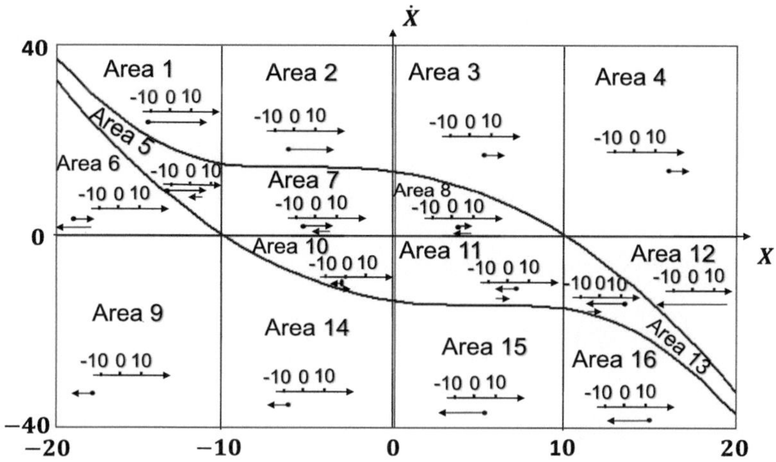

Fig. 4.6 Working areas

4.1.9 *Working Areas*

Figure 4.6 shows each working area. As an example, in unstable area 1, a sufficiently high initial speed causes the material point to pass the point $X = 10$ and move indefinitely along the X-axis. For the initial conditions in area 11, the speed is sufficiently low for the system to reach the equilibrium point(0, 0).

4.2 Autonomous Van Der Pol Oscillator

4.2.1 *System Description*

A mathematical model to describe triode oscillations in an electrical circuit is given by Eq. (4.15) (Balthazar Van Der Pol, 1926). This is a very useful representation for analysing a large number of systems:

$$\ddot{\theta} = (p - \theta^2)\dot{\theta} - \theta. \tag{4.15}$$

Using our usual notations, It can be written as the following set of differential equations:

$$S : \begin{cases} \frac{dX_1}{dt} = X_2 = F_1(X_1, X_2) \\ \frac{dX_2}{dt} = (p - X_1^2)X_2 - X_1 = F_2(X_1, X_2) \end{cases}. \tag{4.16}$$

X_1 and X_2 are the state parameters and p is a system parameter.

4.2.2 Equilibrium Points, Jacobian Matrix and Local Stability

Solving the system S_e yields a unique equilibrium point:

$$(X_{1e}, X_{2e}) = (0, 0). \tag{4.17}$$

The Jacobian matrix at this point is given by:

$$J = \begin{pmatrix} \frac{\partial F_1}{\partial X_1} & \frac{\partial_1}{\partial X_2} \\ \frac{\partial F_2}{\partial X_1} & \frac{\partial F_2}{\partial X_2} \end{pmatrix} = \begin{pmatrix} 0 & 1 \\ -2X_{1e}X_{2e} - 1 & p - X_{1e}^2 \end{pmatrix} = \begin{pmatrix} 0 & 1 \\ -1 & p \end{pmatrix}. \tag{4.18}$$

The characteristic equation is:

$$\det(J - sI) = s^2 - ps + 1 = 0. \tag{4.19}$$

It has two roots

$$s_1 = \frac{p + \sqrt{p^2 - 4}}{2}, s_2 = \frac{p - \sqrt{p^2 - 4}}{2}. \tag{4.20}$$

It is easy to show that the real parts of s_1 and s_2 have the same sign as p. There are equal to zero for $p = 0$. It follows that the Van der pol system is locally stable and tends to $(X_{1e}, X_{2e}) = (0, 0)$ when $p < 0$, and is locally unstable for $p > 0$.

4.2.3 Basin of Attraction for p < 0

Consider the Lyapunov function:

$$V(X_1, X_2) = \frac{1}{2}(X_1^2 + X_2^2). \tag{4.21}$$

$V(X_1, X_2)$ meets the following conditions to be considered a Lyapunov function

- $V(X_{1e}, X_{2e}) = 0$.
- For each $V > 0$ the corresponding region in the phase portrait (X_1, X_2) defined by $V(X_1, X_2) < V$ is a closed surface.
- For each closed surface defined by V_1 and V_2, if $V_1 < V_2$, the surface corresponding to V_1 will be inside the surface corresponding to V_2.

For each $(X_1, X_2) \neq (0, 0)$ with $V(X_1, X_2) > 0$, the calculation of $\dot{V}(X_1, X_2)$ gives:

$$\dot{V}(X_1, X_2) = \frac{\partial V}{\partial X_1}\dot{X}_1 + \frac{\partial V}{\partial X_2}\dot{X}_2 = X_2^2(p - X_1^2). \tag{4.22}$$

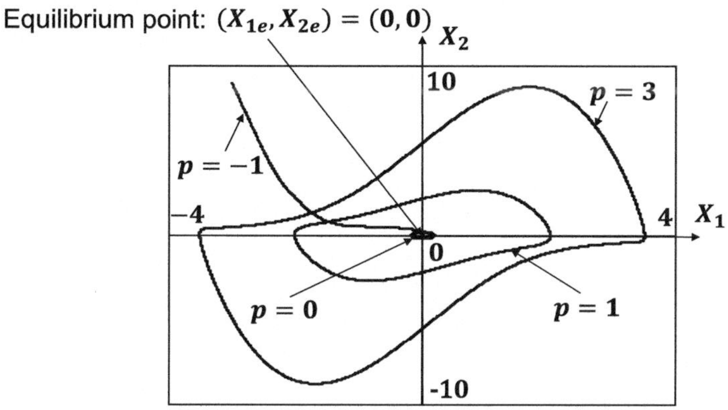

Fig. 4.7 Attractors of a Van Der Pol system for $p = -1, p = 0, p = 1$ and $p = 3$.

We check that $\dot{V}(X_{1e}, X_{2e}) = 0$, and that $\dot{V}(X_1, X_2)$ is always negative when $p < 0$. Thanks to Lyapunov's theory, we can conclude that the basin of attraction of a Vander pol system is $\mathbb{R}^2$ for $p < 0$.

4.2.4 *Periodic Attractors for p ≥ 0*

Numerical simulation shows that the attractor of a Vander pol system is a periodic orbit for $p \geq 0$. Figure 4.7 represents the periodic orbits for $p = 0, p = 1$ and $p = 3$ in steady-state mode. This figure also shows a trajectory converging towards the equilibrium point for a negative value of ($p = -1$).

4.2.5 *Stability of Orbits and Basin of Attraction for p ≥ 0*

For a given value of $p \geq 0$, the stability of the corresponding orbit can be demonstrated using the Floquet matrix. Keeping the same notation as in Para. 2.2.2 the Poincaré map on section $S = \{X_1 < 0, X_2 = 0\}$ is given by:

$$P : X_{10} \rightarrow P(X_{10}) = \vec{\varphi}(X_{10}, T). \tag{4.23}$$

For $p = 2$, the Floquet matrix is the following coefficient:

$$F = \frac{\partial P}{\partial X_{10}}(X_{10}) \simeq -0.045. \tag{4.24}$$

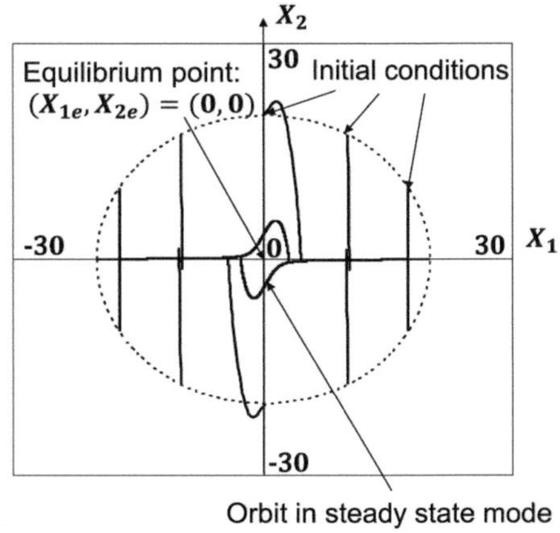

Fig. 4.8 Stability region determination for $p = 2$

Since the absolute value of the only eigenvalue of F is less than one, the corresponding orbit is asymptotically stable. For the same value of p, a stability region can be determined through simulation. Consider a circle with centre $(X_{1e}, X_{2e}) = (0, 0)$ and radius $R = 20$. The results of the simulations, with different initial conditions on this circle, show that each trajectory converges towards the stable orbit (see Fig. 4.8). Therefore, the area inside this circle is a stability region for $p = 2$.

4.3 Hénon System

The Hénon system is a simplified model of the Poincaré section of the Lorenz model (Michel Henon, 1970). It is among the most extensively studied examples of dynamical systems exhibiting chaotic behaviour (Aybar et al., 2013).

4.3.1 System Description

The Hénon system is a second-order nonlinear discrete system defined by the following recursion relations:

$$S : \begin{cases} X_{1(k+1)} = -P_1 X_{1k}^2 + X_{2k} + 1 = F_1(X_{1k}, X_{2k}) \\ X_{2(k+1)} = P_2 X_{1k} = F_2(X_{1k}, X_{2k}) \end{cases} . \tag{4.25}$$

$\vec{X}_k = (X_{1k}, X_{2k})$ and $\vec{P} = (P_1, P_2)$ are, respectively, the discrete state vector ($n = 2$) and the parameter vector ($p = 2$). In this example, P_2 is constant and equal to 0.3. The dynamics of this system will be studied for each value of P_1.

4.3.2 Equilibrium Points

As explained in Chap. 1, the equilibrium points (X_{1ei}, X_{2ei}) are determined by the following S_e system:

$$S_e : \begin{cases} X_{1ei} = -P_1 X_{1ei}^2 + X_{2ei} + 1 \\ X_{2ei} = P_2 X_{1ei} \end{cases}, \tag{4.26}$$

which is equivalent to:

$$S_e : \begin{cases} P_1 X_{1ei}^2 + (1 - P_2) X_{1ei} - 1 = 0 \\ X_{2ei} = P_2 X_{1ei} \end{cases}. \tag{4.27}$$

Three cases need to be considered:

- If $P_1 < \frac{-(1-P_2)^2}{4} = -0.1225$, there is no equilibrium point.
- If $P_1 = \frac{-(1-P_2)^2}{4} = -0.1225$, there is one equilibrium point equal to

$$(X_{1e1}, X_{2e1}) = \left(\frac{2}{1 - P_2}, \frac{2P_2}{1 - P_2} \right) \simeq (2.857, 0.857).$$

- If $P_1 > \frac{-(1-P_2)^2}{4} = -0.1225$ and $P_1 \neq 0$ there are two equilibrium points given by:

$$\begin{aligned} (X_{1e1}, X_{2e1}) &= \left(\frac{-(1 - P_2) + \sqrt{(1 - P_2)^2 + 4P_1}}{2P_1}, \frac{-P_2(1 - P_2) + P_2\sqrt{(1 - P_2)^2 + 4P_1}}{2P_1} \right), \\ (X_{1e2}, X_{2e2}) &= \left(\frac{-(1 - P_2) - \sqrt{(1 - P_2)^2 + 4P_1}}{2P_1}, \frac{-P_2(1 - P_2) - P_2\sqrt{(1 - P_2)^2 + 4P_1}}{2P_1} \right). \end{aligned} \tag{4.28}$$

- If $P_1 = 0$, we obtain only one equilibrium point:

$$(X_{1e1}, X_{2e1}) = \left(\frac{1}{(1 - P_2)}, \frac{P_2}{(1 - P_2)} \right) \simeq (1.429, 0.429). \tag{4.29}$$

This point is also the limit of (X_{1e1}, X_{2e1}) given by Eq. (4.28) as $P_1 \to 0$. Figure 4.9 summarises the evolution of X_{1e1} and X_{1e2} in relation to P_1.

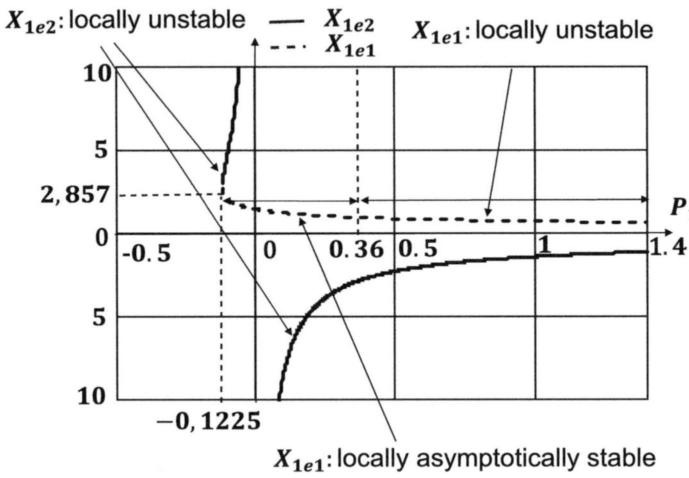

Fig. 4.9 Equilibrium points of the Hénon system

4.3.3 *Jacobian Matrix and Local Stability*

The Jacobian matrix of the Hénon system at the equilibrium point (X_{1ei}, X_{2ei}) is:

$$J = \begin{pmatrix} \frac{\partial X_{1(k+1)}}{\partial X_{1k}} & \frac{\partial X_{1(k+1)}}{\partial X_{2k}} \\ \frac{\partial X_{2(k+1)}}{\partial X_{1k}} & \frac{\partial X_{2(k+1)}}{\partial X_{2k}} \end{pmatrix} = \begin{pmatrix} -2P_1X_{1ei} & 1 \\ P_2 & 0 \end{pmatrix}. \tag{4.30}$$

The type of local stability around each equilibrium point (X_{1ei}, X_{2ei}) is determined by the characteristic equation:

$$\det(J - zI) = -z(-2P_1X_{1ei} - z) - P_2 = z^2 + 2P_1X_{1ei}z - P_2 = 0. \tag{4.31}$$

It follows that:

- If $P_1 < \frac{-(1-P_2)^2}{4} = -0.1225$, there is no equilibrium point, therefore, the system cannot be locally stable.
- If $P_1 = \frac{-(1-P_2)^2}{4} = -0.1225$, the roots of the characteristic equation are 1 and -0.3, therefore, we cannot draw any conclusion about the local stability of the equilibrium point $(2.857, 0.857)$. Numerical simulations show that the Henon system is locally stable for this value of P_1.
- If $P_1 > \frac{-(1-P_2)^2}{4} = -0.1225$, the equilibrium point (X_{1e2}, X_{2e2}) has at least one root of the characteristic equation outside the unit circle. This point is therefore locally unstable, whatever the value of P_1. For the equilibrium point (X_{1e1}, X_{2e1}), the roots are inside the unit circle when P_1 is approximately lower than 0.36, therefore, the system is locally asymptotically stable. For P_1 approximately higher than

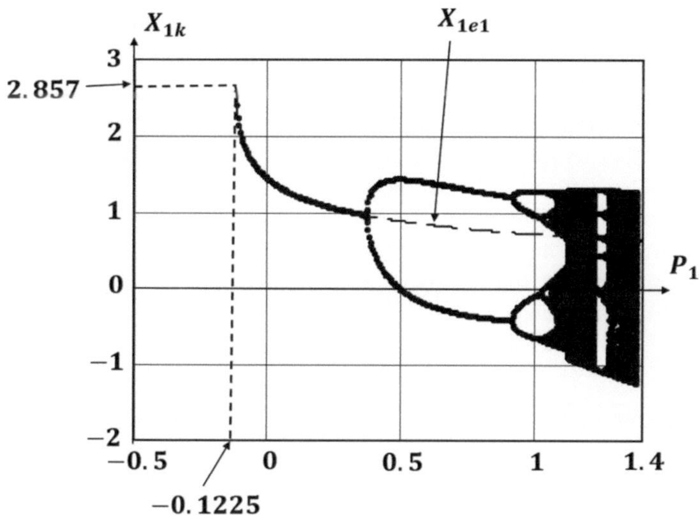

Fig. 4.10 Bifurcation diagram of the Hénon system

0.36, at least one root is outside the unit circle, and consequently the equilibrium point becomes locally unstable.

4.3.4 Bifurcation Diagram

Figure 4.10 represents the bifurcation diagram of the Hénon system. It is obtained by iterating the parameter value P_1 many times. For each value of P_1, the first 155 iterations are discarded to eliminate transient conditions and the latest 155 are plotted. For each simulation, the initial conditions are chosen close to the equilibrium point (X_{1e1}, X_{2e1}).

When $-0.1225 < P_1 < 0.36$, the attractor is (X_{1e1}, X_{2e1}) given by Eq. (4.28). The diagram shows a doubling and a quadrupling of the period as the parameter P_1 increases. For higher P_1 values, the system becomes chaotic.

4.3.5 Stability Areas

As previously mentioned, the Hénon system is always unstable for $P_1 < -0.1225$, whatever the values of the intial conditions. In this paragraph, we will determine the stability region for four representative values of the parameter P_1, corresponding to different dynamical behaviours of the Hénon system.

We will use the technique described in Para. 2.3 and illustrated by the flow chart in Fig. 2.12. For $P_1 = -0.1225$, a stability region is defined by the rectangle shown in Fig. 4.11a. All the trajectories starting from initial conditions on this rectangle converge towards the equilibrium point. For other values of P_1, the stability regions are constructed with circles whose centre are (X_{1e1}, X_{2e1}). For $P_1 = 0.1$,0.5 and 1.4, all trajectories starting on the circles converge towards a stable equilibrium point (Fig. 4.11b), either to an orbit of period two (Fig. 4.12a) or to a strange attractor (Fig. 4.12b). These figures also show unstable trajectories for initial conditions outside these stability regions.

For each value of P_1, an unstable trajectory is also represented. Obvously, these stability regions are not the basins of attraction, which are generally difficult to determine for the Hénon system.

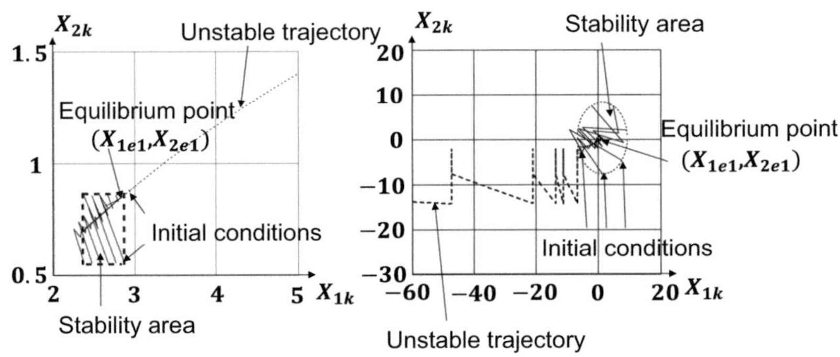

Fig. 4.11 Stability areas for $P_1 = -0.1225$ and $P_1 = 0.1$

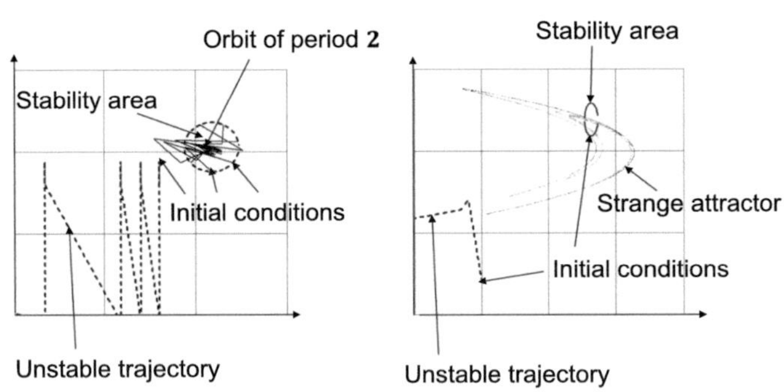

Fig. 4.12 Stability areas for $P_1 = 0.5$ and $P_1 = 1.4$

4.4 Lorenz System

4.4.1 System Description and Equilibrium Points

In this paragraph, we will study further the Lorenz system mentioned in Chap. 3 (Souleres, 2022). The general differential equations of this system are:

$$\begin{cases} \frac{dX_1}{dt} = \sigma(-X_1 + X_2) \\ \frac{dX_2}{dt} = RX_1 - X_2 - X_1X_3 \\ \frac{dX_3}{dt} = -bX_3 + X_1X_2 \end{cases}. \tag{4.32}$$

R, σ and b are system parameters. The system S_e has the following three possible solutions:

$$\begin{aligned} (X_{1e1}, X_{2e1}, X_{3e1}) &= (0, 0, 0), \\ (X_{1e2}, X_{2e2}, X_{3e2}) &= \left(\sqrt{b(R-1)}, \sqrt{b(R-1)}, R-1\right), \\ (X_{1e3}, X_{2e3}, X_{3e3}) &= \left(-\sqrt{b(R-1)}, -\sqrt{b(R-1)}, R-1\right). \end{aligned} \tag{4.33}$$

For $R < 1$, the only one equilibrium point is $(X_{1e1}, X_{2e1}, X_{3e1}) = (0, 0, 0)$, whereas for $R \geq 1$, three equilibrium points exist. Below, we will analyse the influence of the parameter R (for $26 \geq R \geq 0$) on the dynamics of this system. σ and b will be considered constant, equal to 10 and $\frac{8}{3}$, respectively.

4.4.2 Jacobian Matrix and Local Stability

The Jacobian matrix for each equilibrium point $(X_{1ei}, X_{2ei}, X_{3ei})$ is given by:

$$J = \begin{pmatrix} \frac{\partial F_1}{\partial X_1} & \frac{\partial F_1}{\partial X_2} & \frac{\partial F_1}{\partial X_3} \\ \frac{\partial F_2}{\partial X_1} & \frac{\partial F_2}{\partial X_2} & \frac{\partial F_2}{\partial X_3} \\ \frac{\partial F_3}{\partial X_1} & \frac{\partial F_3}{\partial X_2} & \frac{\partial F_3}{\partial X_3} \end{pmatrix} = \begin{pmatrix} -\sigma & \sigma & 0 \\ R - X_{3ei} & -1 & -X_{1ei} \\ X_{2ei} & X_{1ei} & -b \end{pmatrix}. \tag{4.34}$$

The corresponding characteristic equation for the equilibrium point $(X_{1e1}, X_{2e1}, X_{2e1}) = (0, 0, 0)$ is defined by:

$$\det(J - sI) = s^3 + s^2(b + \sigma + 1) + s(b + \sigma + b\sigma - R\sigma) + b\sigma(1 - R) = 0. \tag{4.35}$$

The three eigenvalues are real and given by:

$$s_1 = -b, s_2 = \frac{-\sigma - 1 - \sqrt{(\sigma + 1)^2 + 4\sigma(R - 1)}}{2}, \tag{4.36}$$

$$s_3 = \frac{-\sigma - 1 + \sqrt{(\sigma + 1)^2 + 4\sigma(R - 1)}}{2}. \tag{4.37}$$

When $R < 1$, the real parts of s_1,s_2,s_3 are negative so the origin is locally asymptotically stable. When $R > 1$, the real part of s_3 is positive and the real parts of s_1 and s_2 remains negative, thus, the origin becomes unstable with one-dimensional unstable manifold. When $R > 1$, the characteristic equation is the same for equilibrium points $(X_{1e2}, X_{2e2}, X_{3e2})$ and $(X_{1e3}, X_{2e3}, X_{3e3})$. After some reductions we obtain:

$$\det(J - sI) = s^3 + s^2(b + \sigma + 1) + sb(\sigma + R) + 2b\sigma(R - 1) = 0. \tag{4.38}$$

We can use the Routh stability criterion to determine the stability of the roots based on the value of the coefficient R. With all the coefficients a_i having the same sign (> 0), the first necessary condition to apply this criterion is met. The corresponding Routh array is constructed as follows:

		Col 1	Col 2	Col 3
Row 0	s^3	1	$b(\sigma + R)$	0
Row 1	s^2	$b + \sigma + 1$	$2b\sigma(R - 1)$	0
Row 2	s^1	$\frac{b(b+\sigma+1)\sigma+R-2b\sigma(R-1)}{\sigma+b+1}$	0	
Row 3	s^0	$2\sigma b(R - 1)$		

The Routh criterion states that all the eigenvalues lie in the stable left-half plane if:

$$b(b + \sigma + 1)(\sigma + R) - 2b\sigma(R - 1) \geq 0 \quad \text{and} \quad R \geq 1. \tag{4.39}$$

Thus, if

$$1 < R \leq R_H = \frac{\sigma(b + \sigma + 3))}{\sigma - b - 1} \simeq 24.74, \tag{4.40}$$

the two equilibrium points $(X_{1e2}, X_{2e2}, X_{3e2})$ and $(X_{1e3}, X_{2e3}, X_{3e3})$ are locally asymptotically stable.

To summarise:

- If $1 > R \geq 0$, there is one equilibrium point $(X_{1e1}, X_{2e1}, X_{3e1}) = (0, 0, 0)$. This point is locally asymptotically stable.
- If $R_H > R \geq 1$,there are three equilibrium points $(X_{1e1}, X_{2e1}, X_{3e1})$, $(X_{1e2}, X_{2e2}, X_{3e2})$ and $(X_{1e3}, X_{2e3}, X_{3e3})$. The origin is unstable, whereas the other two points are locally asymptotically stable.

- If $26 > R > R_H$, the three equilibrium points are unstable. We begin to observe chaotic behaviour similar to that originally described in Chap. 3.

Figure 3.13 in Chap. 3 shows a fixed point attractor for $R = 0.5$ and $R = 2$ and a strange attractor for $R = 25$.

4.4.3 Stability Areas

It is possible to prove that the stability region of the origin for $1 > R \geq 0$ is $\mathbb{R}^3$ by considering the following Lyapunov function:

$$V(X_1 X_2, X_3) = X_1^2 + \sigma X_2^2 + \sigma X_3^2. \tag{4.41}$$

We can easily verify that $V(X_1 X_2, X_3)$ meets the usual conditions to be eligible as a Lyapunov function. Moreover, we have:

$$\forall (X_1 X_2, X_3) \in \mathbb{R}^3, \quad V(X_1 X_2, X_3) > 0. \tag{4.42}$$

The time derivative of this Lyapunov function is

$$\dot{V}(X_1, X_2, X_3) = 2\sigma\left[(1+R)X_1 X_2 - X_1^2 - X_2^2 - bX_3^2\right]. \tag{4.43}$$

It is easy to show that

$$(1+R)X_1 X_2 - X_1^2 - X_2^2 < 0 \forall (X_1 X_2) \in \mathbb{R}^2. \tag{4.44}$$

Consequently, we can conclude that

$$\dot{V}(X_1, X_2, X_3) < 0 \forall (X_1 X_2, X_3) \in \mathbb{R}^3. \tag{4.45}$$

Thanks to the Lyapunov theory, we can determine that the basin of attraction for the origin is $\mathbb{R}^3$. Thus, starting from any point away from the origin, $V(X_1 X_2, X_3)$ will decrease and the trajectory will tend towards the equilibrium point $(0, 0, 0)$. Figure 4.13a represents a stability region in the form of a cube centred at

$$(X_{1e2}, X_{2e2}, X_{3e2}) = \left(\sqrt{b(R-1)}, \sqrt{b(R-1)}, R-1\right), \tag{4.46}$$

with edges of length 2 and $R = 2$. This figure illustrates that, for any initial condition located at a corner of the cube, all trajectories converge towards $(X_{1e2}, X_{2e2}, X_{3e2})$. With more initial conditions on the cube, the result will be the same. This cube can be considered a stability region of the attractor $(X_{1e2}, X_{2e2}, X_{3e2})$. Figure 4.13b represents the simulation for the same equilibrium point, but with edge length is equal to 4.

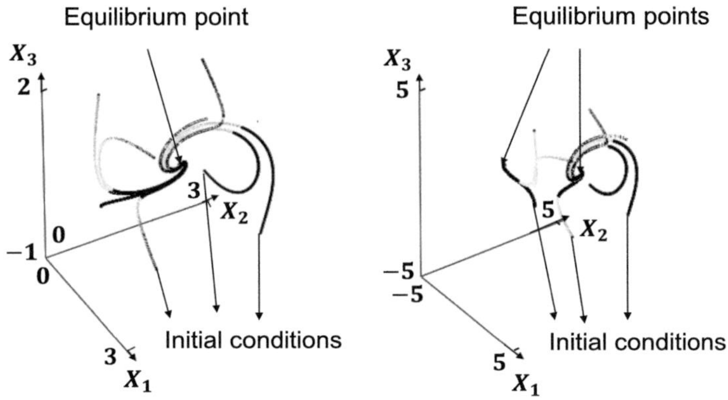

Fig. 4.13 Stability area of $(X_{1e1}, X_{2e1}, X_{3e1})$ for $R = 2$

We can observe that, for some initial conditions on this cube, the trajectories converge towards the second equilibrium point:

$$(X_{1e3}, X_{2e3}, X_{3e3}) = \left(-\sqrt{b(R-1)}, -\sqrt{(R-1)}, R-1\right) \tag{4.47}$$

This phenomenon can also be seen in Fig. 3.12 in Chap. 3, which shows the bifurcation diagram of this system. As the parameter R increases, the attractor undergoes a sudden bifurcation.

4.5 Two-Tank System with Two Valves

4.5.1 System Description

This system (Hiskens, 2001) consists of two tanks and two valves (V_1, V_2) (see Fig. 4.14). The valve V_1 controls the inflow in tank 1, while the valve V_2 serves as a drain for the tank 2.

There is also a constant input flow into the tank 1 (Q_{IN}) and a constant outflow from thank 2 (Q_{OUT}) caused by a pump. The discrete open/close switching control of the valves allows the water levels in both tanks to be kept within limits. The switching strategy for the valves is defined as follows:

- The valve V_1 should open when the level of tank 1 (X_1) falls to -1 and close when the level of tank 2 (X_2) rises to 1.
- The valve V_2 should open when the level of tank 2 rises to 1 and close when the level of tank 2 falls to 0.

After linearisation, the dynamics of this system becomes.

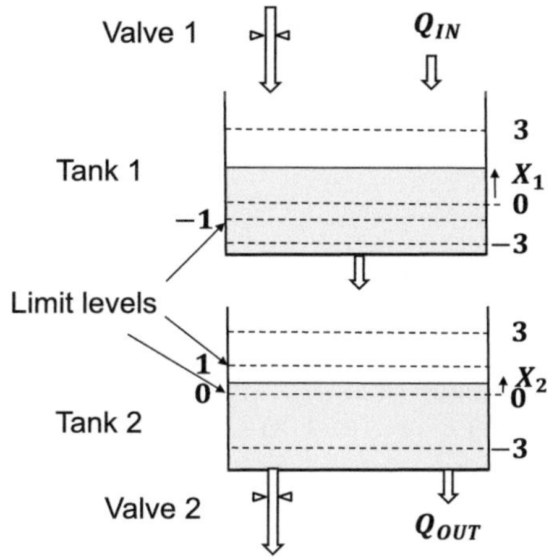

Fig. 4.14 Two-tanks system with two valves

- $\frac{dX_1}{dt} = -X_1 - 2$ when V_1 is closed ($V_1 = 0$),
- $\frac{dX_1}{dt} = -X_1 + 3$ when V_1 is open ($V_1 = 1$),
- $\frac{dX_2}{dt} = X_1$ when V_2 is closed ($V_2 = 0$),
- $\frac{dX_2}{dt} = X_1 - X_2 - 5$ when V_2 is open ($V_2 = 1$).

Using the definitions given in Para. 3.1, this system can be considered an autonomous hybrid system, as shown in Fig. 4.15.

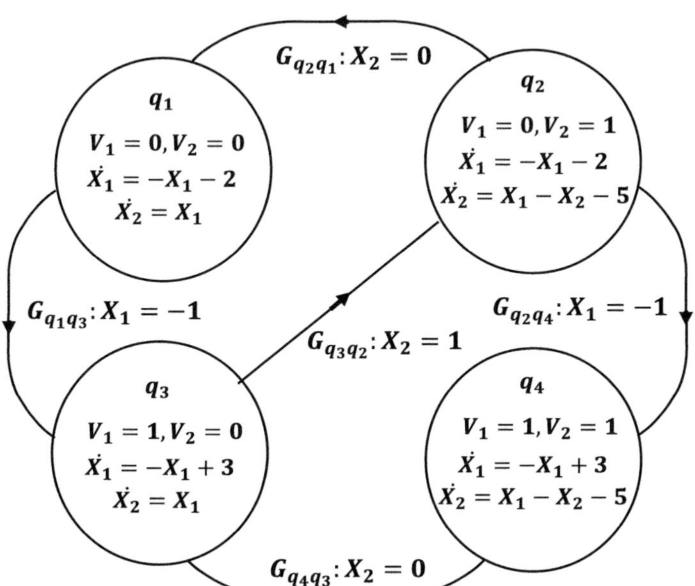

Fig. 4.15 Two-tank system with two valves

The mathematical model of this hybrid system can be represented by the following generic representation:

$$H = (Q, X, \text{Init}, \text{Dom}, F, E, G,) \tag{4.48}$$

where

- Q is a set of 4 discrete states ($n_Q = 4$) $Q = \{q_1, q_2, q_3, q_4\}$, corresponding to the valve states:$\{(V_1, V_2)\} = \{(0,0), (0,1), (1,0), (1,1)\}$,
- $X = \{\vec{X} : \mathbb{R} \to [-3, 3] \times [-3, 3]\}$ is a set of two-dimensional continuous state vectors ($n = 2$),
- $\text{Init} \in Q \times [-3, 3] \times [-3, 3]$ is a set of $1 + n = 3$ initial conditions,
- $\text{Dom} = \left\{D_{q_i}, q_i \in Q\right\}$ is a set of continuous domains:

$$D_{q_1} = D_{q_2} = D_{q_3} = D_{q_4} = [-3, 3] \times [-3, 3],$$

- $F = \left\{\overrightarrow{F_{q_i}}, q_i \in Q\right\}$ is the set of two-dimensional vector fields describing the dynamics of the system. We have:

$$q = q_1 : \begin{cases} \frac{dX_1}{dt} = -X_1 - 2 = F_{q_{1,1}}(X_1, X_2) \\ \frac{dX_2}{dt} = X_1 = F_{q_{1,2}}(X_1, X_2) \end{cases},$$
$$q = q_2 : \begin{cases} \frac{dX_1}{dt} = -X_1 - 2 = F_{q_{2,1}}(X_1, X_2) \\ \frac{dX_2}{dt} = X_1 - X_2 - 5 = F_{q_{2,2}}(X_1, X_2) \end{cases},$$
$$q = q_3 : \begin{cases} \frac{dX_1}{dt} = -X_1 + 3 = F_{q_{3,1}}(X_1, X_2) \\ \frac{dX_2}{dt} = X_1 = F_{q_{3,2}}(X_1, X_2) \end{cases},$$
$$q = q_4 : \begin{cases} \frac{dX_1}{dt} = -X_1 + 3 = F_{q_{4,1}}(X_1, X_2) \\ \frac{dX_2}{dt} = X_1 - X_2 - 5 = F_{q_{4,2}}(X_1, X_2), \end{cases}$$

- $E \subset Q \times Q$ is a collection of discrete transitions

$$E = \{(q_1, q_3), (q_2, q_1), (q_2, q_4), (q_3, q_2), (q_4, q_3)\},$$

- $G = \{G_e, e \in E\}$. is a set of guards that determine when a discrete state transition occurs

$$G_{q_1q_3} = \{X_1 = -1\}, G_{q_2q_1} = \{X_2 = 0\}, G_{q_2q_4} = \{X_1 = -1\},$$
$$G_{q_3q_2} = \{X_2 = 1\}, G_{q_4q_3} = \{X_2 = 0\}.$$

In this example, there is no control vector (autonomous system) and no reset state functions, the state parameters evolve continuously.

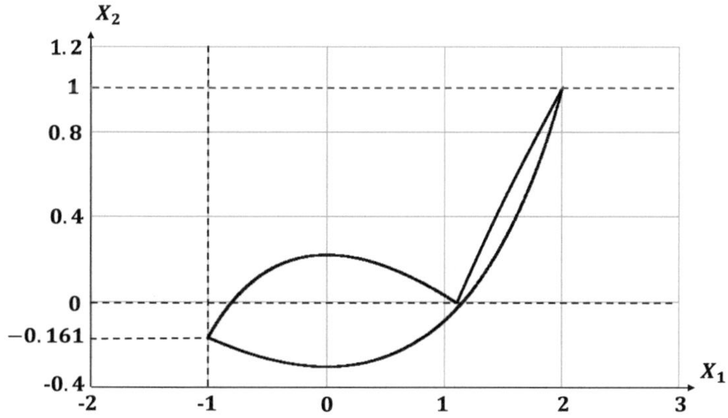

Fig. 4.16 Limit cycle of the two-tank system

4.5.2 Limit Cycle

As is quite common for hybrid systems, after the transition phase, the two-tank system reaches a cycle limit. The corresponding periodic orbit is represented in the phase plane in Fig. 4.16.

4.5.3 Local Stability of the Periodic Orbit

The local stability of this limit cycle can be determined at the point $(X_{O1}, X_{O2}) = (-1, -0.161)$ using the Monodromy matrix (see Fig. 4.16). A small disturbance Δx_{1o} generated at $t = 0$ will produce a difference of $\Delta x_{11}(T)$ and $\Delta x_{21}(T)$ compared with the steady-state mode at $t = T$. In the same way, a small disturbance Δx_{2o} generated at $t = 0$ will generate $\Delta x_{12}(T)$ and $\Delta x_{22}(T)$ at $t = T$. The Monodromy matrix is determined by:

$$M = \begin{pmatrix} \frac{\Delta x_{11}(T)}{\Delta x_{1o}} & \frac{\Delta x_{12}(T)}{\Delta x_{2o}} \\ \frac{\Delta x_{21}(T)}{\Delta x_{1o}} & \frac{\Delta x_{22}(T)}{\Delta x_{2o}} \end{pmatrix} \simeq \begin{pmatrix} 1.6244 & 2.4977 \\ -0.4337 & -0.7348 \end{pmatrix}. \tag{4.49}$$

T is the period of the cycle. This Monodromy matrix has for eigenvalues 1 and -0.1103 showing that the limit cycle is locally asymptotically stable. For a more detailed discussion, refer to Hiskens (2001).

4.5.4 Basin of Attraction of Two-Thank System

Numerical simulations show that the area defined by

$$(X_1, X_2) \in [-3, 3] \times [-3, 3] \tag{4.50}$$

is a stability region of this system. For each initial condition inside this area, and regardless of the discrete state, the trajectories will converge towards the same cycle limit. The phase portrait in the Fig. 4.17 shows the behaviour of the system for the various initial conditions ((3, 3), (3, −3), (−3, −3), (−3, 3)).

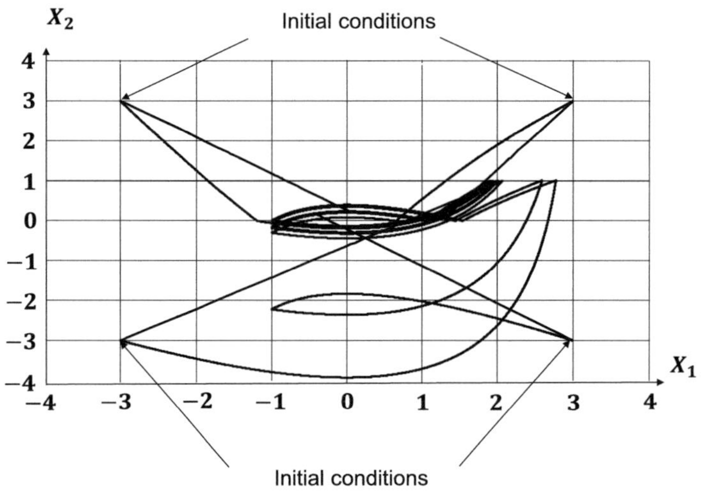

Fig. 4.17 Basin of attraction of the two-tank system

Chapter 5
Basics of Control Theory and Linear Controls

In this chapter, we will review some key points of control theory and the basics of linear control (D'Azzo & Houpis, 2003).

5.1 Basics of Control Theory

As usual, we consider nonlinear systems of the following form:

$$\frac{\mathrm{d}\vec{X}}{\mathrm{d}t} = \vec{F}\left(\vec{X}, \vec{U}\right), \vec{Y} = \vec{G}\left(\vec{X}\right), \quad \vec{X} \in \beta \subseteq \mathbb{R}^n. \tag{5.1}$$

5.1.1 Definition of Controllability

A system is said to be controllable if, for any initial state vector $\vec{X}_I \in \beta$ and any final state vector $\vec{X}_F \in \beta$ at times t_0 and t_F respectively, there exists a control input $\vec{U}$ such that the solution of:

$$\frac{\mathrm{d}\vec{X}}{\mathrm{d}t} = \vec{F}\left(\vec{X}, \vec{U}\right), \vec{X}\,(t = t_0) = \vec{X}_I \tag{5.2}$$

satisfies

$$\vec{X}\,(t = t_F) = \vec{X}_F. \tag{5.3}$$

B. Rivet, *Theories of Nonlinear Systems for Power Converters*,
https://doi.org/10.1007/978-3-032-25062-9_5

5.1.2 Definition of Observability

Indistinguishable state vectors

Two state vectors $\vec{X}_1 \in \beta$ and $\vec{X}_2 \in \beta$ are said to be indistinguishable if, for every admissible control input $\vec{U}(t)$ $(t_0 < t < t_F)$, they produce identical output vectors

$$\vec{Y} = \vec{G}\left(t, \vec{X}_1\right) = \vec{G}\left(t, \vec{X}_2\right) \quad \text{for } t_0 < t < t_F. \tag{5.4}$$

The set of all vectors that are indistinguishable from a given state vector $\vec{X}$ is called its indistinguishability set $T\left(\vec{X}\right)$. A system is observable if:

$$\forall \vec{X} \in \beta T\left(\vec{X}\right) = \vec{X} \tag{5.5}$$

The conditions for controllability and observability of a linear system are given in Para. 5.3.1.

5.1.3 Basics of Control Theory

The fundamental concept in control theory is the feedback control system, as illustrated in Fig. 5.1.
A description of each element in the closed-loop system is provided below:

- The reference signals $\overrightarrow{X_{\mathrm{ref}}}$ are the desired values of the measured outputs.
- The control errors $\vec{e_r}$ are the differences between the reference signals and the transduced outputs.
- The control signals $\vec{U}$ dynamically influence the dynamics of the target system.
- The controller determines the settings of the control signals to achieve the reference signal.
- $\vec{D}$ represents the disturbances that affect the measured outputs.
- $\vec{Y}$ represents the measured outputs.

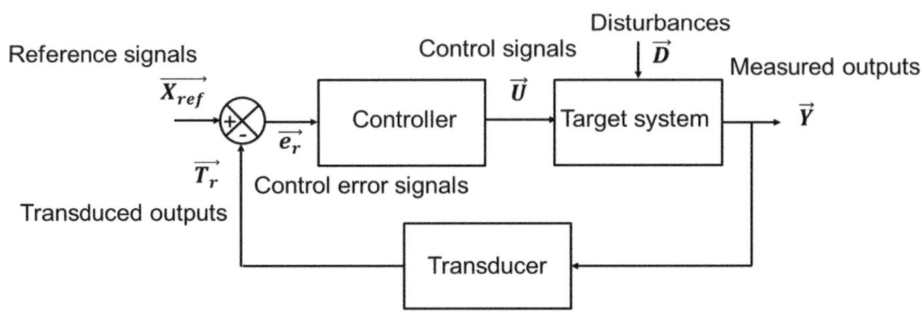

Fig. 5.1 Block diagram of a feedback control system

- The transducer transforms the measured outputs so that they can be compared with the reference signals.
- $\overrightarrow{T_r}$ represents the transduced outputs.

5.1.4 Objectives of the Controller

The aims of the controller are the following:

- Ensure that the transduced outputs are close to the reference signals within the accepted accuracy.
- Ensure the stability of the closed-loop system.
- Guarantee adequate robustness against disturbances and system uncertainties (see Chap. 7).
- Ensure specific performance (fast transient answers, stability during variations of system parameters…).
- Reject high-frequency disturbances.
- In some specific cases, ensure optimal control (see Chap. 6).

5.2 Controller of a Single-Input Single-Output Linear System

5.2.1 Block Diagram of the Feedback Control of a Linear System

For a single-input single-output linear (SISO) system, the diagram in Fig. 5.1 corresponds to that shown in Fig. 5.2. Each block is represented by a Laplace transfer function.

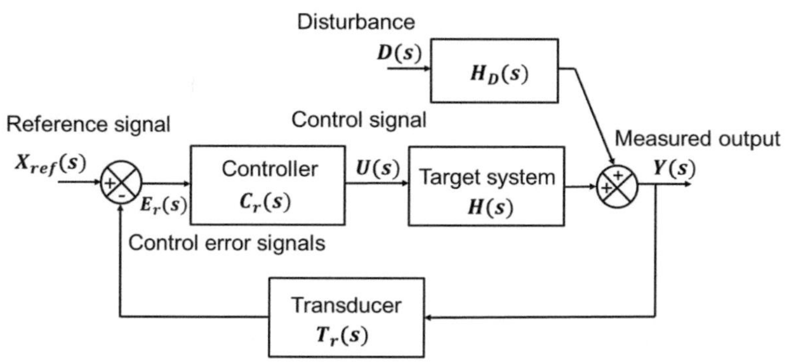

Fig. 5.2 Block diagram of a feedback control for a SISO linear system

5.2.2 Effect of Negative Feedback on $\frac{Y(s)}{X_{ref}(s)}$ and $\frac{Y(s)}{D(s)}$

The transfer function from $X_{\text{ref}}(s)$ to $Y(s)$, when $D(s) = 0$, is given by:

$$\frac{Y(s)}{X_{\text{ref}}(s)} = \frac{H_P(s)C_r(s)}{1 + T_r(s)H_P(s)C_r(s)} = \frac{1}{T_r(s)}\frac{H_{BO}(s)}{1 + H_{BO}(s)} \tag{5.6}$$

where

$$H_{BO}(s) = T_r(s)H_P(s)C_r(s) \tag{5.7}$$

is the open-loop transfer function. If the open-loop gain is large in magnitude, the transfer function becomes:

$$\frac{Y(s)}{X_{\text{ref}}(s)} \simeq \frac{1}{T_r(s)}. \tag{5.8}$$

This result is very interesting because the output depends only on the transfer function $T_r(s)$. It is a key result of negative feedback control. If the reference signal is constant and equal to X_{ref}, the output is:

$$Y = \frac{X_{\text{ref}}}{T_r(0)}\frac{H_{BO}(0)}{1 + H_{BO}(0)} \simeq \frac{X_{\text{ref}}}{T_r(0)}. \tag{5.9}$$

Note that the error is given by:

$$E_r(s) = X_{\text{ref}}(s) - T_r(s)Y(s) = \frac{1}{1 + H_{BO}(s)}X_{\text{ref}}(s). \tag{5.10}$$

The transfer function from $D(s)$ to $Y(s)$, when $X_{\text{ref}}(s) = 0$ is

$$\frac{Y(s)}{D()} = \frac{H_D(s)}{1 + H_{BO}(s)}. \tag{5.11}$$

If $H_{BO}(s)$ is large in amplitude in the frequency range of the disturbance, this disturbance will have little effect on the output. This is also a key result of feedback control.

5.2.3 Analysis of Stability

Ensuring the stability of the controlled system is also a key feature of negative feedback control. It has been shown in Chap. 2 that the stability of a unity feedback loop system can be determined using the Nyquist stability theorem by testing

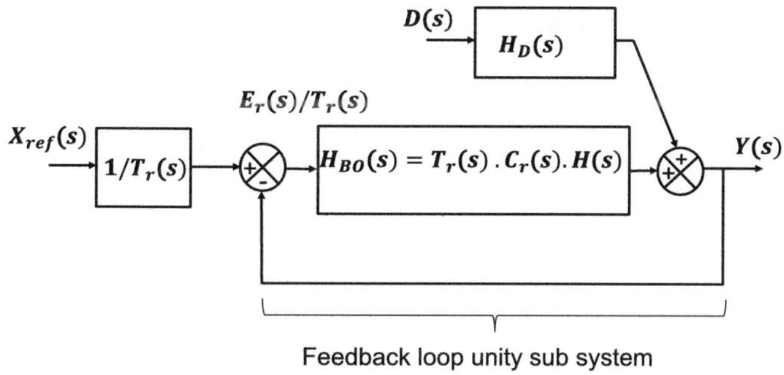

Fig. 5.3 Equivalent diagram with a unity feedback loop

$H_{BO}(s)$. The diagram in Fig. 5.2 is equivalent to the unity feedback loop shown in Fig. 5.3.

Consider now this new representation. If, according to the Nyquist criterion, the unity feedback loop subsystem is stable, then it follows that:

$$\frac{H_{BO}(s)}{1 + H_{BO}(s)} = \frac{T_r(s)H_P(s)C_r(s)}{1 + T_r(s)H_P(s)C_r(s)} \tag{5.12}$$

has no poles with positive real parts. As a consequence, since $T_r(s)$ generally has no zero, it will follow that:

$$\frac{1}{T_r(s)}\frac{H_{BO}(s)}{1 + H_{BO}(s)} = \frac{H_P(s)C_r(s)}{1 + T_r(s)H_P(s)C_r(s)} \tag{5.13}$$

has no poles with positive real parts. Hence, by testing only the unity feedback loop subsystem, we can determine the stability of the complete system. If $H_{BO}(s)$ has no poles in the right half-plane and on the imaginary axis, the simple phase margin test can be used. For complex $H_{BO}(s)$, it is recommended to draw the Nyquist path to test the validity of the phase margin test. The designer has to determine an appropriate phase margin, φ_m. If φ_m is too low, the closed-loop system will be prone to undesirable ringing and overshoots and with $\varphi_m < 0$, the system becomes unstable. If φ_m is too large, the response time will be too slow. In general, a phase margin between 30° and 70° offers a good trade-off between the response time and the stability margin. The example given in Para. 5.2.5 illustrates the design of an optimal phase margin.

5.2.4 Types of Compensators

In this paragraph, we review the basics of PD (proportional-derivative), PI (proportional-integral) and PID (proportional-integral derivative) compensators.

Proportional-derivative (PD) compensator

The main goal of a PD compensator is to increase the phase margin φ_m. The transfer function has a low-frequency zero and multiple high-frequency poles. The transfer function of such a compensator with only one high-frequency pole is given by:

$$C_{r\mathrm{PD}}(jw) = C_{r\mathrm{PD0}} \frac{\left(1 + \frac{jw}{w_z}\right)}{\left(1 + \frac{jw}{w_p}\right)}. \tag{5.14}$$

The corresponding Bode diagram with its asymptotic representation is shown in Fig. 5.4. The maximum phase $\varphi_{\max}$ occurs at the frequency $f_{\varphi\max}$ given by:

$$f_{\varphi\max} = \sqrt{f_z f_p}. \tag{5.15}$$

Generally, the designer aims to increase the phase margin from a given value of $\varphi_{\max}$ at the crossover frequency f_c of $H_{B0}(jw)$. So f_c and $\varphi_{\max}$ can be considered as input parameters. From Eqs. (5.14) and (5.15), we obtain the following equations expressing f_z and f_p in terms of f_c and $\varphi_{\max}$

$$f_z = f_c \sqrt{\frac{1 - \sin\varphi_{\max}}{1 + \sin\varphi_{\max}}}, \quad f_p = f_c \sqrt{\frac{1 + \sin\varphi_{\max}}{1 - \sin\varphi_{\max}}}. \tag{5.16}$$

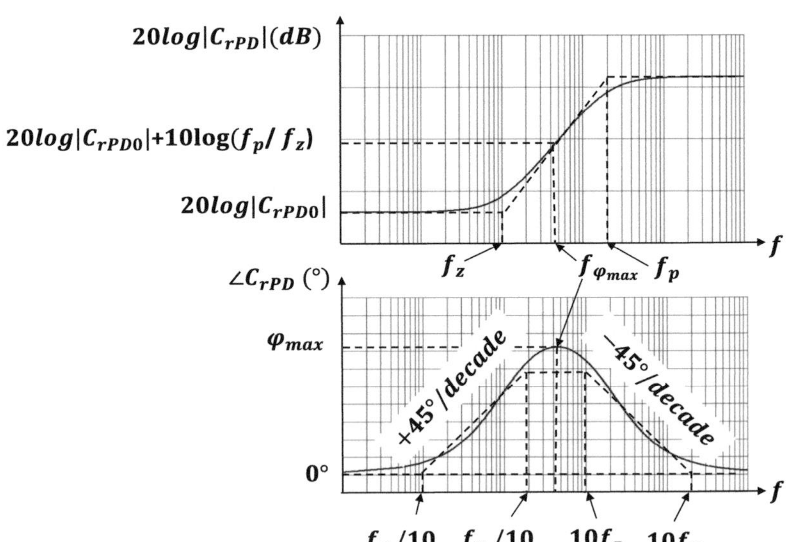

Fig. 5.4 Bode diagram of a PD compensator

To maintain the open-loop gain constant at f_c, the gain of the PD compensator must be equal to:

$$C_{rPD0} = \sqrt{\frac{f_z}{f_p}}. \tag{5.17}$$

If C_{rPD0} is different from this value, the closed-loop crossover frequency will be different from f_c. This type of compensator allows extending the bandwidth of the feedback loop while maintaining the desired phase margin.

Proportional-integral (PI) compensator

The main goal of a PI compensator is to increase the low-frequency open-loop gain in order to improve the regulation of the output at frequencies well below f_c. The transfer function contains an inverted zero at the frequency f_i as shown in the formula below:

$$C_{rPI}(jw) = C_{rPI0}\left(1 + \frac{w_i}{jw}\right). \tag{5.18}$$

Generally, f_i is chosen sufficiently lower than f_c to keep the phase margin unchanged. The Bode diagram of C_{rPI}, with its asymptotic representation, is shown in Fig. 5.5. A PI compensator allows one to obtain an error signal and a transfer function between the output and the disturbance that are close to zero at DC.

Combined (PID) compensator

Generally speaking, to comply with the regulation specifications, it is necessary to combine PI and PD compensators. The resulting compensator, called PID, has the following transfer function:

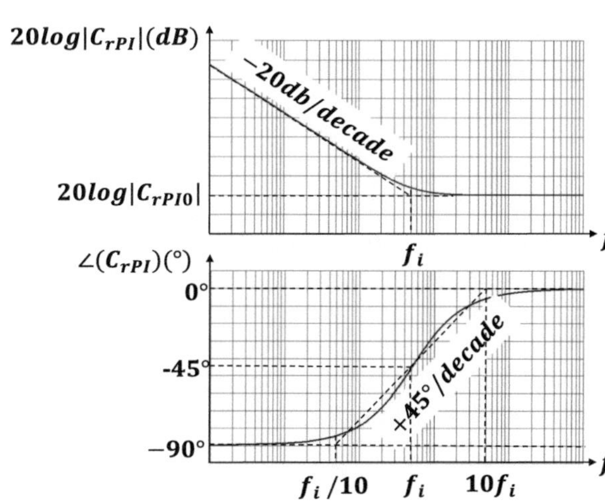

Fig. 5.5 Bode diagram of a PI compensator

$$C_{r\mathrm{PID}}(jw) = C_{r\mathrm{PID0}} \frac{\left(1 + \frac{jw}{w_z}\right)\left(1 + \frac{w_i}{jw}\right)}{\left(1 + \frac{jw}{w_{p1}}\right)\left(1 + \frac{jw}{w_{p2}}\right)}. \tag{5.19}$$

In practical design, a second pole is often added to reject high-frequency noise.

5.2.5 Illustrative Example of a PID Design

As an illustrative example, we consider the system shown in Fig. 5.2 with the following numerical data:

$$X_{ref} = 1, \quad T_r(s) = 0.5, \quad D(s) = 0.1, \quad H_P(s) = \frac{200}{\left(1 + \frac{s}{2\pi.100}\right)\left(1 + \frac{s}{2\pi.700}\right)}. \tag{5.20}$$

Our goal is to design a PID compensator given by Eq. (5.19). The crossover frequency of $T_rH(jw)$ is $f_c = 2.6$ kHz with, a phase margin equal to:

$$180^\circ + \angle T_rH(f_c) = 17.3^\circ. \tag{5.21}$$

We choose $f_i = 50$ Hz and $f_{p2} = 50$ kHz in order, respectively, to reduce the DC error and reject the high-frequency noise. f_i and f_{p2}, not being close to f_c, have no impact on the crossover frequency or the phase margin. In this example, we want to keep the same crossover frequency and increase the phase margin from 17.3° to $\varphi_m = 45^\circ$. Thus, it is possible to use Eqs. (5.16) and (5.17) to determine, respectively, f_z, f_{p1} and $C_{r\mathrm{PID0}}$. We have to increase the phase margin of $\varphi_{\max}$, which is given by:

$$\varphi_{\max} = \varphi_m - 180^\circ - \angle T_rH(f_c) = 27.7^\circ. \tag{5.22}$$

We get:

$$f_z \simeq 1.57 \text{ kHz}, \quad f_{p1} \simeq 4.3 \text{ kHz}, \quad C_{r\mathrm{PID0}} \simeq 0.604. \tag{5.23}$$

Figure 5.6 shows the corresponding Bode diagrams of T_rH_P and $C_{r\mathrm{PID}}$. We can do the same for different values of the phase margin, for example $\varphi_m = 17.3^\circ, 70^\circ, 130^\circ$. Figure 5.7 illustrates the input unit step response of the regulated closed-system for different values of φ_m. We find the same result as that mentioned in Para. 5.2.3. If φ_m is close to zero, the closed-loop system exhibits undesirable ringing, and if φ_m is too large, the response time increases. In this example, $\varphi_m = 70^\circ$ is a good trade-off. With this phase margin, the parameters of the PD compensator are:

$$f_z \simeq 877 \text{ Hz}, \quad f_{p1} \simeq 7.7 \text{ kHz}, \quad C_{r\mathrm{PID0}} \simeq 0.337. \tag{5.24}$$

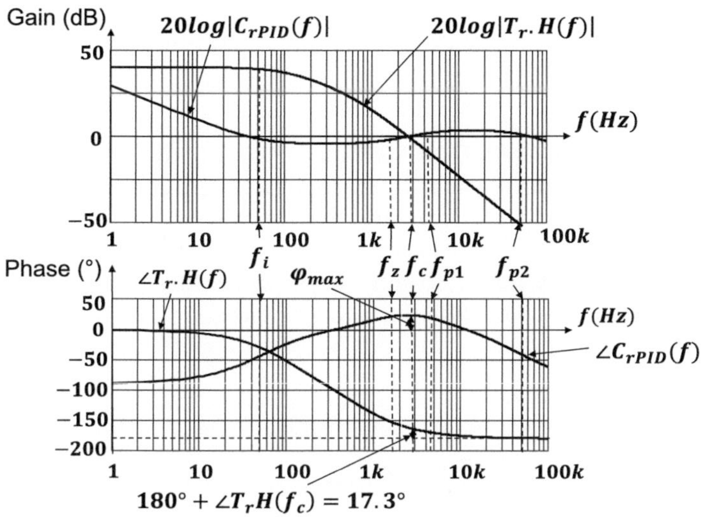

Fig. 5.6 Bode diagrams of T_rH and $C_{r\text{PID}}$

Let us now analyse the influence of the crossover frequency of H_{BO} on the response to the input unit step. We keep the same transfer functions $T_r(s)$ and $H(s)$, and the same frequencies f_i(50 Hz) and f_{p2}(50 kHz). Let us determine f_z, f_{p1} and $C_{r\text{PID}0}$ in order to get a phase margin $\varphi_m = 70°$ for $f_{c1} = 5$ kHz and $f_{c2} = 2.6$ kHz (already determined). Using Eq. (5.16) we get for $f_{c1} = 5$ kHz

$$f_z = 1\ kHz \quad \text{and} \quad f_p = 24.8\ \text{kHz}. \tag{5.25}$$

Now, $C_{r\text{PID}0}$ is calculated to have:

$$20\log|H_{BO}(f_{c1})| = 0. \tag{5.26}$$

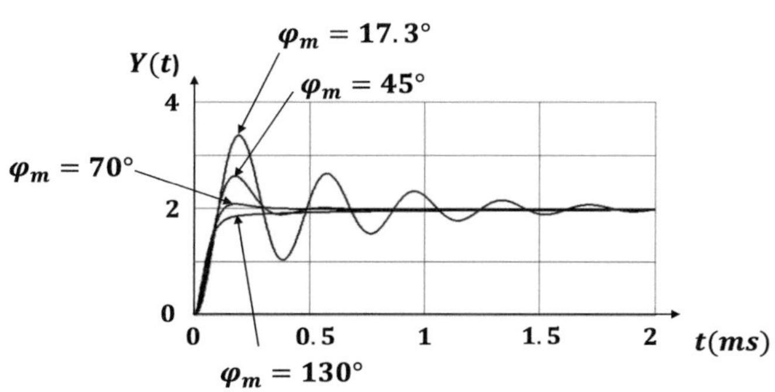

Fig. 5.7 Response to a unit step for different values of φ_m

We find:

$$C_{r\mathrm{PID0}} = 0.732. \tag{5.27}$$

The Bode diagrams of H_{BO} for $f_{c1} = 5$ kHz and $f_{c2} = 2.6$ kHz are shown in Fig. 5.8. The responses to a unit step input for f_{c1} and f_{c2} illustrated in Fig. 5.9. As expected, the response is faster with the higher crossover frequency. A designer is often confronted with conflicting specifications, for example, increasing the crossover frequency while rejecting high-frequency noise. To meet complex specifications, it is sometimes necessary to add zeros and poles to the PID compensator.

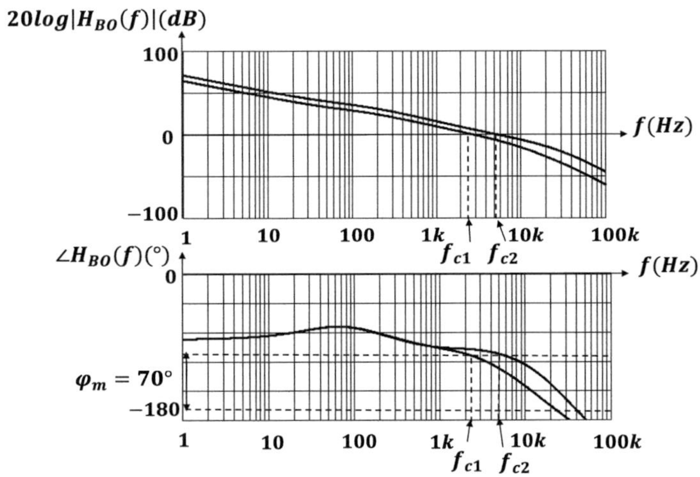

Fig. 5.8 Bode diagrams of H_{BO} for $f_{c1} = 5$ kHz and $f_{c2} = 2.6$ kHz

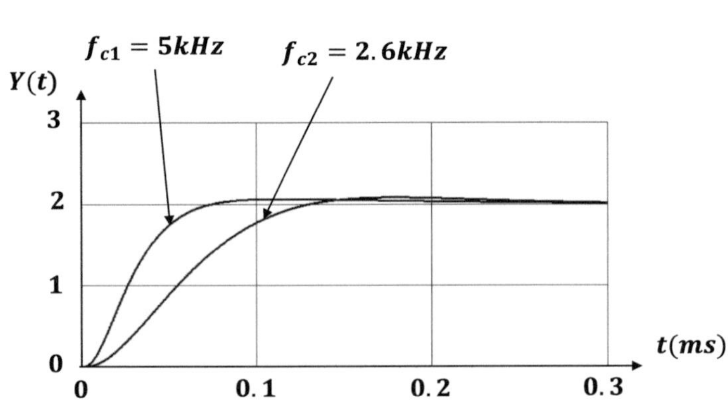

Fig. 5.9 Response to a unit step input for $f_{c1} = 5$ kHz and $f_{c2} = 2.6$ kHz

5.2.6 *Effect of a Right-Half-Plane Zero on the Transfer System Function*

The effect of a Right-Half-Plane Zero (RHPZ) is illustrated using a system having the following transfer function:

$$H_P(s) = \frac{\left(1 - \frac{s}{2\pi f_{\mathrm{RHPZ}}}\right)}{\left(1 + \frac{s}{2\pi f_{p1}}\right)\left(1 + \frac{s}{2\pi f_{p2}}\right)}. \tag{5.28}$$

At low frequencies, the gain

$$H_2(s) = \frac{-\frac{s}{2\pi \cdot f_{\mathrm{RHPZ}}}}{\left(1 + \frac{s}{2\pi f_{p1}}\right)\left(1 + \frac{s}{2\pi f_{p2}}\right)} \tag{5.29}$$

has negligible magnitude, so the output is determined only by:

$$H_1(s) = \frac{1}{\left(1 + \frac{s}{2\pi f_{p1}}\right)\left(1 + \frac{s}{2\pi f_{p2}}\right)}. \tag{5.30}$$

At high frequencies, the gain of $H_P(s)$ is practically equal to the gain of $H_2(s)$. Hence, there is a phase reversal at high frequencies due to the negative sign.
Figure 5.10 shows the response to a unit step input of $H_1(s)\left(Y_{H_1}(t)\right)$, $H_2(s)\left(Y_{H_2}(t)\right)$, and $H_P(s) = H_1(s)+H_2(s)$ $(Y_H(t))$. The simulation is performed using the following parameters:

$$f_{p1} = 500\ \mathrm{Hz}, \quad f_{p2} = 900\ \mathrm{Hz} \quad \text{and} \quad f_{\mathrm{RHPZ}} = 700\ \mathrm{Hz}. \tag{5.31}$$

The output $Y_H(t)$ initially moves in the opposite direction from the final value. It is a classical transient response of a system containing a Right-Half-Plane Zero.

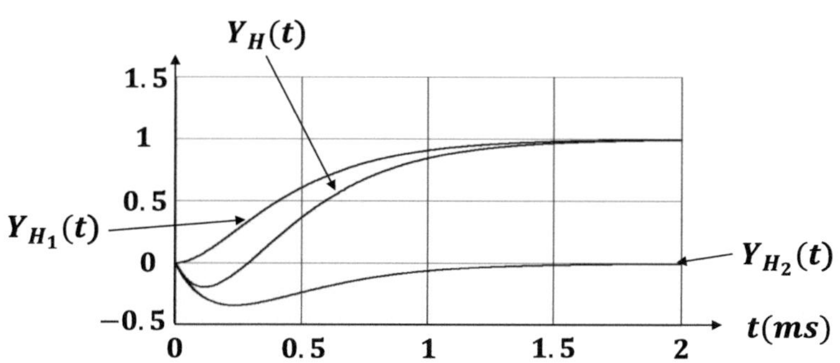

Fig. 5.10 Response to a step input of $H_1(s)$, $H_2(s)$ and $H_P(s) = H_1(s) + H_2(s)$

5.2.7 Example of a Regulator Design for a System Containing a RHPZ

The basic idea consists of designing a compensator with a crossover frequency lower than f_{RHPZ}. Consider the system defined by Eqs. (5.28) and (5.31) with a unit return ($T_r(s) = 1$ and $X_{\mathrm{ref}} = 1$). The regulator is a PID controller with the transfer function given by Eq. (5.19). We choose:

$$f_i = 50\ \mathrm{Hz} \quad \text{and} \quad f_{p2} = 10\ \mathrm{kHz} \tag{5.32}$$

to reduce the DC error and reject high-frequency noise, respectively. f_{p1}, f_z and $C_{r\mathrm{PID0}}$ will be determined so to achieve a phase margin $\varphi_m = 70°$ for the following two crossover frequencies, $f_{c1} = 300$ Hz and $f_{c2} = 500$ Hz.
Using the methodology of the previous paragraph, we obtain for $f_{c1} = 300$ Hz :

$$f_{p11} \simeq 187\ \mathrm{Hz}, \quad f_{z1} \simeq 482\ \mathrm{Hz}, \quad C_{r\mathrm{PID01}} \simeq 1.792, \tag{5.33}$$

and for $f_{c2} = 500$ Hz

$$f_{p12} \simeq 577\ \mathrm{Hz}, \quad f_{z2} \simeq 433\ \mathrm{Hz}, \quad C_{r\mathrm{PID02}} \simeq 1.137. \tag{5.34}$$

It is one of the cases where it is recommended to check the stability margin using the Nyquist diagram. This diagram is shown in Fig. 5.11. We observe that the stability margin ($d_{1\,\mathrm{min}}$, $d_{2\,\mathrm{min}}$) is greatest for the crossover frequency farthest from f_{RHPZ}. The responses to a unit step are shown in Fig. 5.12.
The response obtained with the crossover frequency nearest to f_{RHPZ} has more ringing. Due to the low values of the crossover frequencies, the response times are slow. That is the major control problem of a system with a low frequency f_{RHPZ}.

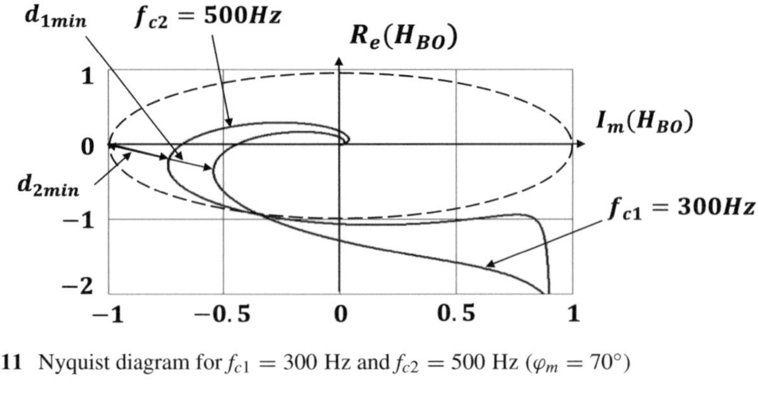

Fig. 5.11 Nyquist diagram for $f_{c1} = 300$ Hz and $f_{c2} = 500$ Hz ($\varphi_m = 70°$)

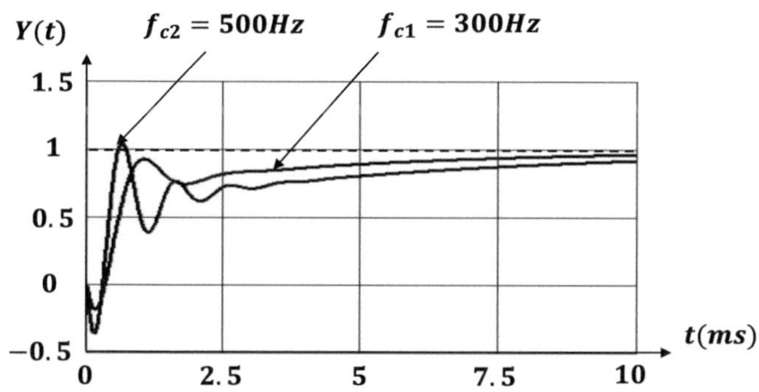

Fig. 5.12 Response to a unit step for $f_{c1} = 300$ Hz and $f_{c2} = 500$ Hz

5.3 Controller of a Multiple-Input and Multiple-Output Linear System

Let us consider, using the usual notation, a linear system of the form:

$$\frac{\mathrm{d}\vec{X}}{\mathrm{d}t} = A\vec{X} + B\vec{U}, \vec{Y} = C\vec{X} \tag{5.35}$$

where A, B, and C are $n \times n$, $n \times u$ and $y \times n$ matrices, respectively (Skogestad & Postlethwaite, 2005).

5.3.1 *Controllability of a MIMO Linear System*

A linear MIMO system is controllable if and only if the following controllability matrix

$$C = \left[BABA^2B \ldots A^{n-1}B\right] \tag{5.36}$$

has full rank, i.e.,

$$\text{rank } C = n. \tag{5.37}$$

5.3.2 *Observability of a MIMO Linear Systems*

A linear MIMO system is observable if and only if the following observability matrix

$$\mathrm{O} = \begin{bmatrix} C \\ CA \\ \vdots \\ CA^{n-1} \end{bmatrix} \tag{5.38}$$

has full rank, i.e.,

$$\text{rank } \mathrm{O} = n. \tag{5.39}$$

5.3.3 *Illustrative Examples of Controllability and Observability*

As an illustration, consider the following example ($n = 2, u = 1, y = 2$):

$$\frac{\mathrm{d}\vec{X}}{\mathrm{d}t} = \begin{pmatrix} 1 & 4 \\ 3 & 2 \end{pmatrix}\vec{X} + \begin{pmatrix} 1 \\ 2 \end{pmatrix}\vec{U},\ \vec{Y} = \begin{pmatrix} 1 & 1 \\ 5 & 0 \end{pmatrix}\vec{X}. \tag{5.40}$$

We find:

$$\text{rank } C = \text{rank}\begin{pmatrix} 1 & 9 \\ 2 & 7 \end{pmatrix} = n = 2 \text{ and rank } \mathrm{O} = \text{rank}\begin{pmatrix} 1 & 1 \\ 5 & 0 \\ 4 & 6 \\ 5 & 20 \end{pmatrix} = n = 2; \tag{5.41}$$

therefore, this system is controllable and observable.

5.3.4 *Transfer Function Matrix*

In order to simplify the notation, we now consider a MIMO system with two inputs and two outputs. The system is defined by the transfer matrix $H_C(s)$ as follows:

$$\begin{pmatrix} Y_1(s) \\ Y_2(s) \end{pmatrix} = \begin{pmatrix} H_{C11}(s) & H_{C12}(s) \\ H_{C21}(s) & H_{C22}(s) \end{pmatrix}\begin{pmatrix} U_1(s) \\ U_2(s) \end{pmatrix} = H_C(s)\begin{pmatrix} U_1(s) \\ U_2(s) \end{pmatrix}. \tag{5.42}$$

The corresponding function block is shown in Fig. 5.13.

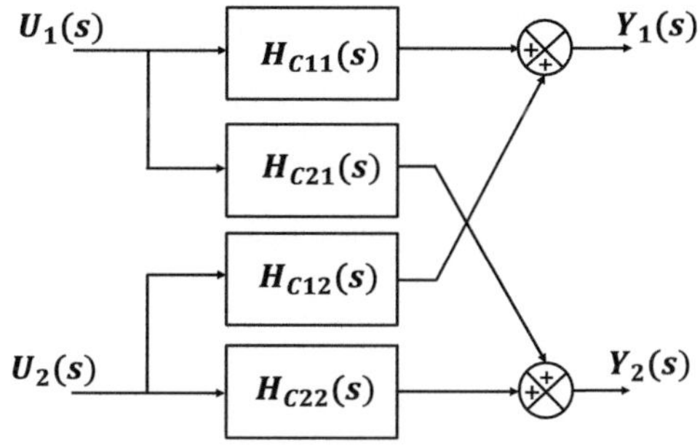

Fig. 5.13 Function block of a two-input, two-output linear system

5.3.5 *Relative Gain Array*

The scale of the cross effect in a MIMO system can be determined using the Relative Gain Array (RGA) matrix. Let U_i and Y_i be a particular input–output pair for the MIMO plan $H_C(s)$. It is possible to define the following two extreme gains between U_i and Y_j.

- The open-loop gain OLG_{ij} when all the other inputs are equal to zero is:

$$\mathrm{OLG}_{ij}(s) = \left(\frac{Y_j(s)}{U_i(s)}\right)_{U_k(s)=0, k \neq i}. \tag{5.43}$$

- The closed-loop gain CLG_{ij} when all the other outputs are equal to zero is:

$$\mathrm{CLG}_{ij}(s) = \left(\frac{Y_j(s)}{U_i(s)}\right)_{Y_k(s)=0, k \neq j}. \tag{5.44}$$

The value $\lambda_{ij}(s)$ of an element in the RGA matrix is:

$$\lambda_{ij}(S) = \frac{\mathrm{OLG}_{ij}(s)}{\mathrm{CLG}_{ij}(s)}. \tag{5.45}$$

For a two-input, two-output system, we can easily calculate $\lambda_{1,1}$. First, consider the case when $U_2(s) = 0$. we have:

$$\mathrm{CLG}_{11}(s) = H_{C11}(s). \tag{5.46}$$

Next, consider the case when the other loop is closed with $Y_2(s) = 0$. Setting $Y_2(s) = 0$ in Eq. (5.42) gives:

$$U_2(s) = -\frac{H_{C21}(s)}{H_{C22}(s)} U_1(s). \tag{5.47}$$

It follows that:

$$Y_1(s) = \left(H_{C11}(s) - \frac{H_{C21}(s)}{H_{C22}(s)} H_{C12}(s)\right) U_1(s). \tag{5.48}$$

The $\lambda_{11}(s)$ RGA element becomes:

$$\lambda_{11}(s) = \frac{1}{1 - \frac{H_{C12}(s)H_{C21}(s)}{H_{C11}(s)H_{C22}(s)}}. \tag{5.49}$$

Applying the same method to calculate $\lambda_{12}(s)$, $\lambda_{21}(s)$, and $\lambda_{22}(s)$, we find

$$\text{RCA}(H(s)) = \begin{pmatrix} \lambda_{11}(s) & \lambda_{12}(s) \\ \lambda_{21}(s) & \lambda_{22}(s) \end{pmatrix} = \begin{pmatrix} \lambda_{11}(s) & 1 - \lambda_{11}(s) \\ 1 - \lambda_{11}(s) & \lambda_{11}(s) \end{pmatrix}. \tag{5.50}$$

The RGA matrix can be used to select a control loop in the following cases:

- If $|\lambda_{1,1}| = 0$, the control of the system can only be achieved by pairing U_1 with Y_2 and U_2 with Y_1. These two pairs will be non-interacting.
- If $|\lambda_{1,1}| = 1$, the control of the system can only be achieved by pairing U_1 with Y_1 and U_2 with Y_2. These two pairs will be non-interacting.
- If $|\lambda_{1,1}| = 0.5$, the two inputs affect the two outputs to the same degree. This is the worst case in terms of control.
- If $|\lambda_{1,1}| < 0.5$, the most suitable pairings will be U_1 with Y_2 and U_2 with Y_1.
- If $0.5 < |\lambda_{1,1}| < 1$, the most suitable pairings will be U_1 with Y_1 and U_2 with Y_2.
- If $|\lambda_{1,1}| > 1$, the pairings U_1 with Y_2 and U_2 with Y_1 are unsuitable because the interaction will cause the controlled outputs to move in a direction away from the one the control is trying to achieve.

The transfer matrix RGA is generally computed as a function of frequency. The previous rules must be applied within the system's frequency range.

5.3.6 Decoupler

In a MIMO system, interactions occur because one input affects more than one controlled output. One possible solution to eliminate these interactions is to design a decoupler. A possible matrix transfer function $D(s)$ for a decoupler is given by

$$\begin{pmatrix} U_1(s) \\ U_2(s) \end{pmatrix} = \begin{pmatrix} 1 & -\frac{H_{C12}(s)}{H_{C11}(s)} \\ -\frac{H_{C21}(s)}{H_{C22}(s)} & 1 \end{pmatrix} \begin{pmatrix} U_1^*(s) \\ U_2^*(s) \end{pmatrix} = D(s) \begin{pmatrix} U_1^*(s) \\ U_2^*(s) \end{pmatrix}. \tag{5.51}$$

The corresponding block diagram is shown in Fig. 5.14.

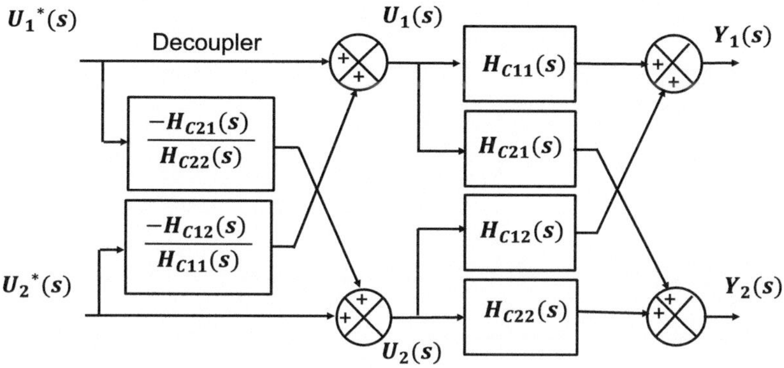

Fig. 5.14 Decoupler associated with a two-input, two-output linear system

Augmenting the transfer matrix $H(s)$ with the decoupler $D(s)$ allows mathe-matically eliminating the effect of the interactions, as the relation below shows:

$$\begin{pmatrix} Y_1(s) \\ Y_2(s) \end{pmatrix} = \begin{pmatrix} H_{C11}(s) - \frac{H_{C12}(s)\cdot H_{C21}(s)}{H_{C22}(s)} & 0 \\ 0 & H_{C22}(s) - \frac{H_{C12}(s)\cdot H_{C21}(s)}{H_{C11}(s)} \end{pmatrix} \begin{pmatrix} U_1^*(s) \\ U_2^*(s) \end{pmatrix}. \quad (5.52)$$

By adding a decoupler, the control design of a MIMO system is reduced to the control design of several independent SISO systems.

5.3.7 *Illustrative Example*

The block diagram of the illustrative example is shown in Fig. 5.15. The transfer matrix function is

$$H_C(s) = \begin{pmatrix} H_{C11}(s) \; H_{C12}(s) \\ H_{C21}(s) \; H_{C22}(s) \end{pmatrix} = \begin{pmatrix} \frac{20}{\left(1+\frac{s}{2\pi\cdot f_{p1}}\right)} & 4 \\ -2 & \frac{5}{\left(1+\frac{s}{2\pi\cdot f_{p2}}\right)} \end{pmatrix}, \quad (5.53)$$

with $f_{p1} = 500$ Hz and $f_{p2} = 1500$ Hz.
The two reference DC signals are $X_{\text{ref1}} = 2$ and $X_{\text{ref2}} = 1$. Using Eq. (5.49), we can plot $|\lambda_{1,1}(f)|$ in relation to f. This curve is shown in Fig. 5.16. At low frequencies, $0.5 < |\lambda_{1,1}| < 1$, so the most suitable pairings will be U_1 with Y_1 and U_2 with Y_2.since $|\lambda_{1,1}|$ is different from 1, we will use a decoupler, as shown in Fig. 5.15, to avoid any interactions between U_1^* and Y_2 and between U_2^* and Y_1.

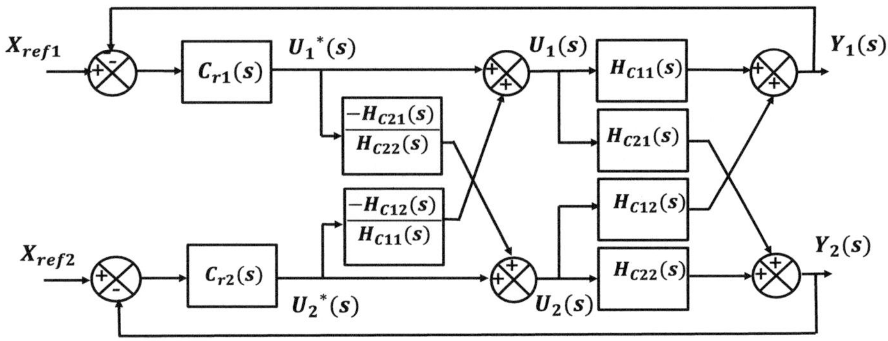

Fig. 5.15 Block diagram of the illustrative example

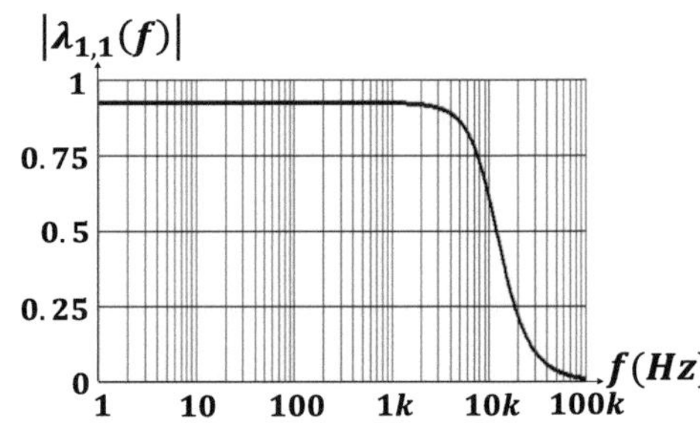

Fig. 5.16 Relation between $|\lambda_{1,1}(f)|$ and f

The control problem is thus reduced to the design of controllers for two SISO systems characterised by the following transfer functions:

$$H_1(s) = H_{C11}(s) - \frac{H_{C12}(s)H_{C21}(s)}{H_{C22}(s)} = \frac{20 + \frac{8}{5}\left(1 + \frac{s}{2\pi f_{p1}}\right)\left(1 + \frac{s}{2\pi f_{p2}}\right)}{\left(1 + \frac{s}{2\pi f_{p1}}\right)} \tag{5.54}$$

and

$$H_2(s) = H_{C22}(s) - \frac{H_{C12}(s)H_{C21}(s)}{H_{C11}(s)} = \frac{5 + \frac{2}{5}\left(1 + \frac{s}{2\pi f_{p1}}\right)\left(1 + \frac{s}{2\pi f_{p2}}\right)}{\left(1 + \frac{s}{2\pi f_{p2}}\right)}. \tag{5.55}$$

The block diagram corresponding to the control output Y_1 is shown in Fig. 5.17.

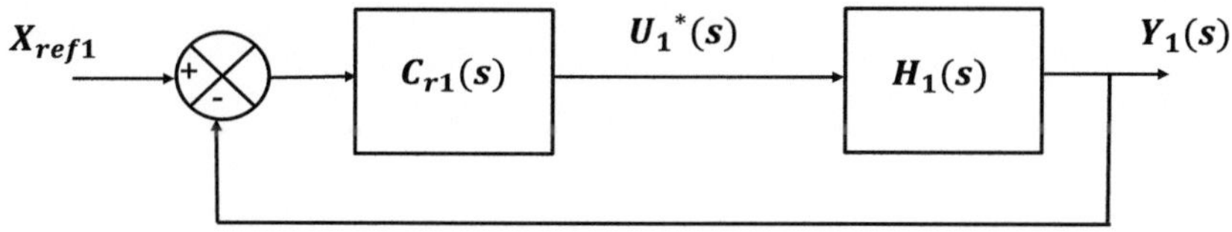

Fig. 5.17 Block diagram of the equivalent SISO system of Y_1

Chapter 6
Optimal Control

6.1 Introduction

6.1.1 *History*

Over the centuries, optimal control has emerged from extensive research in the field of calculus of variations. Renowned mathematicians and physicists including Newton, the Bernoulli family, Euler, Lagrange, Laplace, Hamilton and Hilbert have contributed to the development of this field. In physics, fields such as mechanics, electromagnetism, gravitation, quantum mechanics, string theory and many others can be explored using this theory. Every minimisation problem can be solved using the calculus of variation. In this introduction, we will review the basic variational problem and the associated Euler–Lagrange equation. As an illustration, we will present the resolution to, two classical problems: the shortest plane curve joining two points and the famous brachistochrone problem (Ramana, 2006).

6.1.2 *Variational Problem*

Consider the following integral expression:

$$I(y(x)) = \int_{x_1}^{x_2} f(x, y, y')\mathrm{d}x \quad \text{with } y' = \frac{\mathrm{d}y}{\mathrm{d}x}. \tag{6.1}$$

$I(y(x))$ is called the cost functional. A function $y = y(x)$ that extremises $I(y(x))$ and satisfies the conditions $y(x_1) = y_1$ and $y(x_2) = y_2$ is called an extremal function of $I(y(x))$. A variational problem consists of finding such an extremal function.

B. Rivet, *Theories of Nonlinear Systems for Power Converters*,
https://doi.org/10.1007/978-3-032-25062-9_6

6.1.3 Euler–Lagrange Equation

A necessary and sufficient condition for a function $y(x)$ to extremise $I(y(x))$ is that it satisfies the following Lagrange's equation:

$$\frac{\partial f}{\partial y} - \frac{\mathrm{d}}{\mathrm{d}x}\left(\frac{\partial f}{\partial y'}\right) = 0 \quad \text{with } x_1 < x < x_2. \tag{6.2}$$

6.1.4 Shortest Distance Between Two Points

The length L of a curve C, defined by $y = y(x)$ in the plane (x, y) and passing through the two points $P_1 = (x_1, y_1)$ and $P_2 = (x_2, y_2)$ (see Fig. 6.1), can be expressed as follows:

$$L(y(x)) = \int_{x_1}^{x_2} \sqrt{1 + y'^2}\mathrm{d}x. \tag{6.3}$$

Equation (6.3) is obtained by integrating the infinitesimal element dL which, has the form:

$$dL^2 = dx^2 + dy^2. \tag{6.4}$$

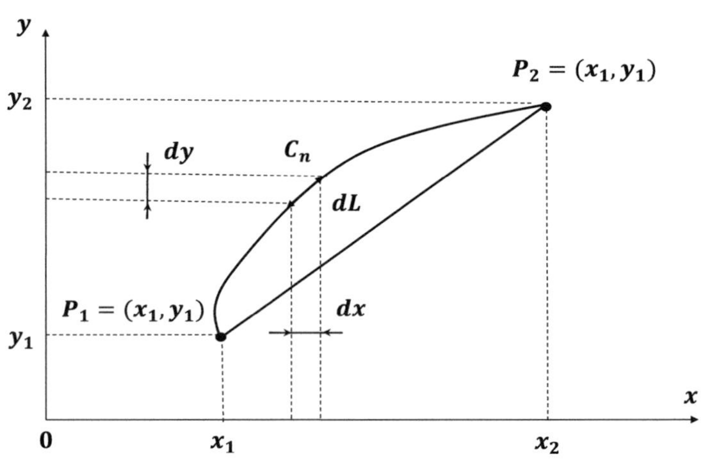

Fig. 6.1 Shortest distance between two points $P_1(x_1, y_1)$ and $P_2(x_2, y_2)$

The problem consists of finding, among the infinite number of curves C_n passing through points P_1 and P_2, the curve, that minimises $L(y(x))$.
In others words, we want to determine the function $y = y(x)$ that minimises the cost functional $L(y(x))$. Here,

$$f\left(x, y, y'\right) = \sqrt{1 + y'^2} \tag{6.5}$$

so

$$\frac{\partial f}{\partial y} = 0 \quad \text{and} \quad \frac{\partial f}{\partial y'} = \frac{y'}{\sqrt{1 + y'^2}}. \tag{6.6}$$

Then, the Euler–Lagrange equation leads to:

$$\frac{\mathrm{d}}{\mathrm{d}x}\left(\frac{y'}{\sqrt{1 + y'^2}}\right) = 0, \quad \text{so } y' = k\sqrt{1 + y'^2}. \tag{6.7}$$

Since k is a constant, we obtain:

$$y' = \pm\sqrt{\frac{k^2}{1 - k^2}} = p. \tag{6.8}$$

Finally, by integrating, we find:

$$y(x) = px + m. \tag{6.9}$$

The two constants p and m can be determined using (x_1, y_1) and (x_2, y_2). Thus, the curve joining points $P_1(x_1, y_1)$ and $P_2(x_2, y_2)$ with the shortest distance is a straight line.

6.1.5 Brachistochrone Problem

This is the most famous variational problem in the calculus of variations. It was posed by the mathematician Johann Bernoulli in 1696. It consists of finding the curve on which a particle slides without friction from point $A_1(x_1, y_1)$ to point $A_2(x_2, y_2)$ in the shortest time. Consider the positive direction of the y-axis to be vertically downward, and let $P(x, y)$ be the position of the particle at any time t, as shown in Fig. 6.2. To simplify the problem, we choose point A_1 as the origin and set the initial speed to zero.

The particle moves without friction, so energy is conserved, and the speed V along the trajectory is given by:

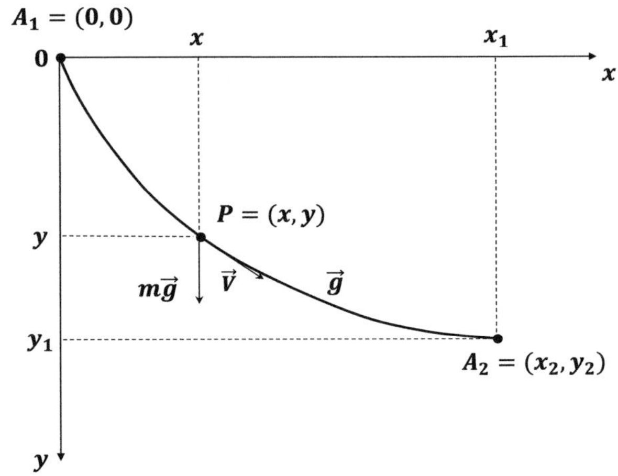

Fig. 6.2 Brachistochrone problem

$$V = \sqrt{2gy} = \sqrt{1 + y'^2}\frac{\mathrm{d}x}{\mathrm{d}t} \tag{6.10}$$

where g is the acceleration due to the gravity. From Eq. (6.10), we can express the time $t(y(x))$ taken by the particle to move from $A_1(0, 0)$ to $A_2(x_2, y_2)$ as follows:

$$t(y(x)) = \frac{1}{\sqrt{2g}} \int_{x_1=0}^{x_2} \frac{\sqrt{1 + y'^2}}{\sqrt{y}} \mathrm{d}x \tag{6.11}$$

Here, the cost functional is $t(y(x))$ and

$$f(x, y, y') = \frac{\sqrt{1 + y'^2}}{\sqrt{y}}. \tag{6.12}$$

A different way to write the Euler–Lagrange equation is

$$\frac{\mathrm{d}}{\mathrm{d}x}\left(f - y'\frac{\partial f}{\partial y'}\right) - \frac{\partial f}{\partial x} = 0, \tag{6.13}$$

which lead to

$$\frac{\mathrm{d}}{\mathrm{d}x}\left(\frac{\sqrt{1 + y'^2}}{\sqrt{y}} - \frac{y'^2}{\sqrt{y(1 + y'^2)}}\right) = 0. \tag{6.14}$$

Integrating, we get

$$y\left(1 + y'^2\right) = k, \tag{6.15}$$

with k being a constant. With the following change of variable: $y' = \cot\theta$, we have:

$$y = \frac{k}{1+\cot^2\theta} = k\sin^2\theta = \frac{k}{2}(1-\cos 2\theta). \tag{6.16}$$

Integrating

$$\mathrm{d}x = \frac{\mathrm{d}y}{y'} = k(1-\cos 2\theta)\mathrm{d}\theta, \tag{6.17}$$

we find:

$$x - k_1 = \frac{k}{2}(2\theta - \sin 2\theta). \tag{6.18}$$

Since $x = y = 0$ for $t = 0$, the constant of integration is $k_1 = 0$. Setting $\varphi = 2\theta$, we finally obtain:

$$x = \frac{k}{2}(\varphi - \sin\varphi),\ y = \frac{k}{2}(1-\cos\varphi). \tag{6.19}$$

The corresponding curve is a cycloid. The value of k can be determined knowing that the curve passes through $A_2(x_2, y_2)$.

6.2 Basic Optimal Problem

6.2.1 Introduction

We begin our discussion by considering the autonomous nonlinear system defined by the following ordinary differential equation

$$\frac{\mathrm{d}\vec{X}}{\mathrm{d}t} = \vec{F}\left(\vec{X}(t), \vec{U}(t)\right). \tag{6.20}$$

$\vec{X}(t) \in \mathbb{R}^n$ is the state vector, $\vec{U}(t) \in A$ is a control vector defined in the set of all admissible controls $A \subset \mathbb{R}^u$ and $\vec{F} : \mathbb{R}^n \times \mathbb{R}^u \rightarrow \mathbb{R}^n$ is, as usual, a nonlinear smooth function. General boundary conditions are given by

$$\vec{X}(0) \in M_0 \quad \text{and} \quad \vec{X}(T) \in M_T, \tag{6.21}$$

where $M_0 \subset \mathbb{R}^n$ and $M_T \subset \mathbb{R}^n$ are smooth submanifolds, respectively at the origin $(t = 0)$ and at the final time $(t = T)$. They can be defined by

$$M_0 = \left\{ \vec{X} \in \mathbb{R}^n / f_{01}\left(\vec{X}\right) = \cdots = f_{0k}\left(\vec{X}\right) = 0 \right\}, \quad (6.22)$$

$$M_T = \left\{ \vec{X} \in \mathbb{R}^n / f_{T1}\left(\vec{X}\right) = \cdots = f_{Tk'}\left(\vec{X}\right) = 0 \right\}, \quad (6.23)$$

where the f_{ij} are functions from $\mathbb{R}^n$ to $\mathbb{R}$. The aim of the optimal control is to determine the best control $\overrightarrow{U^*}(t)$ that allows moving the state vector $\vec{X}(t)$ from $\vec{X}(0)$ to $\vec{X}(T)$ by maximising (or minimising) the following cost functional $J(\vec{U}(t))$:

$$J(\vec{U}(t)) = \int_0^T L\left(\vec{X}, \vec{U}\right) \mathrm{d}t + g(\vec{X}(T)) \quad (6.24)$$

L, defined by

$$L : \mathbb{R}^n \times A \to \mathbb{R}, \quad (6.25)$$

is called the Lagrangian, and

$$g : \mathbb{R}^n \to \mathbb{R}, \quad (6.26)$$

is the terminal cost function. In other words, we want to find the optimal control $\overrightarrow{U^*}(t)$ for which

$$J\left(\overrightarrow{U^*}(t)\right) > J\left(\vec{U}(t)\right), \quad (6.27)$$

for all controls $\vec{U}(t) \in A$.

In this paragraph, we will assume that the optimal control exists. Note that the final time T, $\vec{X}(0)$ and $\vec{X}(T)$ can be partially or completely free or fixed (Evans, 2001; Sethi & Thompson, 2000).

6.2.2 Lagrange, Time Optimal and Mayer Problems

Lagrange's problem is defined by $g(\vec{X}(T)) = 0$. The cost functional is

$$J(\vec{U}(t) = \int_0^T L\left(\vec{X}, \vec{U}\right) \mathrm{d}t. \quad (6.28)$$

The time-optimal problem is a Lagrange problem where $L = -1$, so

$$J(\vec{U}(t) = \int_0^T -\mathrm{d}t = -T. \quad (6.29)$$

The Mayer problem is defined when the cost functional is equal to the terminal cost function:

$$J(\vec{U}(t) = g(\vec{X}(T)). \quad (6.30)$$

6.2.3 Bang-Bang Control

We will consider A to be a cube $[U_{i\min}, U_{i\max}]^u$ in $\mathbb{R}^u$ $(i = 1 \ldots u)$. A control $\vec{U}(t) \in A$ is called bang-bang (Evans, 2001) if $\forall 0 \leq t \leq T$, we have:

$$U_i(t) = U_{i\min} \quad \text{or} \quad U_{i\max}. \quad (6.31)$$

6.3 Pontryagin's Theorem

6.3.1 Hamiltonian Definition

The Hamiltonian function is expressed as

$$H\left(\vec{X}, \vec{p}, \vec{U}\right) = \vec{p}\vec{F}\left(\vec{X}, \vec{U}\right) + L\left(\vec{X}, \vec{U}\right), \quad (6.32)$$

$\vec{p}(t) \in \mathbb{R}^n$ is called the adjoint vector.

6.3.2 Theorem

We will only consider the case in which the state vector $\vec{X}(t) \in \mathbb{R}^n$. Assume $\overrightarrow{U^*}(t)$ is the optimal control for the system defined by:

$$\frac{\mathrm{d}\vec{X}(t)}{\mathrm{d}t} = \vec{F}\left(\vec{X}(t), \vec{U}(t)\right) \quad \text{with } \vec{X}(0) \in M_0 \text{ and } \vec{X}(T) \in M_T. \quad (6.33)$$

The cost functional is given by Eq. (6.24). $\overrightarrow{X^*}(t)$ is the corresponding optimal state vector. Pontryagin's theorem (Evans, 2001; Pontryagin,1962) states that there exists an adjoint vector function:

$$\vec{p} : t \in [0, T] \to \mathbb{R}^n, \tag{6.34}$$

such that

- $\dfrac{\mathrm{d}\vec{X}}{\mathrm{d}t} = \dfrac{\partial H}{\partial \vec{p}}$
- $\dfrac{\mathrm{d}\vec{p}}{\mathrm{d}t} = -\dfrac{\partial H}{\partial \vec{X}}$
- $H\left(\overrightarrow{X^*}, \overrightarrow{p^*}, \overrightarrow{U^*}\right) = \max\limits_{\vec{U}} H\left(\vec{X}, \vec{p}, \vec{U}\right)$ or $H\left(\overrightarrow{X^*}, \overrightarrow{p^*}, \overrightarrow{U^*}\right) = \min\limits_{\vec{U}} H\left(\vec{X}, \vec{p}, \vec{U}\right)$
- Transversality conditions must be satisfied (to be discussed in the next section)

(6.35)

6.3.3 Transversality Conditions

Transversality conditions on $\vec{p}$

The transversality conditions on $\vec{p}$ can be expressed as follows:

- The adjoint vector $\overrightarrow{p^*}(0)$ must be orthogonal to the submanifold M_0, which can be translated by:

$$\exists \lambda_1, \ldots \lambda_k \in \mathbb{R}, \quad \text{such that } \overrightarrow{p^*}(0) = \begin{pmatrix} \lambda_1 \frac{\partial f_{01}}{\partial X_1} + \cdots + \lambda_k \frac{\partial f_{0k}}{\partial X_1} \\ \vdots \\ \lambda_1 \frac{\partial f_{01}}{\partial X_n} + \cdots + \lambda_k \frac{\partial f_{0k}}{\partial X_n} \end{pmatrix}. \tag{6.36}$$

- The adjoint vector $\vec{p}(T)$ minus $\frac{\partial g(\vec{X}(T))}{\partial \vec{X}}$ must be orthogonal to the submain fold M_T, which can be translated by:

$$\exists \alpha_1, \ldots \alpha_{k'} \in \mathbb{R}, \quad \text{such that} \overrightarrow{p^*}(T) = \begin{pmatrix} \alpha_1 \frac{\partial f_{T1}}{\partial X_1} + \cdots + \alpha_{k'} \frac{\partial f_{Tk'}}{\partial X_1} \\ \vdots \\ \alpha_1 \frac{\partial f_{T1}}{\partial X_n} + \cdots + \alpha_{k'} \frac{\partial f_{Tk'}}{\partial X_n} \end{pmatrix} + \begin{pmatrix} \frac{\partial g(\vec{X}(T))}{\partial X_1} \\ \vdots \\ \frac{\partial g(\vec{X}(T))}{\partial X_n} \end{pmatrix}. \tag{6.37}$$

Transversality Conditions on H

For an autonomous system with free final time T, the transversality condition on H is given by:

$$\forall 0 \leq t \leq T, \quad H\left(\overrightarrow{X^*}(t), \overrightarrow{p^*}(t), \overrightarrow{U^*}(t)\right) = 0. \tag{6.38}$$

Particular Cases

- If $T, \vec{X}(0)$ and $\vec{X}(T)$ are fixed, then $\overrightarrow{p^*}(0)$ and $\overrightarrow{p^*}(T)$ are free.
- If T and $\vec{X}(T)$ are fixed and $\vec{X}(0)$ is free, then $\overrightarrow{p^*}(T)$ is free and $\overrightarrow{p^*}(0) = \vec{0}$.
- If T and $\vec{X}(0)$ are fixed and $\vec{X}(T)$ is free, then $\overrightarrow{p^*}(0)$ is free and $\overrightarrow{p^*}(T)$ is given by:

$$\overrightarrow{p^*}(T) = \frac{\partial g\left(\vec{X}(T)\right)}{\partial \vec{X}}. \tag{6.39}$$

- If T and $\vec{X}(0)$ are fixed, then $g\left(\vec{X}(T)\right) = 0$ and

$$M_T = \left\{\vec{X}(X_1, X_2, X_3) \in \mathbb{R}^3 / f_{T1}\left(\vec{X}\right) = X_1 - X_1(T) = f_{T2}\left(\vec{X}\right) = X_2 - X_2(T) = 0\right\}. \tag{6.40}$$

With $X_1(T)$ and $X_2(T)$ fixed, $\overrightarrow{p^*}(T)$ is equal to:

$$\overrightarrow{p^*}(T) = \begin{pmatrix} \alpha_1 \frac{\partial f_{T1}}{\partial X_1} + \alpha_2 \frac{\partial f_{T2}}{\partial X_1} \\ \alpha_1 \frac{\partial f_{T1}}{\partial X_2} + \alpha_2 \frac{\partial f_{T2}}{\partial X_2} \\ \alpha_1 \frac{\partial f_{T1}}{\partial X_3} + \alpha_2 \frac{\partial f_{T2}}{\partial X_3} \end{pmatrix} = \begin{pmatrix} \alpha_1 \\ \alpha_2 \\ 0 \end{pmatrix}. \tag{6.41}$$

The useful result is $p_3^*(T) = 0$.
Through various examples, we will first examine the case in which the optimal control is continuous. Then, the case in which H depends linearly on a bang-bang control we will be considered, including both singular and non-singular bang-bang controls.

6.3.4 *Example of Optimal Continuous Control*

As a first example of optimal control (Evans, 2001), we consider the simple circuit shown in Fig. 6.3. The order of this system is one ($n = 1$), the state parameter is V_C and the control is $I (u = 1)$. We want to minimise the quantity:

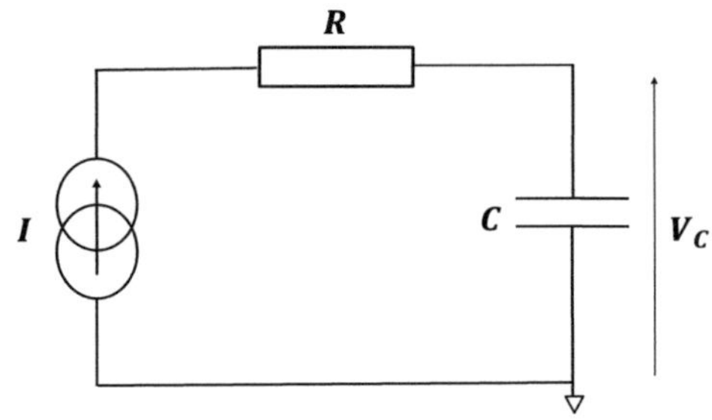

Fig. 6.3 R, C, I circuit

$$\frac{1}{2}\int_{0}^{T}(V_C^2 + R^2 I^2)\mathrm{d}t. \tag{6.42}$$

The initial condition $V_C(0) = V_{C0}$ is fixed, while $V_C(T)$ is free. T is also fixed. The cost functional $J(I(t))$ to minimise is

$$J(I(t)) = \frac{1}{2}\int_{0}^{T}(V_C^2 + R^2 I^2)\mathrm{d}t, \tag{6.43}$$

therefore,

$$L(V_C, I) = \frac{1}{2}(V_C^2 + R^2 I^2) \quad \text{and} \quad g(V_C(T)) = 0. \tag{6.44}$$

The dynamics of this system are defined by the differential equation:

$$\frac{\mathrm{d}V_C}{\mathrm{d}t} = F(V_C, I) = \frac{I}{C}. \tag{6.45}$$

Applying Pontryaging's theorem, we can state that there exists an adjoint parameter $p(t)$ such that:

- $$H(V_C, p, I) = \frac{1}{2}(V_C^2 + R^2 I^2) + p\frac{I}{C}, \tag{6.46}$$
- $$\frac{\mathrm{d}V_C}{\mathrm{d}t} = \frac{\partial H}{\partial p} = \frac{I}{C}, \tag{6.47}$$
- $$\frac{\mathrm{d}p}{\mathrm{d}t} = -\frac{\partial H}{\partial V_C} = -V_C, \tag{6.48}$$
- Since $H\left(V_C^*, p^*, I^*\right) = \min_I H(V_C, p, I)$, we have $\frac{\partial H}{\partial I} = 0$, so $p = -CR^2 I$ (6.49)

Both $V_C(0) = V_{C0}$ and T are fixed, while $V_C(T)$ is free; thus, $p^*(0)$ is free and $p^*(T)$ is given by:

$$p^*(T) = \frac{\partial g(V_C(T))}{\partial V_C} = 0. \tag{6.50}$$

All these conditions amount to solving the system

$$\begin{cases} \frac{\mathrm{d}V_C^*}{\mathrm{d}t} = \frac{I^*}{C} \\ \frac{\mathrm{d}I^*}{\mathrm{d}t} = \frac{V_C^*}{R^2 C} \end{cases} \tag{6.51}$$

with $V_C^*(0) = V_{C0}$ and $I^*(T) = -\frac{p(T)}{CR^2} = 0$. Finally, the optimal control $I^*(t)$ and the corresponding value of $V_C^*(t)$ are expressed by

$$I^*(t) = \frac{V_{C0}}{R}\left(\frac{1}{1+e^{\frac{2T}{RC}}}\right)e^{\frac{t}{RC}} - \frac{V_{C0}}{R}\left(\frac{e^{\frac{2T}{RC}}}{1+e^{\frac{2T}{RC}}}\right)e^{\frac{-t}{RC}}, \tag{6.52}$$

$$V_C^*(t) = V_{C0}\left(\frac{1}{1+e^{\frac{2T}{RC}}}\right)e^{\frac{t}{RC}} + V_{C0}\left(\frac{e^{\frac{2T}{RC}}}{1+e^{\frac{2T}{RC}}}\right)e^{\frac{-t}{RC}}. \tag{6.53}$$

6.3.5 System with a Linear Dependency on H Versus a Bang-Bang Control

Consider systems where the Hamiltonian H depends linearly on a restricted control $\vec{U}$:

$$H\left(\vec{U}\right) = \vec{\varphi}\left(\vec{X}, \vec{p}\right)\vec{U} + \cdots \quad \text{with } U_{i\,\mathrm{min}} < U_i < U_{i\,\mathrm{max}}. \tag{6.54}$$

For each time $0 \le t \le T$ and each index $i = 1 \ldots u$,

$$\vec{\varphi} : \mathbb{R}^n \times \mathbb{R}^n \to \mathbb{R}^u \tag{6.55}$$

is the switching function (Evans, 2001). To maximise (or to minimise) $H\left(\vec{U}\right)$, we must make U_i as large or as small as possible and for each index i, depending on the sign of the $\varphi_i\left(\vec{X}, \vec{p}\right)$. To maximise, it follows that:

$$U_i = \begin{cases} U_{i\,\min} \text{ if } \varphi_i\left(\vec{X}, \vec{p}\right) < 0 \\ U_{i\,\max} \text{ if } \varphi_i\left(\vec{X}, \vec{p}\right) > 0 \\ \text{indefined if } \varphi_i\left(\vec{X}, \vec{p}\right) = 0 \end{cases} . \tag{6.56}$$

If, for each index i, $\varphi_i\left(\vec{X}, \vec{p}\right)$ is never equal to zero or only at discrete times, the control is bang-bang and switches between $U_{i\,\min}$ and $U_{i\,\max}$, according to the sign of each $\varphi_i\left(\vec{X}, \vec{p}\right)$ function.

6.3.6 Singular Bang-Bang Control

When one of the switching functions $\varphi_i\left(\vec{X}, \vec{p}\right)$ remains zero for a finite time interval $(t_1 \leq t \leq t_2)$, the control is called singular. The theory for totally singular and non-singular optimal controls is fairly well developed. The main difficulty with singular control problems arises when both singular and non-singular controls are present. These problems are generally difficult to solve because the maximisation of the Hamiltonian provides no useful information. One way to find a control is to repeatedly differentiate $\frac{\mathrm{d}\varphi_i}{\mathrm{d}t}$ until the control U_i appears in a certain r order time derivative (LAAS-CNRS, 2004). The control often appears in the following form:

$$\frac{\mathrm{d}^r \varphi_i}{\mathrm{d}^r t} = A\left(\vec{X}, \vec{p}\right) + B\left(\vec{X}, \vec{p}\right) U_i\left(\vec{X}, \vec{p}\right), \tag{6.57}$$

with $B\left(\vec{X}, \vec{p}\right) \neq 0$. r is an even number equal to $2q$. q is called the order of the singular control. $U_i\left(\vec{X}, \vec{p}\right)$ will be determined by solving the system:

$$\begin{cases} \frac{\mathrm{d}^j \varphi_i}{\mathrm{d}^j t} = 0 \quad \text{with } j = 0 \ldots r-1 \\ \frac{\mathrm{d}^r \varphi_i}{\mathrm{d}^r t} = A\left(\vec{X}, \vec{p}\right) + B\left(\vec{X}, \vec{p}\right) U_i\left(\vec{X}, \vec{p}\right) = 0 \end{cases} . \tag{6.58}$$

The resulting control will be optimal if it satisfies the Legendre-Clebsch condition:

$$(-1)^q \frac{\partial}{\partial U_i} \left(\frac{d^{2q} \varphi_i}{d^{2q} t} \right) > 0. \tag{6.59}$$

6.3.7 Example of a Non-singular Bang-Bang Control

In this example, we define the optimal control $U^*(t)$ to be applied to a spacecraft to minimise fuel consumption during landing on the surface of the moon (Denis, n.d.; Evans, 2001; LAAS-CNRS, 2004).
Figure 6.4 illustrates this problem with the corresponding notations. This landing is vertical (the spacecraft moves only along the – axis) and soft (at the final time T, $y(T) = 0$ and the velocity $\frac{dy}{dt}(T) = 0$). Minimising consumption means reducing the mass loss $(m(0) - m(T))$ of the spacecraft between 0 and T. The mass variation is expressed as follows:

$$\dot{m} = \frac{dm}{dt} = -\sigma U. \tag{6.60}$$

The control $U(t)$ is bounded by:

$$0 \le U(t) \le 1. \tag{6.61}$$

Two forces contribute to the acceleration, the first is due to the Moon's gravity (mg_m), and the second is generated by the thrust of the rocket $(\alpha\dot{m} = -\alpha\sigma U)$. g_m is the Moon's gravity and α, σ are two constants. The orders of this system and of the control are, respectively, three $(n = 3)$ and one $(u = 1)$. The state vector is defined as:

$$\vec{X}(t) = (X_1, X_2, X_3) = (y, \dot{y}, m) \tag{6.62}$$

The initial and final conditions,

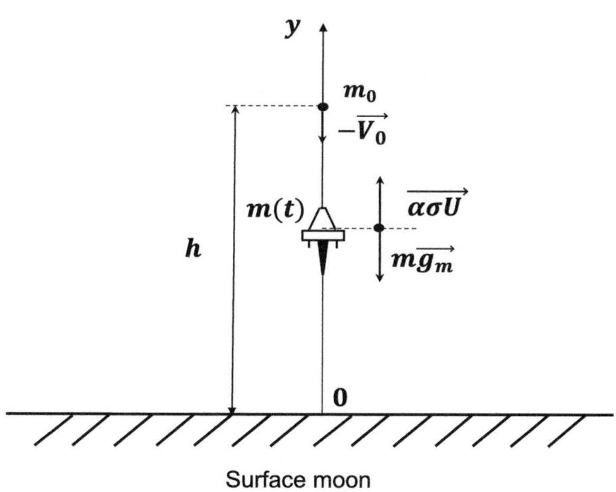

Fig. 6.4 Spacecraft's soft landing with minimal fuel consumption

$$y(0) = h, \quad \dot{y}(0) = -V_0 \ (V_0 > 0), \quad m(0) = m_0, \quad y(T) = 0, \quad \dot{y}(T) = 0, \tag{6.63}$$

are fixed, while

$$T \quad \text{and} \quad m(T) = m_T, \tag{6.64}$$

are free. We want to minimise the mass loss, so the cost functional to be minimise is:

$$J(U(t)) = -\int_0^T \dot{m}\mathrm{d}t = m(0) - m(T) \quad (L = -\dot{m} = \sigma U). \tag{6.65}$$

The dynamics of this system are governed by Newton's law

$$\begin{cases} \frac{\mathrm{d}y}{\mathrm{d}t} = \dot{y} \\ \frac{\mathrm{d}\dot{y}}{\mathrm{d}t} = -g_m + \frac{\alpha\sigma U}{m} \\ \frac{\mathrm{d}m}{\mathrm{d}t} = -\sigma U \end{cases}. \tag{6.66}$$

Pontryagin's theorem indicates that there exits adjoint vector $\vec{p}(t) = (p_1(t), p_2(t), p_3(t))$ such that

$$H\left(\vec{X}, \vec{p}, U\right) = \sigma U + p_1\dot{y} + p_2\left(-g_m + \frac{\alpha\sigma U}{m}\right) - p_3\sigma U, \tag{6.67}$$

$$\frac{\mathrm{d}\vec{X}}{\mathrm{d}t} = \frac{\partial H}{\partial \vec{p}} = \begin{pmatrix} \frac{\partial H}{\partial p_1} \\ \frac{\partial H}{\partial p_2} \\ \frac{\partial H}{\partial p_3} \end{pmatrix} \to \begin{cases} \frac{\mathrm{d}y}{\mathrm{d}t} = \dot{y} \\ \frac{\mathrm{d}\dot{y}}{\mathrm{d}t} = -g_m + \frac{\alpha\sigma U}{m} \\ \frac{\mathrm{d}m}{\mathrm{d}t} = -\sigma U \end{cases}, \tag{6.68}$$

$$\frac{\mathrm{d}\vec{p}}{\mathrm{d}t} = -\frac{\partial H}{\partial \vec{X}} = -\begin{pmatrix} \frac{\partial H}{\partial y} \\ \frac{\partial H}{\partial \dot{y}} \\ \frac{\partial H}{\partial m} \end{pmatrix} \to \begin{cases} \dot{p}_1 = 0 \Rightarrow p_1(t) = a \\ \dot{p}_2 = -p_1 \Rightarrow p_2(t) = -at + b, \\ \dot{p}_3 = p_2\frac{\alpha\sigma U}{m^2} \end{cases} \tag{6.69}$$

where a, b are constants.
Since H depends linearly on the bounded control U, the control is bang-bang. The switching function is

$$\varphi(t) = \sigma + \frac{\alpha\sigma}{m}p_2(t) - \sigma p_3(t). \tag{6.70}$$

The calculation of the time derivative of $\varphi(t)$ yields:

$$\frac{\mathrm{d}\varphi(t)}{\mathrm{d}t} = -\frac{a\alpha\sigma}{m(t)}. \tag{6.71}$$

Since $\frac{\alpha\alpha\sigma}{m(t)}$ always has the same sign, $\varphi(t)$ can switch, at most once at a discret time. Consequently, this control is bang-bang and non-singular. Since the final speed is equal to zero, the only possibility is to begin with a free fall between 0 and t_c $(U(t) = 0)$ and by braking between t_c and $(U(t) = 1)$. It is now possible to calculate the trajectory of the state parameters. We obtain the following:

For $0 \leq t \leq t_c$: $U(t) = 0, y(0) = h, \dot{y}(0) = -V_0, m(0) = m_0$

$$\begin{cases} y = -\frac{g_m}{2}t^2 - V_0 t + h \\ \dot{y} = -g_m t - V_0 \\ m = m_0 \end{cases} \quad \text{at } t = t_c \begin{cases} y(t_c) = -\frac{g_m}{2}t_c^2 - V_0 t_c + h \\ \dot{y}(t_c) = -g_m t_c - V_0 \\ m = m_0 \end{cases}, \tag{6.72}$$

For $t_c \leq t \leq T$: $U(t) = 1, y(T) = 0, \dot{y}(T) = 0$

$$\begin{cases} y = -\frac{g_m}{2}t^2 - V_0 t + h + \frac{\alpha m_o}{\sigma} + \frac{\alpha m_o}{\sigma}\left(1 - \frac{\sigma(t-t_c)}{m_o}\right)\left(\ln\left(1 - \frac{\sigma(t-t_c)}{m_o}\right) - 1\right) \\ \dot{y} = -g_m t - V_0 - \alpha \ln\left(1 - \frac{\sigma(t-t_c)}{m_o}\right) \\ m = m_0 - \sigma(t - t_c) \end{cases}, \tag{6.73}$$

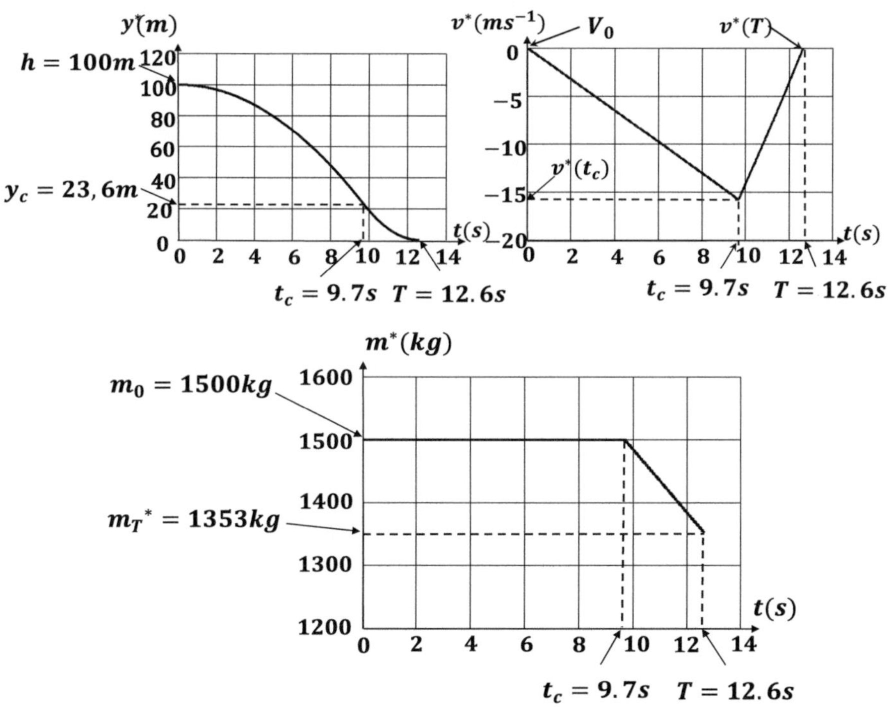

Fig. 6.5 Optimal trajectory of y^*, $v^* = \dot{y}^*$ and m^*

In this system, where $y(T) = 0$, $\dot{y}(T) = 0$ and $m(T) = m_T$, we obtain three equations that allow us to determine the three unknowns: t_c, T and m_T. Using the following numerical values:

$$\begin{aligned} m_0 &= 1500 \text{ kg}, \quad g_m = 1.63 \text{ ms}^{-2}, \quad V_0 = 0 \text{ ms}^{-1}, \\ h &= 100 \text{ m}, \quad \alpha = 200 \text{ ms}^{-1}, \quad \sigma = 50 \text{ kg s}^{-1}, \end{aligned} \tag{6.74}$$

we find:

$$t_c \simeq 9.7 \text{ s}, \quad T \simeq 12.6 \text{ s}, \quad m_T \simeq 1353 \text{ kg}. \tag{6.75}$$

Figure 6.5 shows the optimal trajectories of y^*, $v^* = \dot{y}^*$ and m^* between 0 and T.

6.3.8 Example of a Simple Singular Bang-Bang Control

In this section (Denis, n.d.), we will again consider the basic circuit given in Fig. 6.3. The final time T, as well as the initial and final conditions, are now fixed and have the following value:

$$T = 2s, V_C(0) = 1V, V_C(T) = 0V. \tag{6.76}$$

Moreover, the control $I(t)$ is bounded by:

$$-1A \leq I(t) \leq 1A. \tag{6.77}$$

To simplify the calculation, we will assume $C = 1F$. We want to minimise the quantity:

$$\frac{1}{2} \int_0^T V_C^2(t) \mathrm{d}t. \tag{6.78}$$

The cost functional to maximise is defined by:

$$J(I(t)) = -\frac{1}{2} \int_0^T V_C^2(t) \mathrm{d}t. \tag{6.79}$$

Thanks again to the Pontryaging's theorem, we know that there exists an adjoint parameter $p(t)$ such that:

$$H(V_C, p, I) = -\frac{1}{2}V_C^2 + p\frac{I}{C}, \tag{6.80}$$

$$\frac{\mathrm{d}V_C}{\mathrm{d}t} = \frac{\partial H}{\partial p} = \frac{I}{C}, \tag{6.81}$$

$$\frac{\mathrm{d}p}{\mathrm{d}t} = -\frac{\partial H}{\partial V_C} = V_C, \tag{6.82}$$

With $V_C(0)$, $V_C(T)$ and T being fixed, $p^*(0)$ and $p^*(T)$ will be free.
The maximisation of H with respect to the control I leads to:

$$I = \begin{cases} -1A \text{ if } p(t) < 0 \\ 1A \text{ if } p(t) > 0 \\ \text{undefined if } p = 0 \end{cases}. \tag{6.83}$$

If a singular control exists inside $t \in [t_1, t_2]$, it must satisfy

$$\forall t \in [t_1, t_2]\, p(t) = 0, \quad \text{so } \frac{\mathrm{d}p}{\mathrm{d}t} = V_C = 0. \tag{6.84}$$

Consequently:

$$\frac{\mathrm{d}V_C}{\mathrm{d}t} = \frac{I}{C} = 0. \tag{6.85}$$

In this example, the singular control is straightforward to define. The problem is now reduced to finding $p(0)$ that satisfies all the conditions of the Pontryaging's theorem, with the initial and final conditions given by Eq. (6.76). Suppose $p(0) > 0$, then, that:

$$\frac{\mathrm{d}V_C}{\mathrm{d}t} = \frac{I}{C} = 1 \quad \text{so } V_C = t + 1 \text{ and } p(t) = \frac{t^2}{2} + t + p(0). \tag{6.86}$$

$p(t)$ stays positive between 0 s and 2 s so $I(t)$ will be maintained equal to 1 A. At $t = 2$ s, $V_C(T)$ will be equal to 3 V. It follows that $p^*(0)$ must be negative. The only value of $p^*(0)$ taking into consideration all the conditions, is

$$p^*(0) = -0.5. \tag{6.87}$$

In the next paragraph, we will describe the shooting method, which allows this value to be find more systematically. The optimal control $I^*(t)$, along with the associated state parameter $V_C^*(t)$ and adjoint parameter $p^*(t)$, are given by:

$$I^*(t) = \begin{cases} -1 \text{ A for } t \in [0 \text{ s}, 1 \text{ s}] \\ 0 \text{ A for } t \in [1 \text{ s}, 2 \text{ s}] \end{cases}, \quad V_C^*(t) = \begin{cases} -t + 1 \text{ for } t \in [0 \text{ s}, 1 \text{ s}] \\ 0 \text{for} t \in [1 \text{ s}, 2 \text{ s}] \end{cases},$$

$$p^*(t) = \begin{cases} -\frac{t^2}{2} + t - \frac{1}{2} \text{ for } t \in [0 \text{ s}, 1 \text{ s}] \\ 0 \text{ for } t \in [1 \text{ s}, 2 \text{ s}] \end{cases}. \tag{6.88}$$

We can note that $I^*(t)$ is a singular bang-bang control. The switching function is

$$\varphi(t) = p(t) = 0 \quad \text{for } t \in [1 \text{ s}, 2 \text{ s}]. \tag{6.89}$$

It is easy to check that $V_C^*(0) = 1\text{V}$ and $V_C^*(2) = 0\text{V}$.

6.3.9 *Example of a More Complex Singular Bang-Bang Control*

In this example (Vossen, 2010), the dynamics of the state vector $\vec{X}(t) = (X_1(t), X_2(t), X_3(t))$ are given by the following system:

$$\begin{cases} \dot{X}_1 = X_2 \\ \dot{X}_2 = -X_1 + X_2(1 - X_1^2) + U \\ \dot{X}_3 = 0.5(X_1^2 + X_2^2) \end{cases}. \tag{6.90}$$

The initial conditions and the final time T are fixed and given by:

$$X_1(0) = 0, \quad X_2(0) = 1, \quad X_3(0) = 0, \quad T = 4. \tag{6.91}$$

The control $U(t)$ is bounded by the following limits:

$$-1 \le U(t) \le 1. \tag{6.92}$$

We want to minimise $X_3(T)$, so we have:

$$L = 0 \quad \text{and} \quad g(X_1(T), X_2(T), X_3(T)) = X_3(T). \tag{6.93}$$

The Pontryaging's theorem indicates that there exists an adjoint vector $\vec{p}(t) = (p_1(t), p_2(t), p_3(t))$ such that:

$$H\left(\vec{X}(t), \vec{p}(t), U\right) = p_1 X_2 + p_2\left(-X_1 + X_2(1 - X_1^2)\right) + 0.5 p_3 (X_1^2 + X_2^2) + p_2 U. \tag{6.94}$$

$$\begin{cases} \dot{p}_1 = p_2 + 2p_2 X_1 X_2 - p_3 X_1 \\ \dot{p}_2 = -p_1 - p_2(1 - X_1^2) - p_3 X_2 \\ \dot{p}_3 = 0 \end{cases}. \tag{6.95}$$

Transversality conditions give

$$p_1(T) = \frac{\partial g}{\partial X_1(T)} = 0, \quad p_2(T) = \frac{\partial g}{\partial X_2(T)} = 0, \quad p_3(T) = \frac{\partial g}{\partial X_3(T)} = 1. \quad (6.96)$$

H depends linearly on U, so the control is bang-bang. The switching function is $\varphi(t) = p_2(t)$. The minimisation of H gives:

$$U(t) = \begin{cases} 1 \text{ if } p_2(t) < 0 \\ -1 \text{ if } p_2(t) > 0 \\ \text{undefined if } p_2(t) = 0 \end{cases} . \quad (6.97)$$

Let us consider that there is a singular control for $t_a \leq t \leq t_b$, we then get:

$$\begin{cases} \varphi(t) = p_2(t) = 0 \\ \frac{\mathrm{d}\varphi(t)}{\mathrm{d}t} = \dot{p}_2 = -p_1 - X_2 = 0 \\ \frac{\mathrm{d}^2\varphi(t)}{\mathrm{d}^2 t} = -\dot{p}_1 - \dot{X}_2 = 2X_1 - X_2\left(1 - X_1^2\right) - U = 0 \end{cases} . \quad (6.98)$$

The order of this singular control is $q = 1$. The control U is defined by:

$$U = 2X_1 - X_2\left(1 - X_1^2\right). \quad (6.99)$$

The resulting control will be optimal because the Legendre-Clebsch condition is satisfied:

$$(-1)^q \frac{\partial}{\partial U}\left(\frac{\mathrm{d}^{2q}\varphi(t)}{\mathrm{d}^{2q} t}\right) = (-1)^1(-1) \geq 0. \quad (6.100)$$

Using the shooting method, described in the next section combined with numerical analysis, we can determine the optimal control as follows

$$U^*(t) = \begin{cases} -1 \text{ for } t \in [0, t_1] \\ 1 \text{ for } t \in [t_1, t_2] \\ 2X_1 - X_2\left(1 - X_1^2\right) \text{ for } t \in [t_2, 4] \end{cases} . \quad (6.101)$$

Figure 6.6 shows the graphical representation of the optimal states, adjoint and control parameters which allows us to obtain the minimum value of $X_3(T) = X_{3\,\min}(4)$. The result is:

$$t_1 \simeq 1.63 \text{ s}, \quad t_2 \simeq 2.45 \text{ s}, \quad X_{3\,\min}(4) \simeq 0.76. \quad (6.102)$$

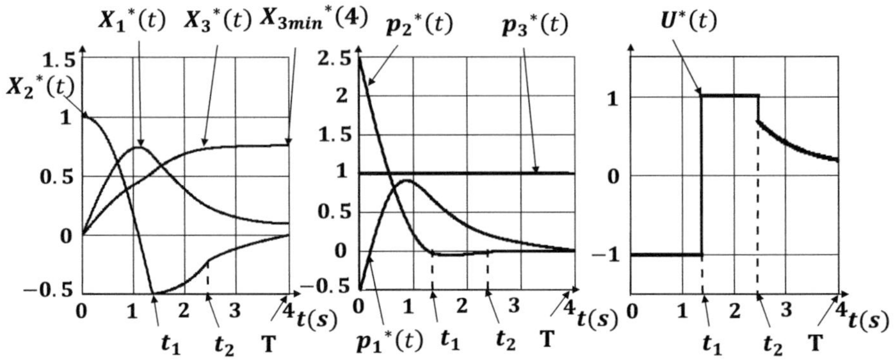

Fig. 6.6 Optimal states, adjoint vector and optimal control

6.3.10 Shooting Method

Globally speaking, the shooting method is a numerical technique for solving boundary value problems by reducing them to the solution of the initial value problems (Martinon, 2005). Such a method, associated with the Pontryaging's theorem, is frequently used to find an optimal control and the corresponding state and adjoint vectors. Consider the $2n$-dimensional vector composed of the state vector $\vec{X}$ and the adjoint vector $\vec{p}\left(\vec{X},\vec{p}\right)$. Generally, n components of this vector are known at $t = 0$ and $t = T$. If we denote by $\vec{S}(T)$ a n-dimensional vector the known components of the vector $\left(\vec{X},\vec{p}\right)$ at $t = T$, the optimal problem consists of finding the n unknown components of $\left(\vec{X}(0),\vec{p}(0)\right)$ in order to reach $\vec{S}(T)$, with the system dynamics defined by the Pontryaging's theorem. With the same dynamics, at each vector $\left(\vec{X}(0),\vec{p}(0)\right)$, we can associate a vector $\vec{G}(T)$ defined by the same n components as $\vec{S}(T)$. Finally, the problem amounts to solving the n-dimensional system:

$$\vec{G}(T) - \vec{S}(T) = \vec{0}, \tag{6.103}$$

where the n unknowns are the n unknown components of $\left(\vec{X}(0),\vec{p}(0)\right)$.

6.3.11 Example of the Use of the Shooting Method

In order to illustrate the use of the shooting method (Martinon, 2005), we will consider the circuit shown in Fig. 6.7. The control parameter is the voltage source U which is limited by:

Fig. 6.7 U, L, I, C circuit

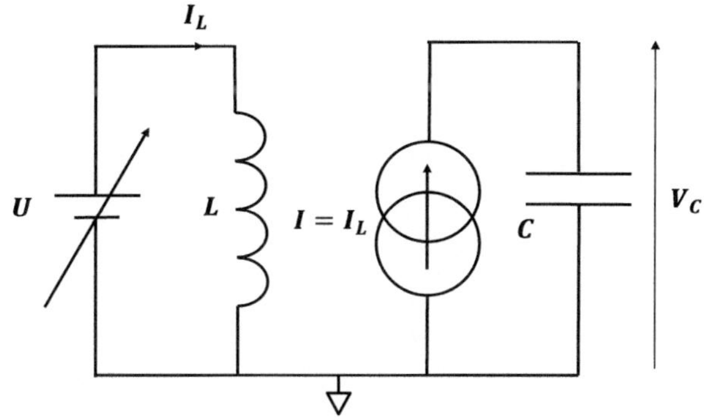

$$-1 \le U(t) \le 1. \tag{6.104}$$

The order of this system is $n = 2$ with the state vector (V_C, I_L). We will consider the numerical values $L = 1H$, $C = 1F$ and $T = 2$ s. The dynamics governing this system are expressed as:

$$\begin{cases} \frac{dV_C}{dt} = \frac{I_L}{C} \\ \frac{dI_L}{dt} = \frac{U}{L} \end{cases}. \tag{6.105}$$

Our purpose is to minimise the cost functional:

$$J(U) = \int_0^T |U(t)| dt. \tag{6.106}$$

Initial and final state vectors are fixed and equal to

$$(V_C(0), I_L(0)) = (0, 0) \quad \text{and} \quad (V_C(2), I_L(2)) = (0.5, 0). \tag{6.107}$$

Pontryaging's theorem leads to:

$$H = |U(t)| + p_1 \frac{I_L}{C} + p_2 \frac{U}{L}. \tag{6.108}$$

$$\begin{cases} \frac{dV_C}{dt} = \frac{I_L}{C} \\ \frac{dI_L}{dt} = \frac{U}{L} \end{cases}. \tag{6.109}$$

$$\begin{cases} \frac{dp_1}{dt} = -\frac{\partial H}{\partial V_C} = 0 \Rightarrow p_1 = a \\ \frac{dp_2}{dt} = -\frac{\partial H}{\partial I_L} = -\frac{p_1}{C} \Rightarrow p_2 = -at + b (C = 1F) \end{cases}, \tag{6.110}$$

where a and b are two constants.
Minimisation of H with respect to U is given by: $(L = 1H, C = 1F)$

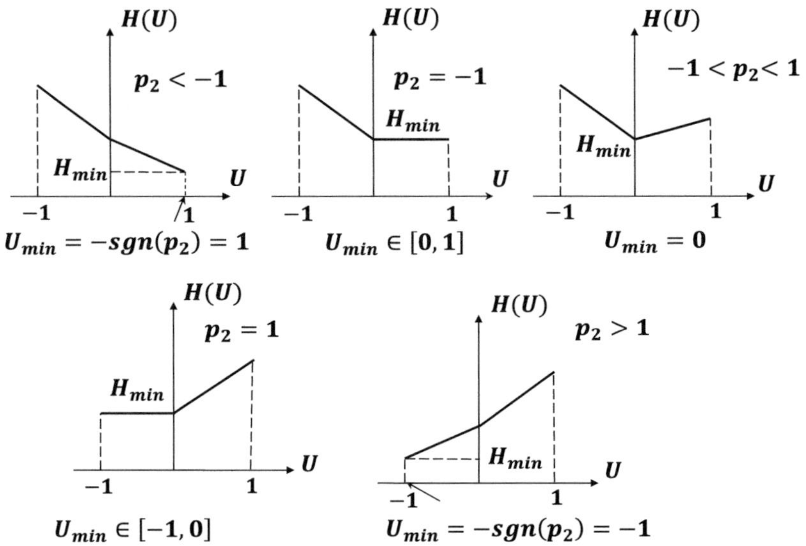

Fig. 6.8 $H_{\min}$ for $p_2 < -1, p_2 = -1, -1 < p_2 < 1, p_2 = 1$ and $p_2 > 1$

$$\text{If } U < 0 \rightarrow |U| = -U \Rightarrow H(U) = p_1 I_L + (p_2 - 1)U \tag{6.111}$$

$$\text{If } U > 0 \rightarrow |U| = U \Rightarrow H(U) = p_1 I_L + (p_2 + 1)U \tag{6.112}$$

Figure 6.8 gives the values of $U_{\min}$ corresponding to $H_{\min}$ for

$$p_2 < -1, \quad p_2 = -1, \quad -1 < p_2 < 1, \quad p_2 = 1 \quad \text{and} \quad p_2 > 1. \tag{6.113}$$

Consequently, the optimal control obtained by minimising $H(U)$ is:

$$\begin{cases} U^* = -\text{sgn}(p_2) \text{ if } |p_2| > 1 \\ U^* = 0 \text{ if } |p_2| < 1 \\ U^* \in [0, 1] \text{ if } p_2 = -1 \\ U^* \in [-1, 0] \text{ if } p_2 = 1 \end{cases}. \tag{6.114}$$

Transversality conditions indicate that $p_1(0), p_2(0)$ and $p_1(T), p_2(T)$ are free. Using the notations defined in Para. 6.3.9 describing the shooting method we get:

$$\left(\vec{X}, \vec{p}\right) = (V_C, I_L, p_1, p_2) \text{ with} \tag{6.115}$$

$$(V_C(0), I_L(0), p_1(0), p_2(0)) = (0, 0, p_1(0), p_2(0)) \text{ and} \tag{6.116}$$

$$(V_C(2), I_L(2), p_1(2), p_2(2)) = (0.5, 0, p_1(2), p_2(2)). \tag{6.117}$$

$$\vec{S}(T) = (V_C(2), I_L(2)) = (0.5, 0). \tag{6.118}$$

$$\vec{G}(T) = (V_C(2), I_L(2)) \quad \text{for a given } (p_1(0), p_2(0)) = (a, b). \tag{6.119}$$

The shooting method consists in using the equations given by the Pontryaging's theorem to find (a, b) by resolving:

$$\vec{G}(T) - \vec{S}(T) = (V_C(2), I_L(2)) - (0.5, 0) = (0, 0) \tag{6.120}$$

Figure 6.9 represents: $V_C(2) - 0.5$ and $I_L(2)$ in relation to $(p_1(0), p_2(0)) = (a, b)$. The numerical analysis of Eq. (6.120) yields:

$$\left(p_1^*(0), p_2^*(0)\right) = (a, b) \simeq (-1.41, -1.41). \tag{6.121}$$

Figure 6.10 shows the change of the optimal parameters between 0 and 2 s.

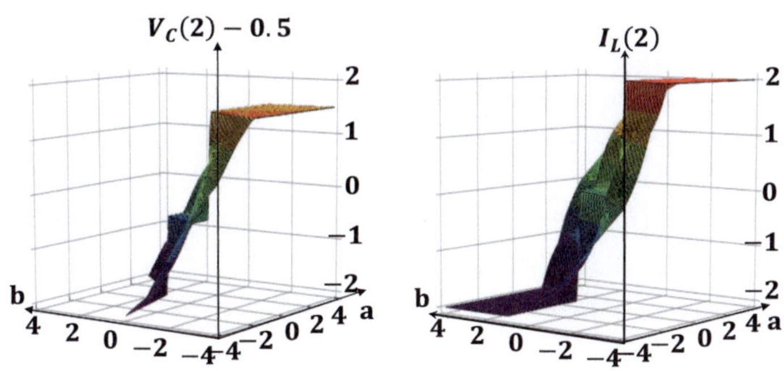

Fig. 6.9 $V_C(2) - 0.5$ and $I_L(2)$, in relation to $(p_1(0), p_2(0)) = (a, b)$

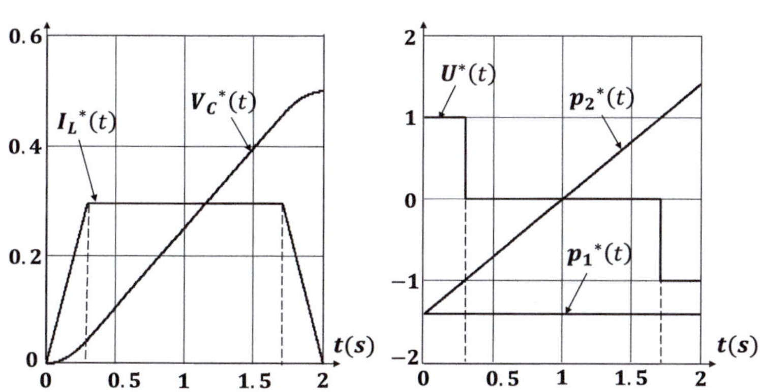

Fig. 6.10 Changes in the optimal parameters

6.4 Bellman Dynamic Programming

6.4.1 Principle

Bellman dynamic programming provides an alternative method to find the optimal control for the basic problem presented in Para. 6.2. This technique was developed by Richard Bellman after World War II (Bellman, 1957; Williams, n.d.). The key idea is to simplify a complicated problem by breaking it down into simpler sub-problems in a recursive manner. As an introduction, it is interesting to consider the problem of a traveller who wants to minimise the length of a journey from towns I to.F. This trip is divided into four stages: from I to A_i, from A_i to B_i,from B_i to C_i and from C_i to.F. Figure 6.11 illustrates these stages and the distance between each town.

The basic principle of dynamic programming, as applied to this example, can be expressed as follows: If an optimal path (with minimal distance) from I to F passes through town M , then the path from M to F is also optimal. By applying this principle starting from the final town F, we get:

- From stage 3 to the town F, we have two possible paths: C_1 to F and C_2 to F, with their corresponding lengths:$L_{C_1F} = 76$ km, $L_{C_2F} = 59$ km.
- From stage 2 to F, the optimal paths are:

$$\begin{aligned}
L_{B_1F} &= \min\{L_{B_1C_1} + L_{C_1F}, L_{B_1C_2} + L_{C_2F}\} = L_{B_1C_2} + L_{C_2F} = 80 \text{ km},\\
L_{B_2F} &= \min\{L_{B_2C_1} + L_{C_1F}, L_{B_2C_2} + L_{C_2F}\} = L_{B_2C_1} + L_{C_1F} = 96 \text{ km},\\
L_{B_3F} &= L_{B_3C_2} + L_{C_2F} = 101 \text{ km},\\
L_{B_4F} &= L_{B_4C_1} + L_{C_2F} = 148 \text{ km}.
\end{aligned}$$

- From stage 1 to F, the optimal paths are:

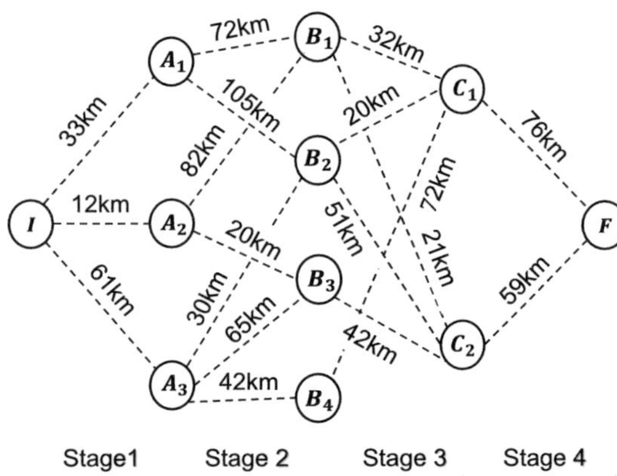

Fig. 6.11 Minimal traveller's journey length between the towns from I to F

$$L_{A_1F} = \min\{L_{A_1B_1} + L_{B_1F}, L_{A_1B_2} + L_{B_2F}\} = L_{A_1B_1} + L_{B_1F} = 152 \text{ km},$$
$$L_{A_2F} = \min\{L_{A_2B_1} + L_{B_1F}, L_{A_2B_3} + L_{B_3F}\} = L_{A_2B_3} + L_{B_3F} = 121 \text{ km},$$
$$L_{A_3F} = \min\{L_{A_3B_2} + L_{B_2F}, L_{A_3B_3} + L_{B_3F}, L_{A_3B_4} + L_{B_4F}\} = L_{A_3B_2} + L_{B_2F} = 126 \text{ km},$$

- And finally, the optimal path from I to F is:

$$L_{IF} = \min\{L_{IA_1} + L_{A_1F}, L_{IA_2} + L_{A_2F}, L_{IA_3} + L_{A_3F}\} = L_{IA_2} + L_{A_2F} = 133 \text{ km}.$$

Thus, the optimal trip, having the minimal length, is:I, A_2, B_3, C_2, F with a minimal length equal to 133 km.

6.4.2 *Hamilton–Jacobi-Bellman Equation (HJB)*

We will keep the same notation as in the section entitled "Basic Optimal Problem". Let us define the optimal value function $V\left(t, \vec{X}(t)\right)$ as the maximum (or minimum) value of the cost functional $J\left(\vec{U}(t)\right)$ that can be obtained, starting at any time $t \in [0, T]$ and from any state vector $\vec{X}(t)$. The mathematical form is:

$$V\left(t, \vec{X}(t)\right) = \max_{\vec{U}}\left(\int_t^T L(\vec{X}, \vec{U})\mathrm{d}t'\right) + g\left(\vec{X}(T)\right). \tag{6.122}$$

Note that:

$$V\left(T, \vec{X}(T)\right) = g\left(\vec{X}(T)\right). \tag{6.123}$$

The principle of optimality states that an optimal control has the property that, regardless of the initial state and initial decision, the remaining decisions must form an optimal control relative to the state resulting from the first decision. This principle allows us to write:

$$V\left(t, \vec{X}(t)\right) = \max_{\vec{U}}\left(\int_t^{t+\Delta t} L(\vec{X}, \vec{U})\mathrm{d}t'\right) + V\left(t + \Delta t, \vec{X}(t + \Delta t)\right). \tag{6.124}$$

By letting Δt tend to zero in this equation, we obtain the following first-order nonlinear equation called the Hamilton–Jacobi-Bellman equation (HJB):

$$-\frac{\partial V\left(t, \vec{X}(t)\right)}{\partial t} = \max_{\vec{U}}\left(L\left(\vec{X}, \vec{U}\right) + \frac{\partial V\left(t, \vec{X}(t)\right)}{\partial \vec{X}}\vec{F}\left(\vec{X}(t), \vec{U}(t)\right)\right). \tag{6.125}$$

In practice, one solves the maximisation (or minisation) problem on the right-hand side of the HJB equation and express $\vec{U}$ as a function of $\vec{X}(t)$ and $V\left(t, \vec{X}(t)\right)$. Then the solution for $\vec{U}$ is substituted back into the HJB equation to obtain a nonlinear first-order partial differential equation, which is to be solved for $V\left(t, \vec{X}(t)\right)$ using the boundary condition: $V\left(T, \vec{X}(T)\right) = g\left(\vec{X}(T)\right)$. Finally, we will deduce the optimal control $\overrightarrow{U^*}$ from $V\left(t, \vec{X}(t)\right)$.

6.4.3 *Optimal Control of the* R, C, I *Circuit Solved Using the HJB Equation*

We consider again the problem from Para. 6.3.4 using the same notations. In this example the optimal value function is:

$$V(t, V_C(t)) = \min_I \left(\frac{1}{2} \int_t^T (V_C^2 + R^2 I^2) \mathrm{d}t' \right). \tag{6.126}$$

By applying the Hamilton–Jacobi-Bellman equation, we can write:

$$-\frac{\partial V(t, V_C)}{\partial t} = \min_I \left(\frac{1}{2}(V_C^2 + R^2 I^2) + \frac{I}{C} \frac{\partial V(t, V_C)}{\partial V_C} \right). \tag{6.127}$$

Minimisation of the right-hand side of the HJB gives.

$$\frac{\partial}{\partial I} \left(\frac{1}{2}(V_C^2 + R^2 I^2) + \frac{I}{C} \frac{\partial V(t, V_C)}{\partial V_C} \right) = 0. \tag{6.128}$$

Thus, I can be expressed by:

$$I = -\frac{1}{R^2 C} \frac{\partial V(t, V_C)}{\partial V_C}. \tag{6.129}$$

Substituting back I into the HJB equation, we obtain the nonlinear first-order partial differential equation:

$$\frac{\partial V(t, V_C)}{\partial t} + \frac{1}{2} V_C^2 - \frac{1}{2R^2 C^2} \left(\frac{\partial V(t, V_C)}{\partial V_C} \right)^2 = 0, \tag{6.130}$$

Subject to the boundary condition:

$$V(T, V_C(T)) = g(V_C(T)) = 0. \tag{6.131}$$

Suppose $V(t, V_C(t))$ is of the form:

$$V(t, V_C(t)) = \varphi(t)V_C^2. \tag{6.132}$$

The nonlinear first-order partial differential equation (6.103) becomes:

$$\frac{d\varphi(t)}{dt} + \frac{1}{2} - \frac{2\varphi^2(t)}{R^2C^2} = 0. \tag{6.133}$$

The resolution of this equation gives, with $\varphi(T) = 0$:

$$\varphi(t) = -\frac{RC}{2}\frac{e^{\frac{2(t-T)}{RC}} - 1}{e^{\frac{2(t-T)}{RC}} + 1}. \tag{6.134}$$

It follows that:

$$\frac{dV_C}{dt} = \frac{I}{C} = -\frac{2}{R^2C^2}\varphi(t)V_C. \tag{6.135}$$

After some transformations, we obtain the same optimal control as in Para. 6.3.4.

Chapter 7
Robust Controls

Physical systems can never be modelled with complete accuracy; there are always inherent modelling errors. Robust control methods aim to achieve robust performance and stability as long as uncertain parameters and disturbances remain within certain limits. These unknown parameters can be either structured or unstructured, but they are always bounded. In this section, the theory of some robust control methods is presented and illustrated with simple examples. We will also address adaptive control, which we consider a form of robust control that includes dynamically estimated values.

7.1 Lyapunov Control

7.1.1 Theory of Lyapunov Control

Since the Lyapunov function allows testing whether a system is stable, it can also be used in control theory (Kokotović & Arcak, 2001; Lyapunov, 1992; Praly & Pietri, 2016). Consider, with the usual notations, the single-input control of the form:

$$\frac{\mathrm{d}\vec{X}(t)}{\mathrm{d}t} = \vec{F}\left(\vec{X}\right) + G\left(\vec{X}\right)\vec{U}\left(\vec{X}\right), \tag{7.1}$$

where $G\left(\vec{X}\right)$ is a $n \times u$ matrix. Our goal is to construct a continuous state feedback law: $\vec{U} = \vec{U}\left(\vec{X}\right)$ so that $\vec{X} = \vec{X}_e$ is a globally stable equilibrium point of the resulting closed-loop system. A Control Lyapunov Function (CLF) $V\left(\vec{X}\right)$ is a continuous, positive definite function such that:

B. Rivet, *Theories of Nonlinear Systems for Power Converters*,
https://doi.org/10.1007/978-3-032-25062-9_7

$$\forall \vec{X} \neq \vec{X}_e \quad \frac{\mathrm{d}V\left(\vec{X}\right)}{\mathrm{d}t} = \frac{\partial V\left(\vec{X}\right)}{\partial \vec{X}} \frac{\mathrm{d}\vec{X}}{\mathrm{d}t} < 0. \tag{7.2}$$

Intuitively, considering $V\left(\vec{X}\right)$ as an energy, if in each state the control $\vec{U}\left(\vec{X}\right)$ can reduce this energy, the system will move towards $\vec{X}_e$. As an illustration, consider the following example ($X_e = 0$):

$$\dot{X} = -X^3 + 2X + 2U. \tag{7.3}$$

Consider the following Lyapunov candidate function:

$$V(X) = \frac{1}{2}X^2. \tag{7.4}$$

We have:

$$V(X) > 0 \; \forall \; X \neq X_e = 0 \text{ and } V(X_e) = 0. \tag{7.5}$$

The time derivative is:

$$\frac{\mathrm{d}V(X)}{\mathrm{d}t} = -X^4 + X(2X + 2U), \tag{7.6}$$

with the state control feedback:

$$U(X) = -X, \tag{7.7}$$

we obtain:

$$\forall \; \vec{X} \in \mathbb{R}^n \; \frac{\mathrm{d}V(X)}{\mathrm{d}t} = -2X^4 < 0. \tag{7.8}$$

$\frac{\mathrm{d}V(X)}{\mathrm{d}t}$ is negative for each $X \neq X_e$ and equal to zero at $X = X_e$, therefore, with this control, the system is globally asymptotically stable at the equilibrium point: $X_e = 0$.

7.1.2 Lyapunov Adaptive Control

Lyapunov control can also be used to stabilise a system with an unknown parameter (Ioannou & Kokotovic, 1984; Kokotović & Arcak, 2001). As an illustration, consider the following system:

$$\dot{X} = 2\theta X + U. \tag{7.9}$$

θ is an unknown time-invariant parameter. We want to stabilise the system towards: $X_e = 0$. We could try to use the Lyapunov candidate function:

$$V(X, \varphi) = \frac{1}{2}X^2 + \frac{1}{2}(\theta - \varphi)^2. \quad (7.10)$$

φ is an estimated value of θ. Initially, we can take $\varphi = 0$. $\dot{V}$ is given by:

$$\dot{V} = X\dot{X} - (\theta - \varphi)\dot{\varphi} = XU + \varphi\dot{\varphi} + \theta\big(2X^2 - \dot{\varphi}\big). \quad (7.11)$$

The aim is to choose a control U and a law φ so that $\dot{V}$ is negative definite. Obviously, neither the control U nor φ can depend on the unknown θ. By taking:

$$\dot{\phi} = 2X^2, \quad (7.12)$$

we can eliminate the term $\theta\big(X^2 - \dot{\varphi}\big)$. With the control:

$$U = -(k + 2\varphi)X, \quad (7.13)$$

we obtain:

$$\dot{V} = -kX^2. \quad (7.14)$$

If k is a positive constant, $\dot{V}$ is negative definite in $\mathbb{R}$, the system will converge towards $X_e = 0$ and φ will converge towards θ. The control also allows us to estimate the unknown parameter θ.

7.2 Backstepping Control

7.2.1 Introduction

Backstepping is a control theory developed by Petar V. Kokotovic (Chen, 1996; Kokotović et al., 1996; Kokotović & Arcak, 2001). It is a recursive procedure that breaks down a design problem for the full system into a sequence of design problems related to lower order systems. The ability of this control to easily account for uncertainties and unknown parameters contributed to its popularity. Unfortunately, this theory is limited to a specific class of dynamic differential equations. However, its success stimulated the development of other recursive methods such as forwarding control.

7.2.2 Theory of Backstepping Control

Consider the simple system:

$$\begin{cases} \dot{X} = X^2 - X^3 + \gamma \\ \dot{\gamma} = U \end{cases}. \tag{7.15}$$

In backstepping control, the state vector is divided into two parts, here X and γ. The design objective is that $X(t) \rightarrow 0$ as $t \rightarrow \infty$. The control law will be defined in two steps. In the first step, γ is regarded as a real control. We use the Lyapunov function candidate:

$$V_1(X) = \frac{1}{2}X^2 \tag{7.16}$$

which gives:

$$\dot{V_1} = X\left(X^2 - X^3 + \gamma\right). \tag{7.17}$$

with the control law:

$$\gamma = -X^2 - k_1 X = \alpha(X), \tag{7.18}$$

it follows that

$$\dot{V_1} = -X^4 - k_1 X^2. \tag{7.19}$$

With $k_1 > 0$ a positive constant, the control objective will be achieved ($\dot{V_1} < 0 \ \forall \vec{X} \in \mathbb{R}^n$). Nevertheless, γ is a state parameter and can not be set to $\alpha(X)$. We define the error variable:

$$z = \gamma - \alpha(X), \tag{7.20}$$

This leads to:

$$\dot{z} = U - (2X + k_1)\left(k_1 X + X^3 - z\right). \tag{7.21}$$

Now, the Lyapunov function candidate is augmented as follows:

$$V_2 = V_1 + \frac{1}{2}z^2. \tag{7.22}$$

Its time derivative is given by:

$$\dot{V}_2 = X\left(-X^3 - k_1X + z\right) + z\left(U - (2X + k_1)\left(k_1X + X^3 - z\right)\right). \tag{7.23}$$

To make $\dot{V}_2$ negative definite, we choose the control law:

$$U = -X + (2X + k_1)\left(k_1X + X^3 - z\right) - k_2z \text{ with } k_2 > 0 \tag{7.24}$$

Then we obtain:

$$\dot{V}_2 = -X^4 - k_1X^2 - k_2z^2, \tag{7.25}$$

which is negative definite in $\mathbb{R}$. This implies that $X(t) \to 0$ and $\gamma \to \alpha(X)$ asymptotically. γ is called a virtual control and $\alpha(X)$ a stabilising function. The backstepping algorithm can be derived from this example. Consider a system of the form:

$$\begin{cases} \dot{X}_1 = F_1\left(X_1, \ldots, X_p\right) \\ \vdots \\ \dot{X}_{p-1} = F_{p-1}\left(X_1, \ldots, X_p\right) \\ \dot{X}_p = F_p\left(X_1, \ldots, X_p\right) + G_1\left(X_1, \ldots, X_p\right)\gamma_1 \\ \dot{\gamma}_1 = F_{p+1}\left(X_1, \ldots, X_p, \gamma_1\right) + G_2\left(X_1, \ldots, X_p, \gamma_1\right)\gamma_2 \\ \vdots \\ \dot{\gamma}_m = F_n\left(X_1, X_2, \ldots, X_p, \gamma_1, \ldots, \gamma_m\right) + G_{m+1}\left(X_1, X_2, \ldots, X_p, \gamma_1, \ldots, \gamma_m\right)U \end{cases}, \tag{7.26}$$

where $\left(X_1, \ldots, X_p, \gamma_1, \ldots, \gamma_m\right)$ are the n state parameters ($n = p + m$). Our goal is that the vector $\vec{X} = \left(X_1, \ldots, X_p\right)$ follows a given reference vector:

$$\overrightarrow{X_r}(t) = \left(X_{r1}(t), \ldots, X_{rp}(t)\right). \tag{7.27}$$

The first step of the backstepping algorithm consists of finding a stabilising function:

$$\gamma_1 = \alpha\left(X_1, \ldots, X_p\right), \tag{7.28}$$

for the $\left(X_1, \ldots, X_p\right)$ subsystem. It can be done using a CLF. A suggestion would be:

$$V\left(X_1, \ldots, X_p\right) = \frac{1}{2}(X_1 - X_{r1})^2 + \cdots \frac{1}{2}\left(X_p - X_{rp}\right)^2. \tag{7.29}$$

The virtual law γ_1 has to be such that $\dot{V} < 0$. To take into account the deviation of the state variable γ_1 from the stabilising function $\alpha\left(X_1, \ldots, X_p\right)$, we consider the error variable:

$$z_1 = \gamma_1 - \alpha(X_1, \ldots, X_p). \tag{7.30}$$

We proceed in the same way as in the first step by augmenting the Lyapunov function:

$$V_1 = V(X_1, \ldots, X_p) + \frac{1}{2}z_1^2 \tag{7.31}$$

and by constructing a stabilising function

$$\gamma_2 = \alpha_1(X_1, \ldots, X_p, z_1) \tag{7.32}$$

such that the time derivative of the Lyapunov function $\dot{V}_1$ is negative definite. This recursive procedure will terminate at the mth step when the U control will be determined.

7.2.3 *Adaptive Backstepping Control*

In this section, adaptive backstepping control is presented through an example (Chen, 1996; Krstić et al., 1995). Describing this technique in a general way is quite complex. The following example illustrates the basic ideas of this control, which can be applied to other dynamical systems. Consider the following system:

$$\begin{cases} \dot{X} = \theta(X^3 - X) + \gamma \\ \dot{\gamma} = U \end{cases}, \tag{7.33}$$

where θ is an unknown constant parameter. Our goal is to regulate X at 0 for all values of θ. The first step involves stabilising the subsystem:

$$\dot{X} = \theta(X^3 - X) + \gamma \tag{7.34}$$

with a virtual control γ. Consider the following Lyapunov function depending on X and the estimation error $\tilde{\theta} = \theta - \hat{\theta}$, where $\hat{\theta}$ is the estimated value of θ:

$$V_1(X, \tilde{\theta}) = \frac{1}{2}X^2 + \frac{1}{2}\tilde{\theta}^2. \tag{7.35}$$

Its time derivative is:

$$\dot{V_1} = X(\theta(X^3 - X) + \gamma) + \tilde{\theta}\dot{\tilde{\theta}}. \tag{7.36}$$

With the virtual control:

$$\gamma = -k_1 X - \hat{\theta}(X^3 - X) = \alpha_1(X, \hat{\theta}) \tag{7.37}$$

and the tuning law chosen as:

$$\dot{\hat{\theta}} = X\left(X^3 - X\right) = \sigma(X) \tag{7.38}$$

$\dot{V}_1$ becomes independent of both θ and $\tilde{\theta}$. We obtain:

$$\dot{V}_1 = -k_1 X^2. \tag{7.39}$$

If the control parameter k_1 is positive, $\dot{V}_1$ is negative definite. Since γ is not a real control, we need to define the deviation of γ from $\alpha_1\left(X, \hat{\theta}\right)$:

$$z = \gamma - \alpha_1\left(X, \hat{\theta}\right). \tag{7.40}$$

With this new parameter, the dynamics of the system can be written as:

$$\begin{cases} \dot{X} = -k_1 X + \tilde{\theta}\left(X^3 - X\right) + z \\ \dot{z} = U - \dot{\alpha}_1 \end{cases}. \tag{7.41}$$

Augment the Lyapunov function by adding the error variable:

$$V_2\left(X, z, \tilde{\theta}\right) = V_1\left(X, \tilde{\theta}\right) + \frac{1}{2} z^2. \tag{7.42}$$

After some transformations, we obtain:

$$\begin{aligned} \dot{V}_2 = &-k_1 X^2 + \tilde{\theta}\left(\sigma(X) - \frac{\partial \alpha_1}{\partial X}\left(X^3 - X\right) z - \dot{\hat{\theta}}\right) \\ &+ z\left(U + X - \frac{\partial \alpha_1}{\partial X}\gamma - \frac{\partial \alpha_1}{\partial \hat{\theta}}\dot{\hat{\theta}} - \hat{\theta}\frac{\partial \alpha_1}{\partial X}\left(X^3 - X\right)\right). \end{aligned} \tag{7.43}$$

To make $\dot{V}_2$ negative and independent of both θ and $\tilde{\theta}$, we can choose:

$$U = -k_2 z - X + \frac{\partial \alpha_1}{\partial X}\left(z - k_1 X\right) + \frac{\partial \alpha_1}{\partial \hat{\theta}}\dot{\hat{\theta}} \quad \text{with } k_2 > 0 \tag{7.44}$$

and

$$\dot{\hat{\theta}} = \sigma(X) - \frac{\partial \alpha_1}{\partial X}\left(X^3 - X\right) z. \tag{7.45}$$

The time derivative $\dot{V}_2$ becomes:

$$\dot{V}_2 = -k_1X^2 - k_1z^2 \leq 0. \tag{7.46}$$

It follows that $X(t) \to 0, \tilde{\theta}(t) \to \theta$ and $z(t) \to 0$ asymptotically. $\dot{\hat{\theta}}$ and U are defined by the state parameters X,γ and the estimated value $\hat{\theta}$.

7.2.4 Robust Backstepping Control

The adaptive backstepping control (Freeman & Kokotović, 1996) can be used for systems with constant unknown parameters. If these parameters vary slowly in time, we can still use this technique. However, if the parameters vary too quickly, the best way to proceed is to ensure these parameters are bounded. It is one of the key ideas of robust control. In this section, robust backstepping control (Chen, 1996; Kokotović & Arcak, 2001) is presented for systems of the form:

$$\begin{cases} \dot{X} = F(X) + W(X, \gamma, t) + \gamma \\ \dot{\gamma} = U \end{cases} \tag{7.47}$$

This control can be extended to higher-system order. The uncertainty $W(X, \gamma, t)$ is bounded by a known limit Δ :

$$|W(X, \gamma, t)| < \Delta. \tag{7.48}$$

Our goal is to stabilise the system around $\vec{X} = \vec{0}$. We choose the robust CLF as:

$$V_1(X) = \frac{1}{2}X^2. \tag{7.49}$$

The virtual control γ will then take into account the worst case of $W(X, \gamma, t)$ by:

$$\gamma = \alpha(X). \tag{7.50}$$

Then, to render $\dot{V}_1$ negative, we look for $\alpha(X)$ which, for $X \neq 0$ and all uncertainties $W(X, \gamma, t)$ bounded by 66 Δ, satisfies:

$$\dot{V}_1\left(\vec{X}\right) = X(F(X) + W(X, \gamma, t) + \alpha(X)) < XF(X) + X\alpha(X) + |X|\Delta < 0. \tag{7.51}$$

A possible choice for the virtual control is:

$$\alpha(X) = -X - F(X) - \Delta\text{sgn}(X) \tag{7.52}$$

which gives:

$$\dot{V}_1(X) < -X^2. \tag{7.53}$$

In order to regulate $\gamma - \alpha(X)$ to zero, we suggest considering:

$$V_2(X, \gamma) = V_1(X) + \frac{1}{2}(\gamma - \alpha(X))^2 \tag{7.54}$$

as a robust CLF. Our objective now is to find a control $U(X)$ so that:

$$\begin{aligned}\dot{V}_2(X) &= X(F(X) + W(X, \gamma, t) + \gamma) \\ &+ (\gamma - \alpha(X))(U - \frac{\partial \alpha}{\partial X}(F(X) + W(X, \gamma, t) + \gamma) < 0.\end{aligned} \tag{7.55}$$

The choice of $\alpha(X)$ implying

$$X(F(X) + W(X, \gamma, t) + \alpha(X)) < -X^2 \tag{7.56}$$

leads to

$$\begin{aligned}\dot{V}_2(X) &< -X^2 + (\gamma - \alpha(X))(U + X - \frac{\partial \alpha}{\partial X}((F(X) \\ &+ W(X, \gamma, t) + \gamma)) < 0\end{aligned} \tag{7.57}$$

with

$$\begin{aligned}U &= -(\gamma - \alpha(X)) - X + \frac{\partial \alpha}{\partial X}(F(X) + \gamma) \\ &- \Delta \mathrm{sgn}\left((\gamma - \alpha(X))\frac{\partial \alpha}{\partial X}\right)\frac{\partial \alpha}{\partial X}.\end{aligned} \tag{7.58}$$

we obtain:

$$\dot{V}_2(X) < -X^2 - (\gamma - \alpha(X))^2. \tag{7.59}$$

Then, this control allows us to stabilise the system at $X = 0$ for all uncertainties $W(X, \gamma, t)$ bounded by Δ.

7.3 Inverse Optimal Control

7.3.1 Theory of Inverse Optimal Control

Inverse optimal control allows us to minimise a meaningful cost function and to stabilise the system towards a given point (Kalman, 1964; Primbs, 1999; Primbs et al., 1999). The great advantage of this technique is that it guarantees optimality without solving the Hamilton–Jacobi-Bellman equation (HJB), which is in general difficult to solve. Bear in mind that the nonlinear system under consideration is given by Eq. (7.1). In this paragraph U is a single input ($U \in \mathbb{R}$), and thus $\vec{G}\left(\vec{X}\right)$ is a n-dimensional vector. Our purpose is to define a state feedback control $U\left(\vec{X}\right)$ such that $\vec{X} = \vec{0}$ is a globally asymptotically stable equilibrium point of the resulting closed-loop system. Moreover, we want this control to minimise a cost functional. In this section, we will consider the following cost functional

$$J = \int_0^{\infty} \left(q(\vec{X}\right) + U^2)\mathrm{d}t. \tag{7.60}$$

The difference with direct optimal control is that the cost functional is not given in principle but is deduced. Only the general form of the cost functional is proposed. For example, in our discussion, the general form of the cost functional is given by Eq. (7.60) but $q\left(\vec{X}\right)$ is unknown. In this approach, the control law is obtained using a known control Lyapunov function $V\left(\vec{X}\right)$. $V\left(\vec{X}\right)$ is, in fact, the value function of the considered optimal problem. Then, $V\left(\vec{X}\right)$ has to be a solution of the following HJB equation, so:

$$\min_U \left(q\left(\vec{X}\right) + U^2 + \frac{\partial V}{\partial \vec{X}} \vec{F}\left(\vec{X}\right) + \frac{\partial V}{\partial \vec{X}} \vec{G}\left(\vec{X}\right) U \right) = 0. \tag{7.61}$$

The minimum is obtained with

$$U\left(\vec{X}\right) = -\frac{1}{2} \frac{\partial V}{\partial \vec{X}} \vec{G}\left(\vec{X}\right). \tag{7.62}$$

It follows that:

$$q\left(\vec{X}\right) = -\frac{\partial V}{\partial \vec{X}} \vec{F}\left(\vec{X}\right) + \frac{1}{4} \left[\left(\frac{\partial V}{\partial \vec{X}} \vec{G}\left(\vec{X}\right) \right) \right]^2. \tag{7.63}$$

Thus, with this function $q\left(\vec{X}\right)$, the control $U\left(\vec{X}\right)$ is optimal. Now, we need to check that this optimal control stabilises the system around the equilibrium point $\vec{X} = \vec{0}$. Consider $V\left(\vec{X}\right)$ a control Lyapunov function such that $V\left(\vec{X}\right)$ is C^1 and positive definite with $V\left(\vec{0}\right) = 0$. We must then have

$$\forall \vec{X} \neq \vec{0} \quad \left(\frac{\partial V}{\partial \vec{X}}\vec{F}\left(\vec{X}\right) + \frac{\partial V}{\partial \vec{X}}\vec{G}\left(\vec{X}\right)U\right) < 0, \tag{7.64}$$

which leads to:

$$\forall \vec{X} \neq \vec{0}, \quad \left(\frac{\partial V}{\partial \vec{X}}\vec{F}\left(\vec{X}\right) - \frac{1}{2}\left[\left(\frac{\partial V}{\partial \vec{X}}\vec{G}\left(\vec{X}\right)\right)\right]^2\right) < 0. \tag{7.65}$$

The initial choice of $V\left(\vec{X}\right)$ is only restricted by:

$$\forall \vec{X} \neq \vec{0} \quad \frac{\partial V}{\partial \vec{X}}\vec{F}\left(\vec{X}\right) < 0. \tag{7.66}$$

It results that $q\left(\vec{X}\right)$ will be positive. Consider the following example:

$$\begin{cases} \dot{X}_1 = -X_1^3 - X_2 \\ \dot{X}_2 = X_1 - X_2 + U \end{cases}. \tag{7.67}$$

It is possible to take

$$V(X_1, X_2) = \frac{1}{2}X_1^2 + \frac{1}{2}X_2^2 \tag{7.68}$$

as a CLF because

$$\begin{aligned} \forall\ (X_1, X_2) \neq (0, 0), \ \frac{\partial V}{\partial \vec{X}}\vec{F}\left(\vec{X}\right) &= \frac{\partial V}{\partial X_1}\left(-X_1^3 - X_2\right) \\ &+ \frac{\partial V}{\partial X_2}(X_1 - X_2) = -X_1^4 - X_2^2 < 0. \end{aligned} \tag{7.69}$$

Thus, the control:

$$U(X_1, X_2) = -\frac{1}{2}\frac{\partial V}{\partial \vec{X}}\vec{G}\left(\vec{X}\right) = -\frac{1}{2}X_2 \tag{7.70}$$

stabilises the system defined by Eq. (7.67) around (0, 0) and minimises the cost functional:

$$J = \int_0^\infty (X_1^4 + \frac{3}{2}X_2^2)\mathrm{d}t. \tag{7.71}$$

7.3.2 Robust Inverse Optimal Control

Sometimes, it can be simpler to translate a nonlinear robust control problem into an optimal control problem that is much easier to solve (El-Farra, 2005; El-Farra & Christofides, 2003). Consider, with the current notations, the nonlinear system:

$$\frac{\mathrm{d}\vec{X}(t)}{\mathrm{d}t} = \vec{F}\left(\vec{X}\right) + \vec{G}\left(\vec{X}\right)U + \vec{H}\left(\vec{X}\right)W(t). \tag{7.72}$$

Once again, U is a single input ($U \in \mathbb{R}$), whereas $\vec{G}\left(\vec{X}\right)$ and $\vec{H}\left(\vec{X}\right)$ are n-dimensional vectors. $W(t)$ is an uncertainty bounded by:

$$|W(t)| < W_B. \tag{7.73}$$

A robust inverse optimal control can be stated as follows: Find a feedback control $U\left(\vec{X}\right)$ such that $\vec{X} = \vec{0}$ is a globally asymptotically stable equilibrium point of the resulting closed-loop system and minimises a cost functional for all uncertainties $W(t)$. In this paragraph, we revisit the cost functional given by Eq. (7.60). As in the previous section, the control law is obtained using a known control Lyapunov function $V\left(\vec{X}\right)$, which must be a solution to the following HJB equation accounting for uncertainty $W(t)$. It can be formulated as follows:

$$\min_U \sup_W \left(q\left(\vec{X}\right) + U^2 + \frac{\partial V}{\partial \vec{X}}\vec{F}\left(\vec{X}\right) + \frac{\partial V}{\partial \vec{X}}\vec{G}\left(\vec{X}\right)U + \frac{\partial V}{\partial \vec{X}}\vec{H}\left(\vec{X}\right)W \right) = 0. \tag{7.74}$$

The minimum is always achieved with:

$$U\left(\vec{X}\right) = -\frac{1}{2}\frac{\partial V}{\partial \vec{X}}\vec{G}\left(\vec{X}\right). \tag{7.75}$$

The worst-case uncertainty is obtained by replacing $\frac{\partial V}{\partial \vec{X}}\vec{H}\left(\vec{X}\right)W$ with:

$$\left|\frac{\partial V}{\partial \vec{X}}\vec{H}\left(\vec{X}\right)\right| W_B. \tag{7.76}$$

It follows that:

$$q\left(\vec{X}\right) = -\frac{\partial V}{\partial \vec{X}}\vec{F}\left(\vec{X}\right) + \frac{1}{4}\left[\left(\frac{\partial V}{\partial \vec{X}}\vec{G}\left(\vec{X}\right)\right)\right]^2 - \left|\frac{\partial V}{\partial \vec{X}}\vec{H}\left(\vec{X}\right)\right| W_B. \tag{7.77}$$

To regulate the system around $\vec{X} = \vec{0}$, we need the time derivative $\dot{V}$ to be negative for all $\vec{X} \neq \vec{0}$, leading to a restriction in the choice of $V\left(\vec{X}\right)$:

$$\forall\, \vec{X} \neq \vec{0}, \quad \frac{\partial V}{\partial \vec{X}} \cdot \vec{F}\left(\vec{X}\right) + \left|\frac{\partial V}{\partial \vec{X}}\vec{H}\left(\vec{X}\right)\right| W_B < 0. \tag{7.78}$$

By adding uncertainty to the previous example, we have:

$$\begin{cases} \dot{X}_1 = -X_1^3 - X_2 \\ \dot{X}_2 = X_1 - X_2 + X_2 W(t) + U \end{cases}, \tag{7.79}$$

where $W(t)$ is bounded by:

$$|W(t)| < W_B = 1. \tag{7.80}$$

Consider again the same CLF given by Eq. (7.68), we obtain for any $(X_1, X_2) \neq (0, 0)$

$$\begin{aligned} \frac{\partial V}{\partial \vec{X}}\vec{F}\left(\vec{X}\right) + \left|\frac{\partial V}{\partial \vec{X}}\vec{H}\left(\vec{X}\right)\right| W_B &= \frac{\partial V}{\partial X_1}(-X_1^3 - X_2) \\ &+ \frac{\partial V}{\partial X_2}(X_1 - X_2) + |X_2 X_2| W_B = -X_1^4 < 0. \end{aligned} \tag{7.81}$$

Therefore, this CLF is a suitable candidate. Consequently, the control

$$U(X_1, X_2) = -\frac{1}{2}\, X_2 \tag{7.82}$$

stabilises the system defined by Eq. (7.79) around (0, 0) and is optimal for the cost functional:

$$J = \int_0^\infty (X_1^4 + \frac{1}{2}X_2^2)\mathrm{d}t. \tag{7.83}$$

7.4 Sliding Mode Control

7.4.1 Generalities

The sliding mode control applies to variable structure systems (Edwards & Spurgeon, 1998; Ferrara et al., 2008; Utkin, 1977). These systems, which belong to the class of hybrid systems, consist of a set of continuous subsystems with a switching logic defined by a feedback control vector $\vec{U}\left(\vec{X}\right)$ and sliding surfaces $\vec{\sigma}\left(\vec{X}\right)$. This control is known to be very robust, even for high-order nonlinear systems. The motion on the sliding surfaces is insensitive to various types of perturbations and uncertainties. Therefore, designers do not need an exact model of the system to implement the control. We consider again the system defined by Eq. (7.1). Each component U_i of the discontinuous feedback control $\vec{U}\left(\vec{X}\right)$ is given by:

$$U_i\left(\vec{X}\right) = \begin{cases} U_i\left(\vec{X}\right)^+ & \text{if } \sigma_i\left(\vec{X}\right) > 0 \\ U_i\left(\vec{X}\right)^- & \text{if } \sigma_i\left(\vec{X}\right) < 0 \end{cases} \quad \text{with } i \in (1, \ldots, u), \tag{7.84}$$

where the condition $\sigma_i\left(\vec{X}\right) = 0$ defines the ith sliding surface. It follows that:

$$\vec{\sigma}\left(\vec{X}\right) = \left(\sigma_1\left(\vec{X}\right), \ldots, \sigma_u\left(\vec{X}\right)\right) = \vec{0}. \tag{7.85}$$

The closed-loop system behaves on $\vec{\sigma}\left(\vec{X}\right) = \vec{0}$ as a reduced $(n-u)$ order system.

7.4.2 Control Problem

The goal of the control is to develop continuous functions $U_i\left(\vec{X}\right)^+$, $U_i\left(\vec{X}\right)^-$ with the corresponding sliding surface $\sigma_i\left(\vec{X}\right) = 0$ so that the closed-loop system slides on the surface $\vec{\sigma}\left(\vec{X}\right) = \vec{0}$ determined by the intersection of u, $(n-1)$ dimensional surfaces $\sigma_i\left(\vec{X}\right) = 0$. These surfaces are designed so that the system exhibits the desired behaviour, such as reaching a desired equilibrium point or following a reference signal. Generally, the design of control begins by constructing sliding surfaces to achieve the desired behaviour, followed by defining the control law that forces the trajectory to reach the surface $\vec{\sigma}\left(\vec{X}\right) = \vec{0}$ and, once reached, keeps the system on it.

7.4.3 Existence of the Sliding Mode

Consider a Lyapunov function candidate

$$V\left(\vec{\sigma}\left(\vec{X}\right)\right) = \frac{1}{2}\vec{\sigma}\left(\vec{X}\right)\vec{\sigma}\left(\vec{X}\right) = \frac{1}{2}\left|\vec{\sigma}\left(\vec{X}\right)\right|^2 > 0, \tag{7.86}$$

where $\left|\vec{\sigma}\left(\vec{X}\right)\right|$ is the distance from $\vec{\sigma}\left(\vec{X}\right)$ to $\vec{\sigma}\left(\vec{X}\right) = \vec{0}$. The solution $\vec{\sigma}\left(\vec{X}\right) = \vec{0}$ will be stable if the time derivative of the Lyapunov function is negative definite:

$$\frac{\mathrm{d}V}{\mathrm{d}t} = \frac{\partial V}{\partial \vec{\sigma}}\frac{\mathrm{d}\vec{\sigma}}{\mathrm{d}t} = \vec{\sigma}\,\dot{\vec{\sigma}} < 0 \tag{7.87}$$

A sufficient condition to achieve this is that, for each $i \in (1, \ldots, u)$, the feedback control law $U_i\left(\vec{X}\right)$ is such that $\sigma_i\left(\vec{X}\right)$ and $\frac{\mathrm{d}\sigma_i(\vec{X})}{\mathrm{d}t}$ have opposite signs. Therefore,$(U_i\left(\vec{X}\right)^+$ must make $\frac{\mathrm{d}\sigma_i(\vec{X})}{\mathrm{d}t}$ negative when $\sigma_i\left(\vec{X}\right)$ is positive and $(U_i\left(\vec{X}\right)^-$ must make $\frac{\mathrm{d}\sigma_i(\vec{X})}{\mathrm{d}t}$ positive when $\sigma_i\left(\vec{X}\right)$ is negative. The condition given by Eq. (7.87) ensures that the sliding surfaces defined by $\vec{\sigma}\left(\vec{X}\right) = \vec{0}$ are reached asymptotically.

7.4.4 Region of Attraction

The region of attraction R_A is the set of all initial conditions $\vec{X}_0$ for which the closed-loop system reaches the sliding surface $\vec{\sigma}\left(\vec{X}\right) = \vec{0}$. Mathematically, it can be expressed as

$$R_A = \left\{\vec{X} \in \mathbb{R}^n : \vec{\sigma}\left(\vec{X}\right)\dot{\vec{\sigma}}\left(\vec{X}\right) < 0\right\}. \tag{7.88}$$

7.4.5 Sliding Motion on the Sliding Surface

Assume that a control law $\vec{U}\left(\vec{X}\right)$ has been defined to achieve a sliding mode on the surface $\vec{\sigma}\left(\vec{X}\right) = \vec{0}$. For each initial condition belonging to R_A, there exists a finite time t_i at which the trajectory of the closed-loop system intercepts the surface $\vec{\sigma}\left(\vec{X}\right) = \vec{0}$. With the sliding mode existing for all instants $t > t_i$, we have:

$$\vec{\sigma}\left(\vec{X}\right) = \vec{0} \text{ and } \frac{\mathrm{d}\vec{\sigma}\left(\vec{X}\right)}{\mathrm{d}t} = \vec{0}. \tag{7.89}$$

Calling $\overrightarrow{U_{\mathrm{eq}}}\left(\vec{X}\right)$ a continuous equivalent control acting on the sliding surface, by differentiating $\vec{\sigma}\left(\vec{X}\right)$ with respect to time, we obtain:

$$\frac{\mathrm{d}\vec{\sigma}\left(\vec{X}\right)}{\mathrm{d}t} = \frac{\partial\vec{\sigma}\left(\vec{X}\right)}{\partial\vec{X}}\frac{\mathrm{d}\vec{X}(t)}{\mathrm{d}t} = \frac{\partial\vec{\sigma}\left(\vec{X}\right)}{\partial\vec{X}}\left(\vec{F}\left(\vec{X}\right) + G\left(\vec{X}\right)\overrightarrow{U_{\mathrm{eq}}}\left(\vec{X}\right)\right) = \vec{0}. \tag{7.90}$$

Thus, the equivalent control is defined as:

$$\overrightarrow{U_{\mathrm{eq}}}\left(\vec{X}\right) = -\left[\frac{\partial\vec{\sigma}\left(\vec{X}\right)}{\partial\vec{X}}\vec{G}\left(\vec{X}\right)\right]^{-1}\frac{\partial\vec{\sigma}\left(\vec{X}\right)}{\partial\vec{X}}\vec{F}\left(\vec{X}\right). \tag{7.91}$$

By substituting this expression into Eq. (7.1), it follows that:

$$\frac{\mathrm{d}\vec{X}(t)}{\mathrm{d}t} = \left[I - \vec{G}\left(\vec{X}\right)\left[\frac{\partial\vec{\sigma}\left(\vec{X}\right)}{\partial\vec{X}}\vec{G}\left(\vec{X}\right)\right]^{-1}\frac{\partial\vec{\sigma}\left(\vec{X}\right)}{\partial\vec{X}}\right]\vec{F}\left(\vec{X}\right). \tag{7.92}$$

It is interesting to note that this dynamic does not depend on the control $\vec{U}\left(\vec{X}\right)$. That is why, in control design, we can start by constructing the sliding surface to achieve the desired behaviour without initially considering the control.

7.4.6 *Gathering Effect*

In actual design, due to the inertia of the actuators and sensors, the control vector cannot switch at infinity frequencies from $\vec{U}\left(\vec{X}\right)^{+}$ to $\vec{U}\left(\vec{X}\right)^{-}$. This implies that the sliding mode control occurs in a region close to the sliding surface. The finite switching frequency of a real control is called the Gathering effect. It is one of the major drawbacks of this control method. This phenomenon can lead to undesirable effects, such as instability. Gathering will be reviewed in Para. 14.5.

7.4.7 Example of One-Dimensional Sliding Control

In this first example, we consider a system with the following dynamics

$$\begin{cases} \dot{X}_1 = X_2 \\ \dot{X}_2 = X_1 - X_2^2 + U \end{cases}. \tag{7.93}$$

We have:

$$\vec{F}(\vec{X}) = \begin{pmatrix} X_2 \\ X_1 - X_2^2 \end{pmatrix} \text{ and } \vec{G}(\vec{X}) = \begin{pmatrix} 0 \\ 1 \end{pmatrix}. \tag{7.94}$$

Our goal is to stabilise this system at the equilibrium point $(X_{1e}, X_{2e}) = (0, 0)$ using sliding mode control. We have: $n = 2$ and $u = 1$. The first step consists of finding a sliding surface $\sigma(X_1, X_2) = 0$ of dimension $n - 1 = 1$ that allows us to achieve our objective. Consider the sliding line

$$\sigma(X_1, X_2) = X_1 + X_2 = 0. \tag{7.95}$$

On this line, the motion is defined by:

$$\sigma(X_1, X_2) = 0 \text{ and } \frac{\mathrm{d}\sigma}{\mathrm{d}t} = 0. \tag{7.96}$$

This implies that:

$$\frac{\mathrm{d}\sigma}{\mathrm{d}t} = \frac{\partial\sigma}{\partial\vec{X}}\frac{\mathrm{d}\vec{X}(t)}{\mathrm{d}t} = \begin{pmatrix} \frac{\partial\sigma}{\partial X_1} \\ \frac{\partial\sigma}{\partial X_2} \end{pmatrix}\left[\begin{pmatrix} X_2 \\ X_1 - X_2^2 \end{pmatrix} + \begin{pmatrix} 0 \\ 1 \end{pmatrix} U_{\text{eq}}\right] = 0 \tag{7.97}$$

gives the equivalent control

$$U_{\text{eq}} = X_2^2 - X_2 - X_1. \tag{7.98}$$

It follows that the dynamics on this line are given by:

$$\begin{cases} \dot{X}_1 = X_2 \\ \dot{X}_2 = -X_2 \end{cases}. \tag{7.99}$$

We obtain:

$$X_2(t) = X_{t_i} e^{-(t-t_i)}, \quad X_1(t) = -X_{t_i} e^{-(t-t_i)}. \tag{7.100}$$

t_i is the time at which the trajectory of the closed-loop system intercepts the line $\sigma = 0$. By Staying on the sliding line, the system will move towards the equilibrium

$(X_{1e}, X_{2e}) = (0, 0)$. Now, we have to find U, which forces the trajectory to reach $\sigma = 0$ and maintain the system on this line. The control must provide:

$$\forall\ (X_1, X_2)\ \ X_1 \neq -X_2 \Rightarrow\ \sigma(X_1, X_2)\dot{\sigma}(X_1, X_2) < 0. \tag{7.101}$$

From this condition, we can deduce the following control law:
when $\sigma > 0$, $(X_1 + X_2 > 0)$, we want to have:

$$\dot{\sigma} = \dot{X}_1 + \dot{X}_2 = X_2 + X_1 - X_2^2 + U < 0, \tag{7.102}$$

thus, $U = U^+ = -X_2 - X_1 + X_2^2 - 1$ is a possibility.
When $\sigma < 0$, $(X_1 + X_2 < 0)$, we want to have:

$$\dot{\sigma} = \dot{X}_1 + \dot{X}_2 = X_2 + X_1 - X_2^2 + U > 0, \tag{7.103}$$

thus, $U = U^- = -X_2 - X_1 + X_2^2 + 1$ is a possibility.
Figure 7.1 shows the closed-loop system trajectory under this control for different initial conditions. The region of attraction R_A for this example is $\mathbb{R}^2$.

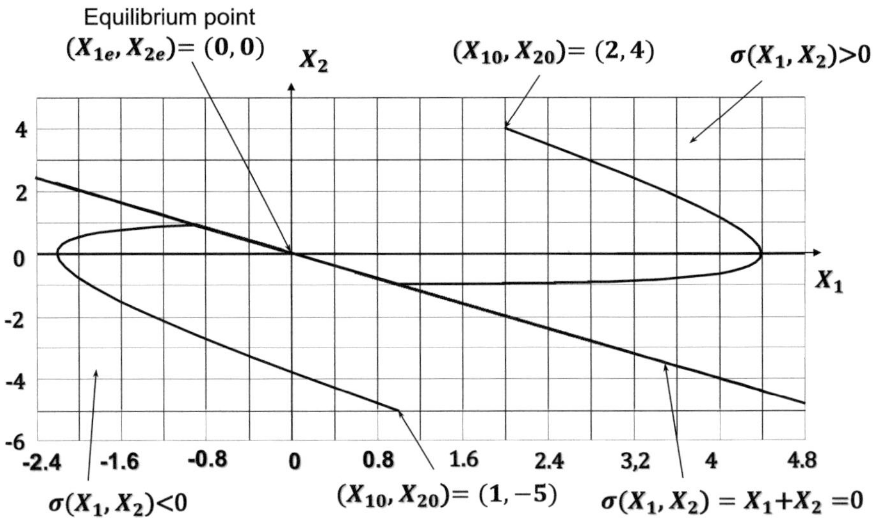

Fig. 7.1 Example of one-dimensional sliding mode control

7.4.8 Example of a Two-Dimensional Sliding Control

The following dynamics illustrate two-dimensional sliding control ($u = 2$):

$$\begin{cases} \dot{X}_1 = X_2 \\ \dot{X}_2 = X_3 + U_1 \\ \dot{X}_3 = X_1 - X_3^2 + U_2 \end{cases}. \tag{7.104}$$

Here we have:

$$\vec{F}\left(\vec{X}\right) = \begin{pmatrix} F_1(X_1, X_2, X_3) \\ F_2(X_1, X_2, X_3) \\ F_3(X_1, X_2, X_3) \end{pmatrix} = \begin{pmatrix} X_2 \\ X_3 \\ X_1 - X_3^2 \end{pmatrix},$$

$$G\left(\vec{X}\right) = \begin{pmatrix} 0\,0 \\ 1\,0 \\ 0\,1 \end{pmatrix}, \quad \vec{U}\left(\vec{X}\right) = \begin{pmatrix} U_1 \\ U_2 \end{pmatrix}. \tag{7.105}$$

Our goal is to stabilise the system to $(X_{1e}, X_{2e}, X_{3e}) = (0, 0, 0)$. Consider the sliding surface defined as:

$$\vec{\sigma}\left(\vec{X}\right) = \begin{pmatrix} \sigma_1(X_1, X_2, X_3) \\ \sigma_2(X_1, X_2, X_3) \end{pmatrix} = \begin{pmatrix} X_1 + X_2 \\ X_2 + X_3 \end{pmatrix} = \begin{pmatrix} 0 \\ 0 \end{pmatrix}. \tag{7.106}$$

On this surface, the dynamics are described by:

$$\frac{d\vec{\sigma}\left(\vec{X}\right)}{dt} = \frac{\partial\vec{\sigma}\left(\vec{X}\right)}{\partial\vec{X}}\frac{d\vec{X}(t)}{dt} = \begin{pmatrix} \frac{\partial\sigma_1}{\partial X_1} & \frac{\partial\sigma_1}{\partial X_2} & \frac{\partial\sigma_1}{\partial X_3} \\ \frac{\partial\sigma_2}{\partial X_1} & \frac{\partial\sigma_2}{\partial X_2} & \frac{\partial\sigma_2}{\partial X_3} \end{pmatrix} \begin{pmatrix} X_2 \\ X_3 + U_1 \\ X_1 - X_3^2 + U_2 \end{pmatrix} = \begin{pmatrix} 0 \\ 0 \end{pmatrix}. \tag{7.107}$$

Thus, the equivalent control $\overrightarrow{U_{\text{eq}}}\left(\vec{X}\right)$ is

$$\overrightarrow{U_{\text{eq}}}\left(\vec{X}\right) = \begin{pmatrix} U_{\text{eq1}} \\ U_{\text{eq2}} \end{pmatrix} = \begin{pmatrix} -X_2 - X_3 \\ X_2 - X_1 + X_3^2 \end{pmatrix}. \tag{7.108}$$

On the surface $\vec{\sigma}\left(\vec{X}\right) = \vec{0}$ (in this example, it is a line because $n - u = 1$) the motion is given by:

$$\begin{cases} \dot{X}_1 = X_2 \\ \dot{X}_2 = -X_2 \\ \dot{X}_3 = X_2 \end{cases}. \tag{7.109}$$

It is easy to deduce that the system moving along this line will reach the equilibrium point. The control $\vec{U}\left(\vec{X}\right)$ must force the trajectory to reach $\vec{\sigma} = \vec{0}$ and maintain the system on this surface. The control must provide:

$$\forall(X_1, X_2, X_3),\quad \vec{\sigma} \neq \vec{0} \Rightarrow \vec{\sigma}\frac{\mathrm{d}\vec{\sigma}}{\mathrm{d}t} < 0. \tag{7.110}$$

From this condition, we can deduce the following control law:
When $\sigma_1 > 0$, $(X_1 + X_2 > 0)$ we want to have:

$$\dot{\sigma_1} = \dot{X}_1 + \dot{X}_2 = X_2 + X_3 + U_1 < 0, \tag{7.111}$$

thus, $U_1 = U_1^+ = -X_2 - X_3 - 1$ is a possibility.
When $\sigma_1 < 0$, $(X_1 + X_2 < 0)$ we want to have:

$$\dot{\sigma_1} = \dot{X}_1 + \dot{X}_2 = X_2 + X_3 + U_1 > 0, \tag{7.112}$$

Thus, $U_1 = U_1^- = -X_2 - X_3 + 1$ is a possibility.
When $\sigma_2 > 0$, $(X_2 + X_3 > 0)$ we want to have:

$$\dot{\sigma_2} = \dot{X}_2 + \dot{X}_3 = X_3 + U_1 + X_1 - X_3^2 + U_2 < 0, \tag{7.113}$$

thus, $U_2 = U_2^+ = -X_3 - U_1 - X_1 + X_3^2 - 1$ is a possibility.
When $\sigma_2 < 0$, $(X_2 + X_3 < 0)$ we want to have:

$$\dot{\sigma_2} = \dot{X}_2 + \dot{X}_3 = X_3 + U_1 + X_1 - X_3^2 + U_2 > 0, \tag{7.114}$$

Thus, $U_2 = U_2^- = -X_3 - U_1 - X_1 + X_3^2 + 1$ is a possibility.
Figure 7.2 shows that the trajectory of the system starts with the initial conditions $(X_{10}, X_{20}, X_{30}) = (3, 2, 2)$ then reaches the first sliding surface $\sigma_2 = 0$. It then

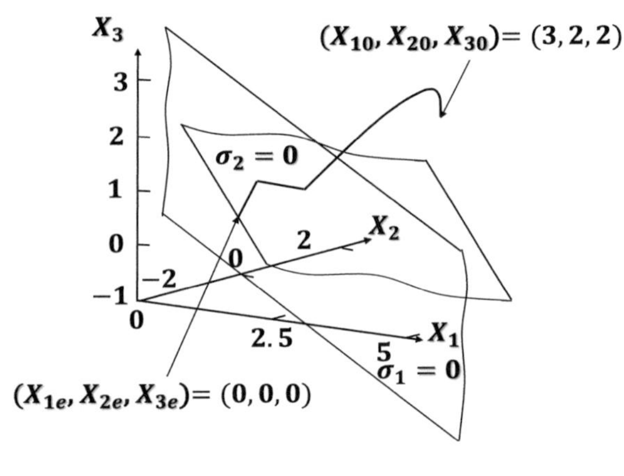

Fig. 7.2 Example of two-dimensional sliding mode control

slides along this surface until it reaches the second sliding surface $\sigma_1 = 0$. Finally, it establishes itself at the equilibrium point $(X_{1e}, X_{2e}, X_{3e}) = (0, 0, 0)$. The region of attraction R_A for this example is $\mathbb{R}^3$.

7.4.9 Example of Robust Sliding Control

The robustness of the sliding control (Utkin, 1977) is illustrated with the following example:

$$\begin{cases} \dot{X}_1 = -X_1 - X_2 \\ \dot{X}_2 = -X_3 \\ \dot{X}_3 = U + W(X_1, X_1, X_1, t) \end{cases} . \tag{7.115}$$

The uncertainty $W(X_1, X_1, X_1, t)$ is bounded by the positive parameter Δ:

$$|W(X_1, X_1, X_1, t)| < \Delta = 1 \tag{7.116}$$

Consider the sliding line:

$$\sigma(X_1, X_2) = -X_2 + X_3 = 0. \tag{7.117}$$

On this line, the motion is defined by:

$$\sigma(X_1, X_2) = 0 \text{ and } \frac{\mathrm{d}\sigma}{\mathrm{d}t} = 0, \tag{7.118}$$

thus,

$$\frac{\mathrm{d}\sigma}{\mathrm{d}t} = \frac{\partial\sigma}{\partial\vec{X}} \frac{\mathrm{d}\vec{X}(t)}{\mathrm{d}t} = \begin{pmatrix} 0 \\ -1 \\ 1 \end{pmatrix} \begin{pmatrix} -X_1 - X_2 \\ -X_3 \\ U_{\text{eq}} - X_1 + W(X_1, X_1, X_1, t) \end{pmatrix} = 0. \tag{7.119}$$

It follows that:

$$U_{\text{eq}} = X_1 - X_3 - W(X_1, X_2, X_3, t). \tag{7.120}$$

On the sliding line $X_2 = X_3$, the motion does not depend on the uncertainty and is described by:

$$\begin{cases} \dot{X}_1 = -X_1 - X_2 \\ \dot{X}_2 = -X_2 \\ \dot{X}_3 = -X_3 \end{cases} . \tag{7.121}$$

The controlled system converges towards $(X_{1e}, X_{2e}, X_{3e}) = (0, 0, 0)$. Now we have to find U that forces the trajectory to reach $\sigma = 0$ and maintain the system on this line for all uncertainties bounded by Δ. The control must provide:

$$\forall (X_1, X_2, X_3)\, X_2 \neq X_3 \Rightarrow \sigma\dot{\sigma} < 0. \tag{7.122}$$

From this condition, we can deduce the following control:
When $\sigma > 0$, $(-X_2 + X_3 > 0)$ we want to have:

$$\dot{\sigma} = -\dot{X}_2 + \dot{X}_3 = X_3 + U + W(X_1, X_1, X_1, t) < 0, \tag{7.123}$$

Thus, $U = U^+ = -X_3 - 2$ is a possibility.
When $\sigma < 0$, $(-X_2 + X_3 < 0)$ we want to have:

$$\dot{\sigma} = -\dot{X}_2 + \dot{X}_3 = X_3 + U + W(X_1, X_1, X_1, t) > 0, \tag{7.124}$$

Thus, $U = U^- = -X_3 + 2$ is a possibility.
With this control, for all initial conditions in the specified region $R_A = \mathbb{R}^2$ and all uncertainties bounded by Δ, the system will reach the equilibrium point $(X_{1e}, X_{2e}, X_{3e}) = (0, 0, 0)$.

7.5 Model Predictive Control (MPC)

7.5.1 Generalities

Model predictive control theory (also known as receding horizon control) was introduced in the 1970s (Fernández Camacho & Bordons Alba, 1999; Pekař, 2010; Qin & Badgwell, 1998). MPC is particularly well suited for problems with multiple objectives and many inputs and outputs with constraints. This control also has the advantage of compensating for disturbances and model inaccuracies. Due to its flexibility, it is now a common solution in various industrial problems. MPC designates a specific control strategy that uses a model of a process to obtain control action by minimising a cost function. This is a special case of optimal control theory. MPC consists of an optimisation problem at each control interval. In most cases, the goal of the control is for the system outputs to follow the predefined reference signals.

7.5.2 Theory of Predictive Control

A graphical illustration of this control is shown in Fig. 7.3. The system to be controlled is usually described (using the usual notations) by recurrence relations of the form:

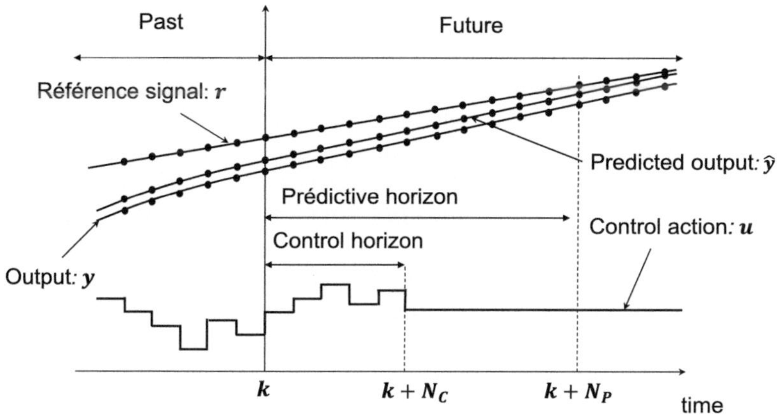

Fig. 7.3 Graphical illustration of model predictive control

$$\begin{cases} \vec{X}_{k+1} = \vec{F}\left(\vec{X}_k, \vec{U}_k\right) \\ \vec{Y}_k = \vec{C}\left(\vec{X}_k\right) \end{cases} . \tag{7.125}$$

We consider the general case where the functions $\vec{F}$ and $\vec{C}$ are nonlinear. The state and control vectors are required to satisfy the following constraints:

$$\vec{X}_k \in X \quad \text{and} \quad \vec{U} \in U \tag{7.126}$$

X and ***U*** are subsets of $\mathbb{R}^n$ and $\mathbb{R}^u$, respectively. The current control vector is obtained by solving a finite-horizon open-loop optimal control problem minimising a cost functional. A common cost functional is given by

$$J(k) = \sum_{i=0}^{N_P} \left(\widehat{\vec{Y}}_{k+i} - \vec{r}_{k+i}\right)^T Q\left(\widehat{\vec{Y}}_{k+i} - \vec{r}_{k+i}\right) + \vec{U}_{k+i}^T R \vec{U}_{k+i}. \tag{7.127}$$

where N_P is the predictive horizon, $\widehat{\vec{Y}}_k$ is the predicted process output vector, $\vec{r}_k$ the reference signal vector, Q and R are the symmetric positive definite weighting matrices for the output and control, respectively. The initial conditions of this problem are the values of the state variables measured at each discrete time k :

$$\widehat{\vec{Y}}_k = \vec{Y}_k \tag{7.128}$$

It is also possible to consider a control horizon N_C, which is the number of samples within the predictive horizon where the MPC controller can affect the control action.

At each interval, an algorithm determines an optimal sequence of control actions minimising $J(k)$. Only the first element of the computed sequence is applied to the system. At each iteration, the predictive horizon shifts forward in time, and the MPC controller predicts the system output again. One of the main challenges of MPC control is ensuring the stability of the controlled system. The closed-loop stability is generally achieved by tuning control parameters such as the length of the control horizon and the choice of weighting matrices, or by adding a terminal penalty at the end of the control horizon.

MPC control provides the possibility of achieving robust control. In general, a distinction is made between the inherent robustness properties of MPC (where the controller can cope with uncertainties without explicit accounting for them) and robust MPC designs which explicitly take uncertainties and disturbances into account. Most robust MPC approaches that take uncertainty and disturbance into account are based on a min–max formulation. In this case, the minimisation of the cost functional is carried out under uncertainties and disturbances corresponding to the worst-case scenario (i.e. the maximum value of $J(k)$).

7.5.3 Illustrative Example

Consider, as illustration, the following hybrid system:

$$\begin{pmatrix} X_{1(k+1)} \\ X_{2(k+1)} \end{pmatrix} = \begin{pmatrix} 1 & -0.011 \\ 0.2 & 0.96 \end{pmatrix} \begin{pmatrix} X_{1k} \\ X_{2k} \end{pmatrix} + \begin{pmatrix} 0.11(1+\text{tol}) \\ 0 \end{pmatrix} u_k, \tag{7.129}$$

with the output

$$\begin{pmatrix} Y_{1k} \\ Y_{2k} \end{pmatrix} = \begin{pmatrix} X_{1k} \\ X_{2k} \end{pmatrix}. \tag{7.130}$$

u_k and tol are, respectively, a scalar Boolean control and a tolerance parameter. Model inaccuracy is defined as tol $\in \{0, 05\}$. There is a constraint on X_{1k} defined by:

$$\text{if } X_{1k} > 1.2 \quad \text{then } u_k = 0. \tag{7.131}$$

The goal is for the system outputs to follow the predefined reference vector:

$$\begin{pmatrix} r_{1k} \\ r_{2k} \end{pmatrix} = \begin{pmatrix} 1 \\ 5 \end{pmatrix}. \tag{7.132}$$

The cost functional is given by:

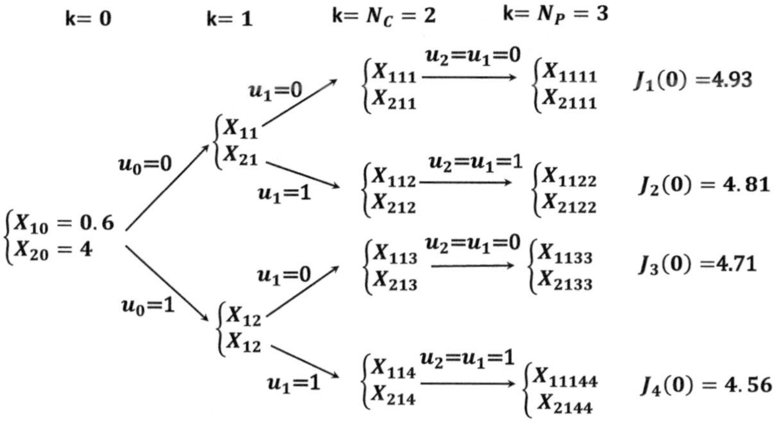

Fig. 7.4 Determination of u_0

$$J(k) = \sum_{i=0}^{N_P} 0.5\left(\widehat{Y_1}_{(k+i)} - r_{1(k+i)}\right)^2 + \left(\widehat{Y_2}_{(k+i)} - r_{2(k+i)}\right)^2. \tag{7.133}$$

Thus, in this example: $Q = (0.5, 1)$ and $R = 0$. Moreover, we take $N_P = 3$ and $N_c = 2$. Predicted outputs $\widehat{Y_1}$ and $\widehat{Y_2}$ are calculated using a predictive model for which tol $= 0$. The following simulation is carried out assuming that the real system is defined with tol $= 0.5$. Figure 7.4 illustrates the calculation of the first control input u_0.

The measured state values at $k = 0$ are:

$$X_{10} = 0.6 \text{ and } X_{20} = 4 \tag{7.134}$$

The four possible sequences (u_0, u_1, u_2) (see Fig. 7.4) are:

$$(0, 0, 0), (0, 1, 1), (1, 0, 0), (1, 1, 1) \tag{7.135}$$

which correspond to:

$$J_1(0), J_2(0), J_3(0), J_4(0). \tag{7.136}$$

As shown in Fig. 7.4, the minimum cost value is:

$$J_4(0) = 4.56 \tag{7.137}$$

Thus, the first control input of this sequence is:

$$u_0 = 1. \tag{7.138}$$

The fact that X_{10} is lower than 1.2 the control value will be confirmed. The next initial conditions will be defined using the real system (tol $=$ 0.5) and the same process will be applied. Figure 7.5 represents the evolution of X_{1k}, $X_{2k,}u_k$ and $J(k)$. Naturally X_{1k}, X_{2k} and $J(k)$ tend, respectively, to r_{1k} and r_{2k} and 0. When

$$X_{1k} > 1.2, \tag{7.139}$$

we can note that

$$u_k = 0 \tag{7.140}$$

This example highlights the inherent robustness properties of MPC.

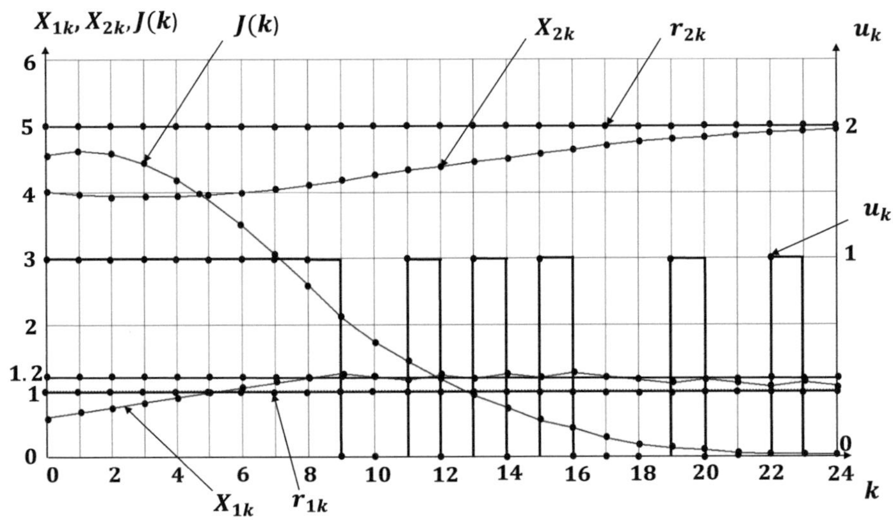

Fig. 7.5 Evolution of X_{1k}, X_{2k}, u_k and $J(k)$

Chapter 8
Control of Chaotic Systems

In a system, chaotic behaviour is usually undesirable. That is why chaos control has attracted considerable attention across various fields. Various control concepts developed for chaotic systems are presented in this section. Examples are provided to demonstrate the effectiveness of these control techniques. We will use the same illustrative examples (the Hénon system for discrete systems, and the Chua circuit and the Rössler system for continuous systems) to show that multiple control approaches can be applied to stabilise a given chaotic system.

8.1 System Parameter Variation Technique

8.1.1 Principle

This is the simplest method to prevent chaotic oscillations. It consists of keeping the system parameters within a region where the system does not exhibit chaotic behaviour (Auteur(s), 1989; Ogorzalek, 1993*)*. As shown in Para. 3.2, this region can be identified using a bifurcation diagram.

8.1.2 Illustrative Example

Consider again the first-order discrete system studied in Para. 3.2.4:

$$X_{k+1} = rX_k(1 - X_k), \tag{8.1}$$

where r is the only system parameter. The bifurcation diagram shown in Fig. 3.11 indicates the values of r to avoid so that the system remains non-chaotic.

B. Rivet, *Theories of Nonlinear Systems for Power Converters*,
https://doi.org/10.1007/978-3-032-25062-9_8

8.2 Control by Adding an Absorber

8.2.1 Principle

The principle of this method is to add a dynamic vibration absorber to the chaotic system without altering its structure. The suppression of chaotic behaviour is achieved without the use of feedback and control signals. With this method, a new stable orbit appears in a neighbourhood of the original attractor.

8.2.2 Illustrative Example

We consider here the Chua circuit shown in Fig. 8.1 (Kapitaniak et al., 1993; Ogorzalek, 1993).
The numerical values of the parameters are as follows:

$$\begin{aligned} &C_1 = 100\,\text{nF}, \quad C_2 = 10\,\text{nF}, \quad L = 22\,\text{mH}, \quad R_L = 10\,\Omega, \quad R = 1900\,\Omega, \\ &m_0 = -0.41\,\text{mS}, \quad m_1 = -0.757\,\text{mS}, \quad B_P = 1.7\,\text{V}. \end{aligned} \tag{8.2}$$

Figure 8.2 illustrates the chaotic behaviour of this system.
The absorber shown in Fig. 8.3 consists of a parallel circuit (L_a, C_a, R_a) coupled to the original Chua circuit via a resistor: R_{ax}. Depending on the choice of the coupling parameter R_{ax}, various stable orbits can be obtained. With the parameters of the absorber given below:

$$L_a = 22\,\text{mH}, \quad C_a = 100\,\text{nF}, \quad R_a = 1\,\text{k}\Omega, \quad R_{ax} = 5\,\text{k}\Omega, \tag{8.3}$$

the result is a stable period-one trajectory.
This controlled orbit is shown in Fig. 8.4.

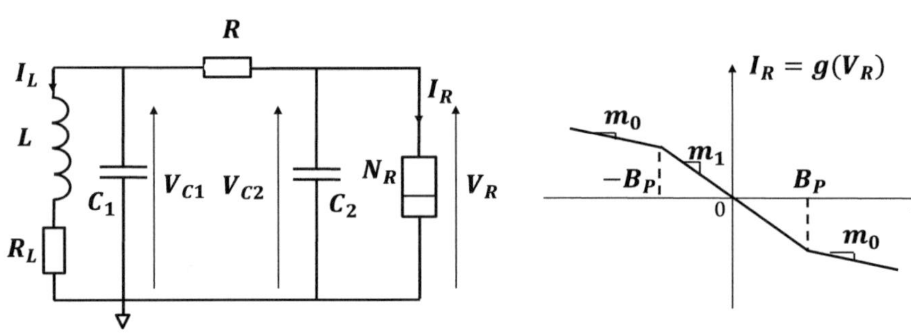

Fig. 8.1 Chua circuit

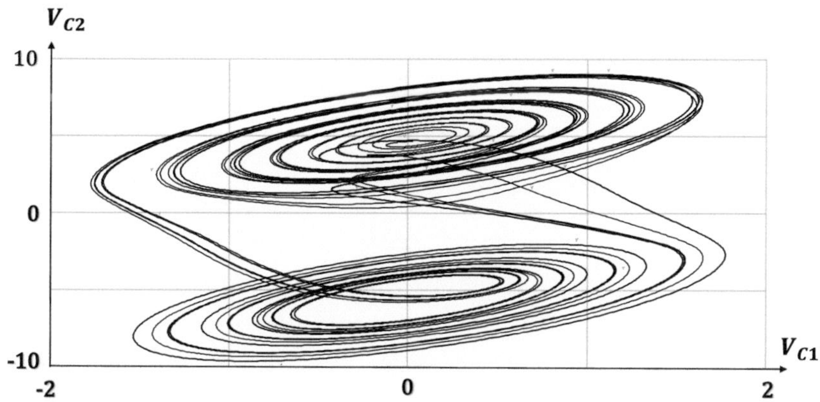

Fig. 8.2 Chaotic behaviour of a Chua's circuit

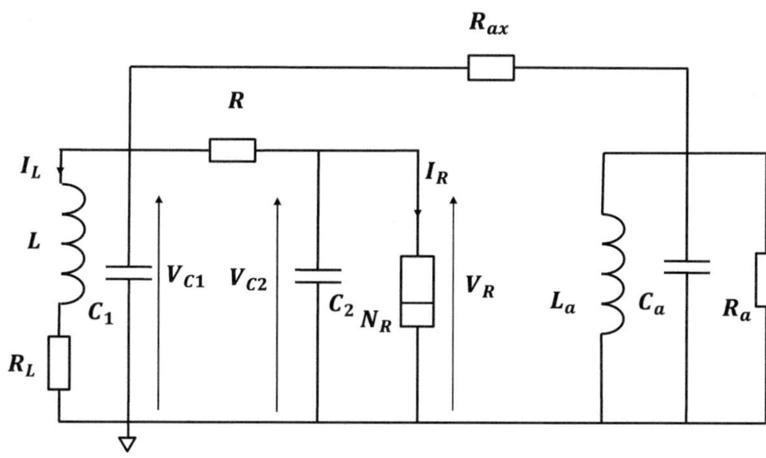

Fig. 8.3 Chua's circuit coupled with its absorber

8.3 Control with a Predictive Model

8.3.1 Principle

With the usual notations, consider a n-dimensional chaotic system associated with a n-dimensional control vector $\overrightarrow{U_k}$:

$$\vec{X}_{k+1} = \vec{F}\left(\vec{X}_{k+1}, \vec{U}_k\right). \tag{8.4}$$

Suppose we have a predictive model of the chaotic system without control:

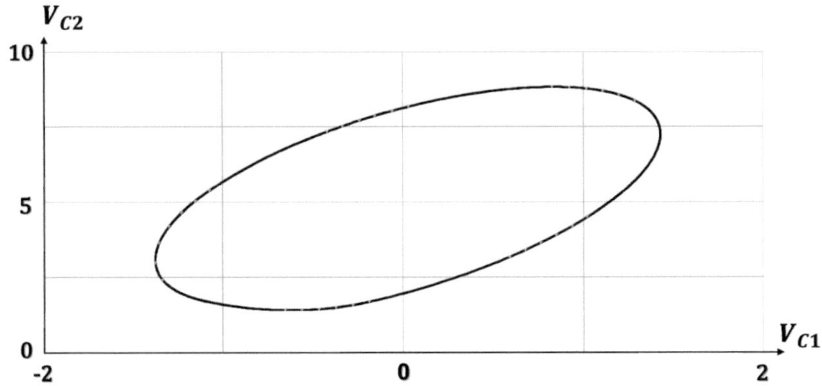

Fig. 8.4 Stable period-one orbit of the Chua's circuit with an absorber

$$\overrightarrow{X_{Pk+1}} = \vec{F}\left(\vec{X}_k, \vec{0}\right). \tag{8.5}$$

The principle of controlling a chaotic system with a predictive model is illustrated in Fig. 8.5 (Boukabou & Mansouri, n.d.; Ott et al., 1990).

Our objective is to stabilise the system around an unstable equilibrium point $\overrightarrow{X_e}$ defined by:

$$\vec{X}_e = \vec{F}\left(\vec{X}_e, \vec{0}\right). \tag{8.6}$$

For this purpose, we will use the following control:

$$\vec{U}_k = K\left(\overrightarrow{X_{Pk+1}} - \vec{X}_k\right) \text{ if } |X_{1k} - X_{1e}| < \varepsilon_1, \ldots, |X_{nk} - X_{ne}| < \varepsilon_n \tag{8.7}$$

and

Fig. 8.5 Principle of chaos control using a predictive model

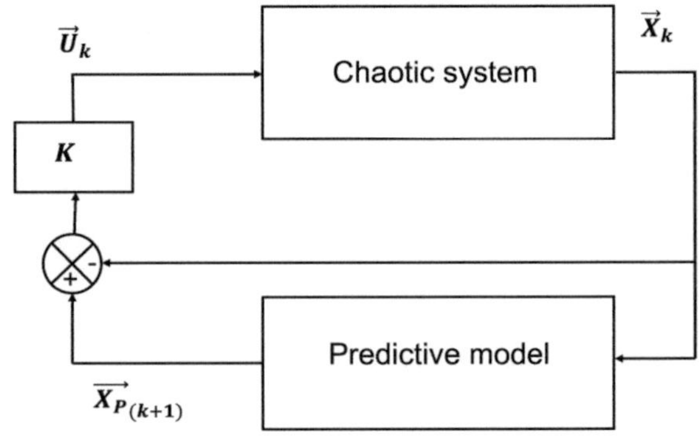

$$\vec{U}_k = \vec{0} \quad \text{otherwise.} \tag{8.8}$$

$K = \text{diag}(k_1, \ldots, k_n)$ is a $n \times n$ diagonal matrix and $\varepsilon_1, \ldots, \varepsilon_n$ are small positive coefficients. In other words, the control is disabled when the state vector is far from the equilibrium point under consideration. When the system approaches $\vec{X}_e$, the control is activated. The gains $(k_1, \ldots, k_n)$ will be determined by the linearisation technique to stabilise the system around the equilibrium point. If $n \neq u$, K becomes a $u \times n$ matrix. An illustrative example is provided in the following section.

8.3.2 Illustrative Example

Consider the Hénon system studied in Para. 4.3. We know that this system is chaotic for $P_1 = 1.4$ and $P_2 = 0.3$. In order to stabilise this system at the unstable equilibrium point:

$$(X_{1e}, X_{2e}) = (0.6313, 0.1894), \tag{8.9}$$

we add a control vector:

$$\vec{U}_k = \begin{pmatrix} U_{1k} \\ U_{2k} \end{pmatrix}. \tag{8.10}$$

The dynamics of the controlled Hénon system are now given by:

$$\begin{cases} X_{1(k+1)} = -1.4X_{1k}^2 + X_{2k} + 1 + U_{1k} \\ X_{2(k+1)} = 0.3X_{1k} + U_{2k} \end{cases}. \tag{8.11}$$

The control is defined by:

$$\vec{U}_k = \begin{pmatrix} k_1\left(-1.4X_{1k}^2 + X_{2k} + 1 - X_{1k}\right) \\ k_2(0.3X_{1k} - X_{2k}) \end{pmatrix} \quad \text{if } |X_{1k} - X_{e1}| < \varepsilon_1 \text{ and } |X_{2k} - X_{e2}| < \varepsilon_2 \tag{8.12}$$

and

$$\vec{U}_k = \begin{pmatrix} 0 \\ 0 \end{pmatrix} \quad \text{otherwise} \tag{8.13}$$

We will take:

$$(\varepsilon_1, \varepsilon_2) = (0.05, 0.05). \tag{8.14}$$

So, when the system gets close to (X_{e1}, X_{e2}), the dynamics become

$$\begin{cases} X_{1(k+1)} = -1.4X_{1k}^2 + X_{2k} + 1 + k_1\left(-1.4X_{1k}^2 + X_{2k} + 1 - X_{1k}\right) \\ X_{2(k+1)} = 0.3X_{1k} + k_2(0.3X_{1k} - X_{2k}) \end{cases}. \tag{8.15}$$

The next step consists in linearising this system around (X_{e1}, X_{e2}). The corresponding Jacobian matrix is:

$$J = \begin{pmatrix} -2.8X_{1e}(1+k_1) - k_1 & (1+k_1) \\ 0.3(1+_2) & -k_2 \end{pmatrix}. \tag{8.16}$$

The characteristic equation is given by:

$$z^2 + z(k_2 + 2.8X_{1e}(1+k_1) + k_1) + k_2(2.8X_{1e}(1+k_1) + k_1) \\ - 0.3(1+k_1)(1+k_2) = 0 \tag{8.17}$$

with the gains

$$k_1 = -0.5 \quad \text{and} \quad k_2 = 0.3. \tag{8.18}$$

The two roots $(-0.786, 0.102)$ are inside the unit circle, so the corresponding control will stabilise the Hénon system towards (X_{1e}, X_{2e}). Figure 8.6 illustrates the behaviour of the controlled Hénon system. The system starts with initial conditions $(X_{10}, X_{20}) = (0.8, 0.3)$ far from (X_{1e}, X_{2e}), so the control is disabled. At $k = 49$, for the first time

$$|X_{1k} - X_{e1}| < \varepsilon_1 \quad \text{and} \quad |X_{2k} - X_{e2}| < \varepsilon_2, \tag{8.19}$$

so the control is activated and stabilises the system around the desired equilibrium point.

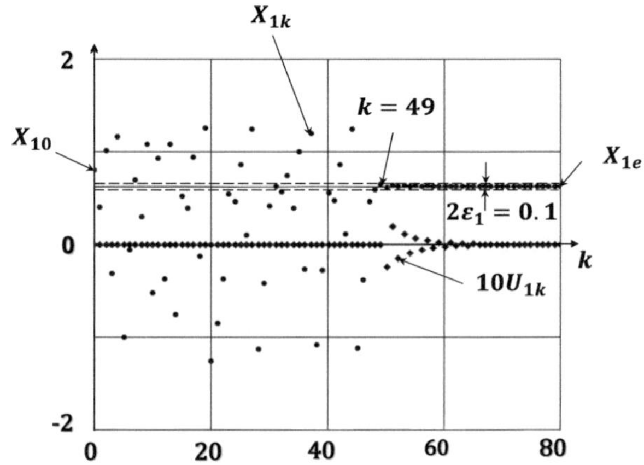

Fig. 8.6 Stabilisation of the Hénon system using a predictive model

8.4 Entrainment Control

8.4.1 Principle

Let us consider again the n-dimensional system:

$$\frac{\mathrm{d}\vec{X}}{\mathrm{d}t} = \vec{F}\left(\vec{X}\right). \tag{8.20}$$

Entrainment control consists in suppressing chaos by entraining the system to an arbitrarily chosen goal orbit $\vec{g}(t)$, which is not a particular solution of Eq. (8.20) (Htischer & Htibler, 1989; Jackson, 1991; Ogorzalek, 1993). To reach this objective, we apply the control:

$$\frac{\mathrm{d}\vec{g}}{\mathrm{d}t} - \vec{F}(\vec{g}) \tag{8.21}$$

to the right-hand side of Eq. (8.20). The dynamics of the new system become:

$$\frac{\mathrm{d}\vec{X}}{\mathrm{d}t} = \vec{F}\left(\vec{X}\right) + \frac{\mathrm{d}\vec{g}}{\mathrm{d}t} - \vec{F}(\vec{g}) \tag{8.22}$$

No particular solution of system can be proposed for this control because with

$$\frac{\mathrm{d}\vec{g}}{\mathrm{d}t} - \vec{F}(\vec{g}) = 0, \tag{8.23}$$

there is no possibility of control. The goal dynamics $\vec{g}$ must be confined to the convergent region of phase space C_r ($\vec{g} \in C_r$) defined by

$$C_r = \left\{\vec{X} : \forall\lambda \in \mathbb{C}\left|J\left(\vec{X}\right) - \lambda I_n\right| = 0 \Rightarrow R_e(\lambda) < 0\right\}, \tag{8.24}$$

where J is the Jacobian matrix of system (8.20). As explained in Para. 2.12, the Routh criteria can be used to determine the convergent region without explicitly calculating the roots λ. If $\vec{g} \in C_r$ we get:

$$\lim_{t\to\infty}\left(\vec{X}(t) - \vec{g}(t)\right) = 0. \tag{8.25}$$

8.4.2 Illustrative Example

As an illustrative example, we choose the Rössler system. The dynamics of this system are given by:

$$\begin{cases} \frac{dX_1}{dt} = -X_2 - X_3 \\ \frac{dX_2}{dt} = X_1 + 0.15X_2 \\ \frac{dX_3}{dt} = 0.2 + X_1X_3 - 10X_3 \end{cases} . \quad (8.26)$$

With the chosen parameters, the Rössler system exhibits chaotic behaviour. To determine the convergent region, the first step is to determine the characteristic equation. It is given by:

$$\begin{vmatrix} -\lambda & -1 & -1 \\ 1 & 0.15-\lambda & 0 \\ X_3 & 0 & X_1 - 10 - \lambda \end{vmatrix} = 0, \quad (8.27)$$

leading to:

$$\lambda^3 + \lambda^2(9.85 - X_1) + \lambda(-0.5 + 0.15X_1 + X_3) - X_1 - 0.15X_3 + 10 = 0. \quad (8.28)$$

The convergent region of the phase space is the area C_r in which all the roots of the characteristic equation have negative real parts. Using the Routh criteria, C_r is determined by:

$$C_r : \begin{cases} X_1 < 9.85 \\ X_3 > \frac{0.15X_1^2 - 2.9775X_1 + 14.925}{10 - X_1} = W(X_1) \\ X_3 < \frac{10 - X_1}{0.15} = U(X_1) \end{cases} . \quad (8.29)$$

The definition of C_r does not depend on X_2. The convergent region is shown in Fig. 8.7. We first begin by stabilising the Rössler system to the following fixed point belonging to C_r :

$$(g_1, g_2, g_3) = (-10, 0, 20). \quad (8.30)$$

The modified system is given by:

$$\begin{cases} \frac{dX_1}{dt} = -X_2 - X_3 + 20 \\ \frac{dX_2}{dt} = X_1 + 0.15X_2 + 10 \\ \frac{dX_3}{dt} = X_1X_3 - 10X_3 + 400 \end{cases} . \quad (8.31)$$

Figure 8.7 shows the entrainment to the fixed point. Secondly, we want to entrain the Rössler system to the desired goal orbit:

Fig. 8.7 Entrainment of the Rössler system to fixed point (10, 0, 20)

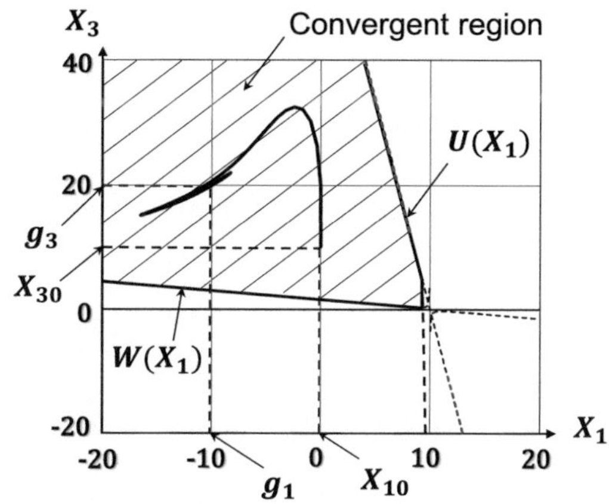

$$(g_1(t), g_2(t), g_3(t)) = (-10 + \sin(t), 0, 20 + \cos(t)). \tag{8.32}$$

The modified system becomes:

$$\begin{cases} \frac{dX_1}{dt} = -X_2 - X_3 + 20 + 2\cos(t) \\ \frac{dX_2}{dt} = X_1 + 0.15X_2 + 10 - \sin(t) \\ \frac{dX_3}{dt} = X_1X_3 - 10X_3 + (20 - \sin(t))(20 + \cos(t)) - \sin(t) \end{cases}. \tag{8.33}$$

The temporal representation of the entrainment to the orbit is shown in Fig. 8.8.

Fig. 8.8 Entrainment of the Rössler system to a periodic orbit

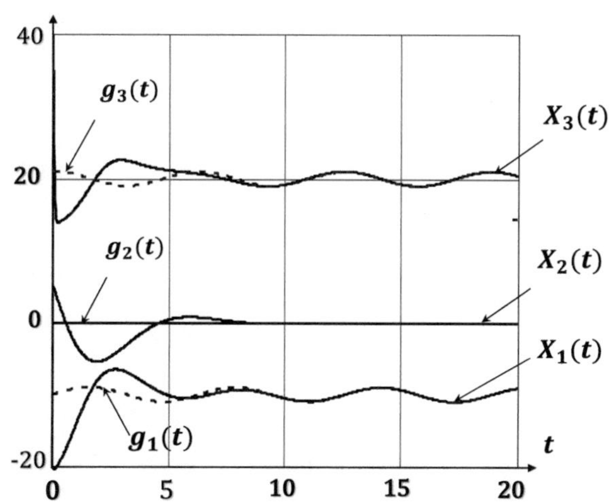

8.5 Linear Feedback Control

8.5.1 Principle

Let us consider again a n-dimensional chaotic system associated with a n-dimensional control vector $\vec{U}$:

$$\frac{d\vec{X}}{dt} = \vec{F}\left(\vec{X}\right) + \vec{U}. \tag{8.34}$$

The principle of linear control is based on the idea of stabilising an unstable orbit (or an unstable equilibrium point) embedded within a strange attractor (Auteur(s), 1993; Chen & Dong, 1992; Ogorzalek, 1993). This orbit (or this point) is represented by the reference vector $\overrightarrow{X_{\text{ref}}}(t)$. Due to the infinite number of different orbits embedded within a strange attractor, an infinite choice of $\overrightarrow{X_{\text{ref}}}(t)$ is available. In practice, the main difficulty lies in the design of a specialised periodic oscillator. The block diagram of this method is shown in Fig. 8.9.

K is a $n \times n$ matrix of the form:

$$K = \begin{pmatrix} k_{11} & \cdots & k_{1n} \\ \vdots & \ddots & \vdots \\ k_{n1} & \cdots & k_{nn} \end{pmatrix}. \tag{8.35}$$

The control vector is defined by:

$$\vec{U} = K\left(\overrightarrow{X_{\text{ref}}} - \vec{X}\right). \tag{8.36}$$

The matrix K will be determined in order to stabilise the controlled system to the reference vector $\overrightarrow{X_{\text{ref}}}(t)$. To do that, we will use the usual linear techniques, as illustrated in the next paragraph.

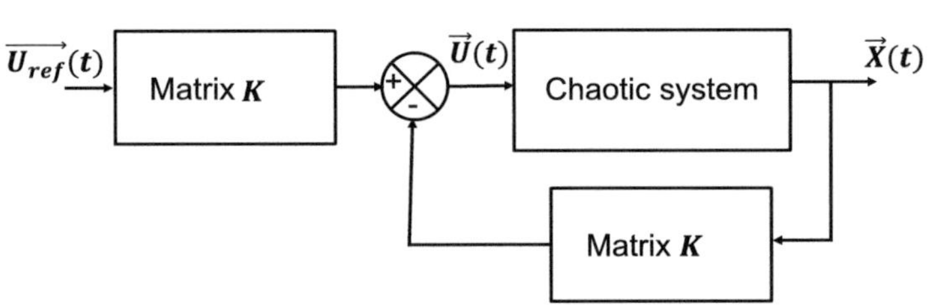

Fig. 8.9 Block diagram of linear feedback control

8.5.2 Illustrative Example

Consider again the Rössler system with the control:

$$\vec{U} = \begin{pmatrix} U_1 \\ 0 \\ 0 \end{pmatrix}. \tag{8.37}$$

In this example, our goal is to stabilise the chaotic system to the unstable equilibrium point:

$$(X_{\text{ref}1}, X_{\text{ref}2}, X_{\text{ref}3}) = (X_{e1}, X_{e2}, X_{e3}) = (0.003, -0.02, 0.02). \tag{8.38}$$

K is given by:

$$K = \begin{pmatrix} k_{11} & 0 & 0 \\ 0 & 0 & 0 \\ 0 & 0 & 0 \end{pmatrix}. \tag{8.39}$$

The dynamics of the controlled system become:

$$\begin{cases} \frac{dX_1}{dt} = -X_2 - X_3 + k_{11}(X_{e1} - X_1) \\ \frac{dX_2}{dt} = X_1 + 0.15X_2 \\ \frac{dX_3}{dt} = 0.2 + X_1X_3 - 10X_3 \end{cases}. \tag{8.40}$$

It is easy to note that $\vec{X}_e = (X_{e1}, X_{e2}, X_{e3})$ is also an equilibrium point of this new system. To do that, we have to calculate k_{11} to stabilise the controlled system around the equilibrium point $\vec{X}_e$. We begin by determining the Jacobian matrix at $\vec{X}_e$, we get:

$$J = \begin{pmatrix} -k_{11} & -1 & -1 \\ 1 & 0.15 & 0 \\ X_{3e} & 0 & X_{1e} - 10 \end{pmatrix}. \tag{8.41}$$

The characteristic equation is:

$$\lambda^3 + a_1\lambda^2 + a_2\lambda + a_3 = 0, \tag{8.42}$$

with

$$a_1 = -X_{1e} + k_{11} + 9.85,\ a_2 = -0.5 + 0.15X_{1e} + X_{3e} + 9.85k_{11} - X_{1e}k_{11},$$
$$a_3 = -X_{1e} - 0.15X_{3e} - 1.5k_{11} + 0.15X_{1e}k_{11} + 10. \tag{8.43}$$

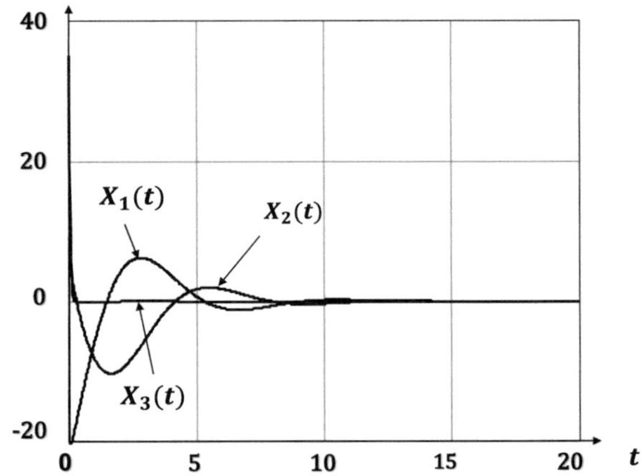

Fig. 8.10 Stabilisation of the Rössler system using linear feedback control

According to the Routh criterion, to stabilise the system around the equilibrium point we have to choose k_{11} such that:

$$a_1 > 0,\ a_1a_2 - a_3 > 0,\ a_3 > 0. \tag{8.44}$$

These three conditions are satisfied with

$$k_{11} = 1. \tag{8.45}$$

Figure 8.10 shows the temporal representation of the state parameters and the stabilisation around the equilibrium point, simulated with $k_{11} = 1$.

8.6 Time-Delayed Feedback Control

8.6.1 Principle

The time-delayed feedback control, developed by K. Pyragas, is a practical approach for stabilising a periodic orbit embedded in a chaotic system (Pyragas, 1992). The key idea is to introduce a control $\vec{U}$ that is proportional to the deviation between the output vector at time t $\left(\vec{Y}(t)\right)$ and the output vector at time $t-\tau\left(\vec{Y}(t-\tau)\right)$.

$$\vec{U} = \begin{pmatrix} k_1(Y_1(t-\tau) - Y_1(t)) \\ \vdots \\ k_y(Y_y(t-\tau) - Y_y(t)) \end{pmatrix}. \tag{8.46}$$

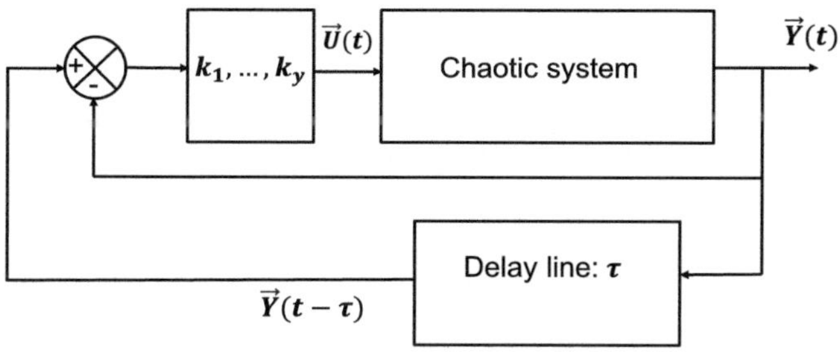

Fig. 8.11 Block diagram of time-delayed feedback control

The block diagram of this control is shown in Fig. 8.11.

The time delay τ has to coincide with the period of the unstable periodic orbit we intend to stabilise. Consider the chaotic system given by Eq. (8.34). Note that this control does not change the solution of the differential equation. The coefficients $k_1, \ldots, k_y$ should be determined to stabilise the desired orbit. The great advantage of this method is its simplicity. We do not need to know the dynamics, and only a simple delay line is required for the control. In practice,τ, $k_1, \ldots, k_y$ can be adjusted by experimentation. In some cases, amplitudes U_i can become too large, and the controlled system can reach an undesirable orbit. This negative effect can be eliminated by adding a clamping circuit for each U_i component. The clamping transfer function:

$$U_{\text{clampi}} = f(U_i), \tag{8.47}$$

of the component U_i is shown in Fig. 8.12.

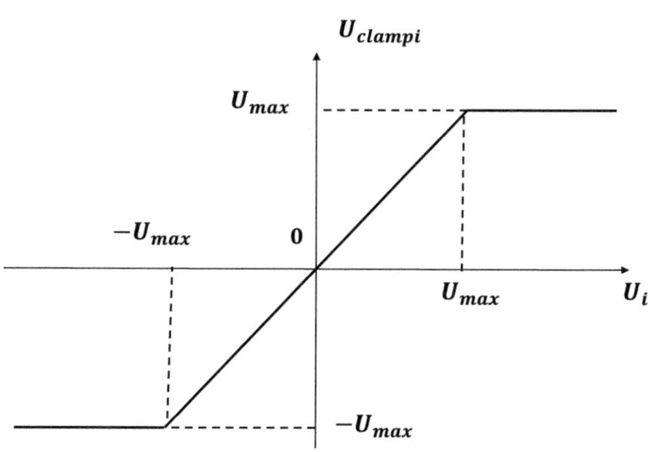

Fig. 8.12 Clamping transfer function

The stability of the desired periodic orbit can be analysed by the sign of the maximum Lyapounov exponent $\lambda_{\max}$. The negative value of $\lambda_{\max}(k_i)$ determines the interval of each k_i corresponding to the stabilised periodic orbit. Another way to study the stability of the desired orbit is to determine the average value of $\vec{U}(t)^2$ once the transient phase is over:

$$S = \vec{U}(t)^2. \tag{8.48}$$

When the system moves along this stabilised orbit, S has to be extremely low so that the stabilisation of the desired orbit can be achieved by a small-time continuous control. The k_i coefficients can be determined to minimise the value of S. If the delay time differs from the period, the dynamics of the system can be chaotic or periodic with a period different from the desired orbit. Time-delayed feedback control can also be used to stabilise an unstable fixed point.

8.6.2 *Illustrative Example with the Rössler System*

As an illustrative example (Chen & Dong, 1994; Pyragas, 1992), we consider a Rössler system where the output vector is one-dimensional and equal to $\vec{Y} = Y = X_1$. The dynamics of the controlled system are given by:

$$\begin{cases} \frac{dX_1}{dt} = -X_2(t) - X_3(t) + k_1(X_1(t-\tau) - X_1(t)) \\ \frac{dX_2}{dt} = X_1(t) + 0.2X_2(t) \\ \frac{dX_3}{dt} = 0.2 + X_1(t)X_3(t) - 5.7X_3(t) \end{cases}. \tag{8.49}$$

Our goal is to stabilise the unstable periodic orbit of period $\tau = 5.9$.

Figure 8.13 shows the variation of S in relation to k_1. For k_1 belonging to $[0.23, 2.3]$, S is close to zero. We will choose $k_1 = 1$. Figure 8.14, which shows the trajectory of $Y = X_1$ in relation to X_2 after the transient phase, illustrates the effectiveness of this control. Figure 8.15 shows the trajectories for $\tau = 5$ and for $\tau = 7$.

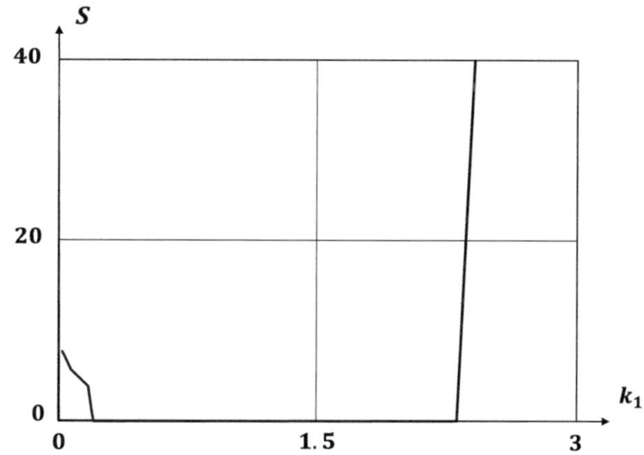

Fig. 8.13 Variation of S with respect to k_1 in the Rössler system

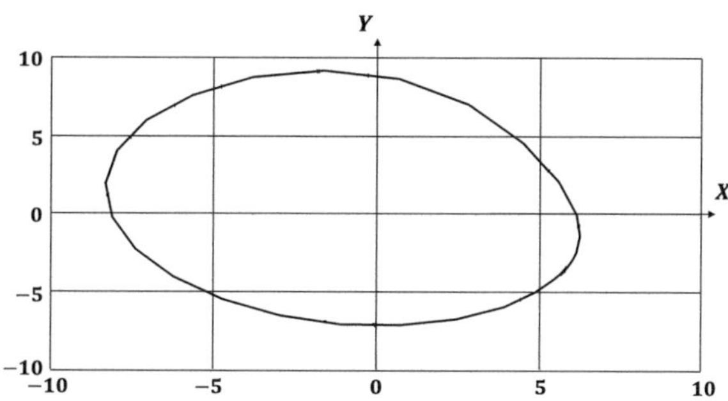

Fig. 8.14 Stabilisation of unstable periodic orbit of period $\tau = 5.9$ using $k_1 = 1$

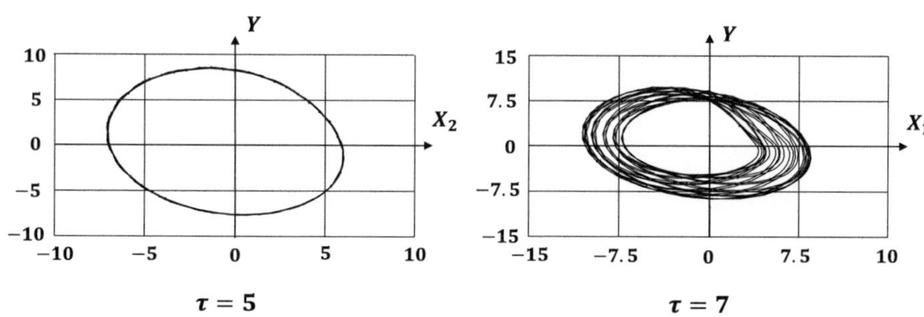

Fig. 8.15 (Y, X_2) trajectories for $\tau = 5$ and $\tau = 7$ using $k_1 = 1$

8.6.3 *Illustrative Example with the Hénon System*

Time-delayed feedback control can also be applied to a discrete system. Consider again the Hénon system, where the output vector is one-dimensional and equal to $\vec{Y}_k = Y_k = X_{1k}$. Our objective is to stabilise this system at the unstable equilibrium point:

$$(X_{e1}, X_{e2}) = (0.6313, 0.1894). \tag{8.50}$$

The time delay corresponds to one iteration. The dynamics of the controlled system become:

$$\begin{cases} X_{1(k+1)} = -1.4X_{1k}^2 + X_{2k} + 1 + k_1\left(X_{1(k-1)} - X_{1k}\right) \\ X_{2(k+1)} = 0.3X_{1k} \end{cases}. \tag{8.51}$$

Figure 8.16 shows the variation of S as a function of k_1. From this figure, we can deduce that $k_1 = -0.7$ is a good coefficient candidate to stabilise this system. Figure 8.17, representing the time variation of Y_k, confirms that the system meets the target

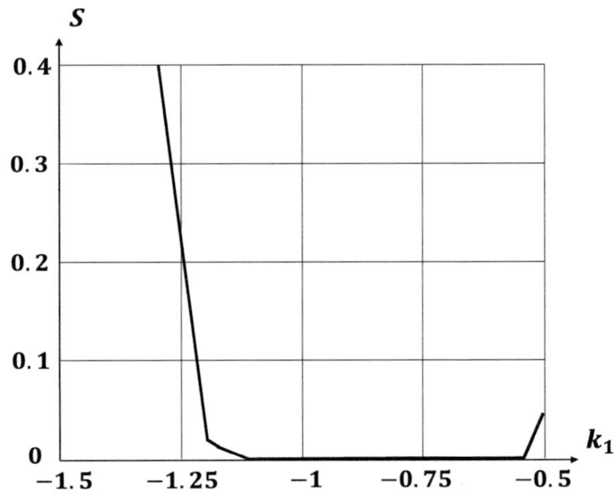

Fig. 8.16 Variation of S as a function of k_1 for the Hénon system

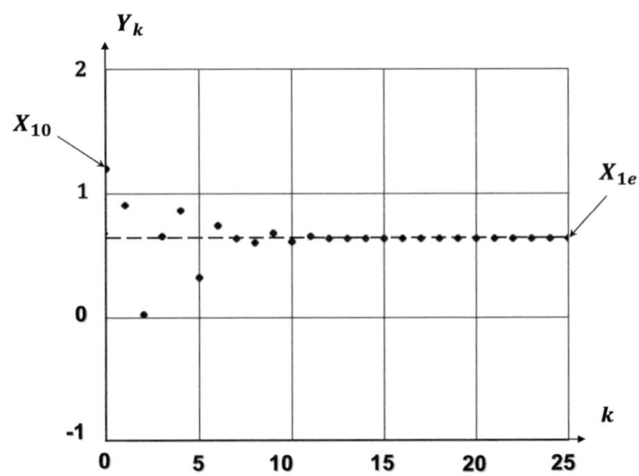

Fig. 8.17 Hénon system stabilisation by time-delayed feedback control

8.7 OGY Control

The OGY control was developed by Ott, Grebogi and Yorke in 1990 (Ott, Grebogi & Yorke, 1990). This method can stabilise an unstable periodic orbit and requires no analytical model of the system. It only requires determining the fixed point and the stable and unstable directions. We begin this section by recalling some concepts from linear algebra.

8.7.1 A Reminder of Linear Algebra Concepts

Let

$$\beta = (\vec{i}, \vec{j}),\ \beta\prime = \left(\overrightarrow{i'}, \overrightarrow{j'}\right), \tag{8.52}$$

be two bases of $\mathbb{R}^2$. We denote $\vec{X}_\beta$,$\vec{X}_{\beta'}$ the $\vec{X}$ vector, respectively, in the basis β and β'. The transition matrix P from β to β' is given by:

$$P = \begin{pmatrix} a_i & b_i \\ a_j & b_j \end{pmatrix} \text{ with } \overrightarrow{i'} = a_i\vec{i} + a_j\vec{j} \text{ and } \overrightarrow{j'} = b_i\vec{i} + b_j\vec{j}. \tag{8.53}$$

This leads to:

$$\vec{X}_\beta = P\vec{X}_{\beta'} \text{ and } \vec{X}_{\beta'} = P^{-1}\vec{X}_\beta. \tag{8.54}$$

Let L be a linear function from $\mathbb{R}^2$ to $\mathbb{R}^2$, defined by the matrices A_β and $A_{\beta'}$ in the bases β and β', respectively. In the basis β, we get:

$$L\left(\vec{X}_\beta\right) = A_\beta\vec{X}_\beta. \tag{8.55}$$

The theory of linear algebra gives

$$A_{\beta'} = P^{-1}A_\beta P \text{ and } A_\beta = PA_{\beta'}P^{-1}. \tag{8.56}$$

P^{-1} is determined by:

$$P^{-1} = \frac{\begin{pmatrix} b_j & -b_i \\ -a_j & a_i \end{pmatrix}}{a_ib_j - a_jb_i}. \tag{8.57}$$

8.7.2 OGY Control Theory

OGY control is only applicable to discrete-time systems. For a continuous-time dynamical system, it should first be converted into a discrete-time system by using a Poincaré map. To simplify explanations, we consider only second-order discrete systems r (corresponding to third-order continuous systems) with one-dimensional control. The following considerations can be extended to higher-order systems. Consider the following discrete chaotic system in the basis $\beta = \left(\vec{i}, \vec{j}\right)$:

$$\begin{cases} X_{1(k+1)} = F_1(X_{1k}, X_{2k}, U_k) \\ X_{2(k+1)} = F_2(X_{1k}, X_{2k}, U_k) \end{cases}. \tag{8.58}$$

Here, the one-dimensional control U_k is an accessible system parameter with the nominal value U_0. Assume the unstable orbit that we wish to stabilise corresponds to the point (X_{1e}, X_{2e}, U_0). The first step of this method involves linearising the system (8.58) around (X_{1e}, X_{2e}, U_0) in the basis β. We get:

$$\vec{x}_{(k+1)\beta} = A_\beta \vec{x}_{k\beta} + \vec{B}_\beta u_k \tag{8.59}$$

where

$$\vec{x}_{k\beta} = \vec{X}_{k\beta} - \vec{X}_{e\beta}, \quad u_k = U_k - U_0, \quad A_\beta = \begin{pmatrix} \frac{\partial F_1}{\partial X_{1k}} & \frac{\partial F_1}{\partial X_{2k}} \\ \frac{\partial F_2}{\partial X_{1k}} & \frac{\partial F_2}{\partial X_{2k}} \end{pmatrix} \text{ and } \vec{B}_\beta = \begin{pmatrix} \frac{\partial F_1}{\partial U_k} \\ \frac{\partial F_2}{\partial U_k} \end{pmatrix}. \tag{8.60}$$

Assume that the matrix A_β has one stable eigenvalue λ_s ($|\lambda_s| < 1$) and one unstable eigenvalue λ_u ($|\lambda_u| > 1$), associated with the eigenvectors $\overrightarrow{e_s}$ and $\overrightarrow{e_u}$, respectively. Let

$$\beta' = \left(\overrightarrow{e_s}, \overrightarrow{e_u}\right) \tag{8.61}$$

Be a new basis of $\mathbb{R}^2$. The transition matrix P from β to β' is given by:

$$P = \begin{pmatrix} e_{s1} & e_{u1} \\ e_{s2} & e_{u2} \end{pmatrix} \text{ where } \overrightarrow{e_s} = e_{s1}\vec{i} + e_{s2}\vec{j} \text{ and } \overrightarrow{e_u} = e_{u1}\vec{i} + e_{u2}\vec{j}. \tag{8.62}$$

The transition matrix from β' to β is

$$P^{-1} = \frac{\begin{pmatrix} e_{u2} & -e_{u1} \\ -e_{s2} & e_{s1} \end{pmatrix}}{e_{s1}e_{u2} - e_{s2}e_{u1}}. \tag{8.63}$$

$\vec{x}_{(k+1)}$ in the basis β' is given by:

$$\vec{x}_{(k+1)\beta'} = A_{\beta'}\vec{x}_{k\beta'} + \vec{B}_{\beta'}u_k \tag{8.64}$$

with

$$A_{\beta'} = P^{-1}A_\beta P, \quad \vec{B}_{\beta'} = P^{-1}\vec{B}_\beta, \quad \vec{x}_{k\beta'} = P^{-1}\vec{x}_{k\beta}. \tag{8.65}$$

This leads to

$$\vec{x}_{(k+1)\beta'} = \begin{pmatrix} x_{1(k+1)} \\ x_{2(k+1)} \end{pmatrix}_{\beta'} = P^{-1}A_\beta\vec{x}_{k\beta} + P^{-1}\vec{B}_\beta u_k. \tag{8.66}$$

The principle of OGY control is to determine the control u_k in order to cancel the unstable component. Mathematically speaking, it can be expressed by:

$$x_{2(k+1)\beta'} = 0. \tag{8.67}$$

We get the following expression for the control:

$$u_k = \frac{\left(e_{s1}\frac{\partial F_2}{\partial X_{1k}} - e_{s2}\frac{\partial F_1}{\partial X_{1k}}\right)x_{1k\beta} + \left(e_{s1}\frac{\partial F_2}{\partial X_{2k}} - e_{s2}\frac{\partial F_1}{\partial X_{2k}}\right)x_{2k\beta}}{e_{s2}\frac{\partial F_1}{\partial u_k} - e_{s1}\frac{\partial F_2}{\partial u_k}}. \tag{8.68}$$

Due to the nature of the chaotic system, a trajectory that starts from an arbitrary initial value inside the basin of attraction will necessarily fall into the neighbourhood of the desired fixed point $\vec{X}_e$. When the system approaches $\vec{X}_e$ within a circle of radius ε, the control is activated. Obviously, there is a trade-off between the size of this control region and the mean time before reaching this region.

8.7.3 Illustrative Example

We consider once again the example of the Hénon system (Ott, Grebogi, & Yorke, 1990; Witvoet, 2005). We seek to stabilise it around

$$(X_{e1}, X_{e2}) = (0.631, 0.189). \tag{8.69}$$

The system parameter p_1 is now our control variable. We thus obtain

$$u_k = U_k - p_1 = U_k - U_0. \tag{8.70}$$

We take $U_0 = p_1 = 1.4$ and $p_2 = 0.3$. The control is activated when the state parameters enter a circle of radius $\varepsilon = 0.05$ and of centre (X_{e1}, X_{e2}). When the system enters this region, the dynamics become

$$\begin{cases} X_{1(k+1)} = -U_{1k}^2 + X_{2k} + 1 \\ X_{2(k+1)} = 0.3X_{1k} \end{cases}. \tag{8.71}$$

Outside this region, we have:

$$U_0 = p_1(u_k = 0). \tag{8.72}$$

By applying the formulas given in Para. 8.7.1, we obtain:

$$A_\beta = \begin{pmatrix} -1.263 & 1 \\ 0.3 & 0 \end{pmatrix}, \vec{B}_\beta = \begin{pmatrix} -0.399 \\ 0 \end{pmatrix}, \lambda_s = 0.2, \lambda_u = -1.467 \tag{8.73}$$

and

$$\vec{e_s} = 0.563\vec{i} + 0.826\vec{j}, \ \vec{e_u} = -0.979\vec{i} + 0.2\vec{j}. \tag{8.74}$$

Thanks to Eq. (8.68), we can say that the control is

$$u_k = -3.681x_{1k\beta} + 2.509x_{2k\beta}. \tag{8.75}$$

Figure 8.18 shows the simulation results with the same initial conditions as those given in Para. 8.3.2. We note that in this example, the reaction time of the OGY control is very fast.

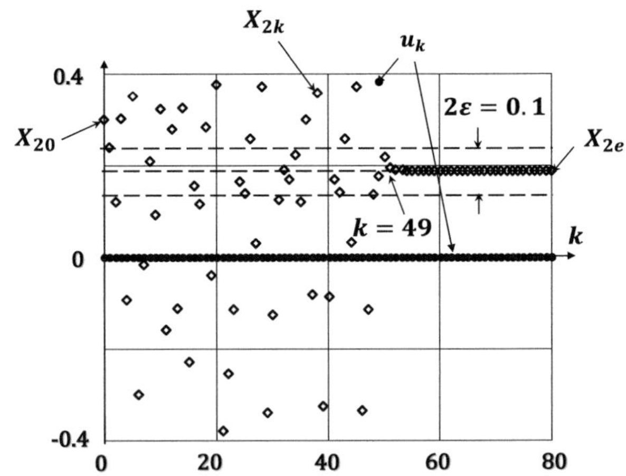

Fig. 8.18 Stabilisation of the Hénon system using OGY control

Part II
Applications of the Theories of Nonlinear Systems to Power Converters

Chapter 9
Nonlinear Dynamic Models of Power Converters

Switching converters inherently comprise multiple nonlinear subsystems; therefore, the most accurate modelling approach naturally relies on nonlinear mathematical models. We will begin studying the dynamics of switching power converters using averaged models (Erickson & Maksimovic, 1996; Erickson & Maksimovic, 2001; Erickson & Maksimović, 2020; Middlebrook & Ćuk, 1976). Both discrete and hybrid models will also be utilised. We will mainly use the DC-to-DC buck converter as an illustration. The results obtained for this converter can easily be extrapolated to other converters.

9.1 Averaged Model of CCM Converters with Voltage-Mode Control

9.1.1 Basic Electrical Laws with Averaged Values

The basic idea of the averaged model is to eliminate the switching ripple in the state parameters (inductor current and capacitor voltage) by averaging over one switching period T_S. Figure 9.1 shows the classical current and voltage waveforms of the inductor and capacitor used in a switching power converter. The averaged values of the inductor current $I_L(t)$ and capacitor voltage $V_C(t)$ at any time t are given by:

$$I_L(t) = \frac{1}{T_S}\int_{t}^{t+T_S} i_L(t')dt', \quad V_C(t) = \frac{1}{T_S}\int_{t}^{t+T_S} v_C(t')dt'. \tag{9.1}$$

Consider now the averaged value of $V_L(t)$, we get:

B. Rivet, *Theories of Nonlinear Systems for Power Converters*,
https://doi.org/10.1007/978-3-032-25062-9_9

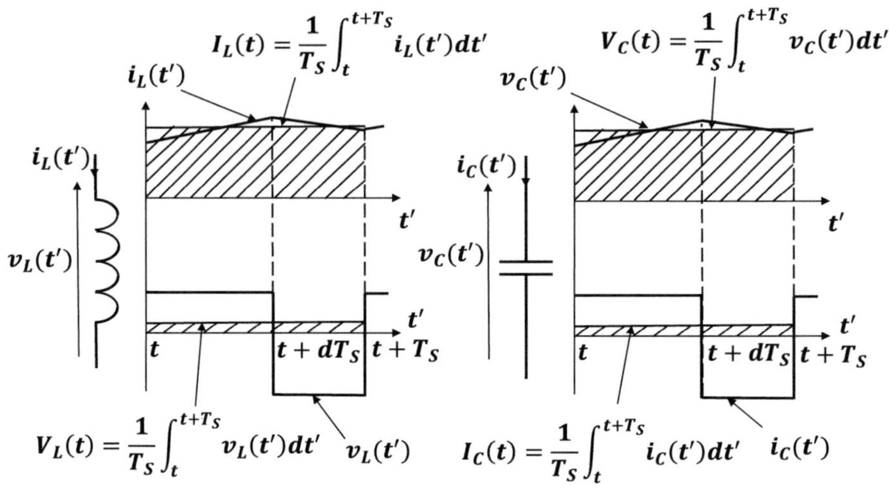

Fig. 9.1 Current and voltage waveforms of the inductor and capacitor in a switching converter operating in CCM

$$V_L(t) = \frac{1}{T_S}\int_t^{t+T_S} v_L(t')dt' = \frac{1}{T_S}\int_t^{t+T_S} L\frac{di_L(t')}{dt'}dt'$$
$$= \frac{L}{T_S}(i_L(t+T_S) - i_L(t)). \tag{9.2}$$

Further, we can write

$$\frac{dI_L(t)}{dt} = \frac{d}{dt}\left(\frac{1}{T_S}\int_t^{t+T_S} i_L(t')dt'\right) = \frac{(i_L(t+T_S) - i_L(t))}{T_S}. \tag{9.3}$$

From Eqs. (9.2) and (9.3), we get the following very important result:

$$V_L(t) = L\frac{dI_L(t)}{dt}. \tag{9.4}$$

In other words, the fundamental electrical law

$$v_L(t) = L\frac{di_L(t)}{dt}, \tag{9.5}$$

is applicable to averaged values. Similarly, for the capacitor we have:

$$I_C(t) = C\frac{dV_C(t)}{dt}. \tag{9.6}$$

9.1.2 Averaged Model of a CCM Converters with Voltage-Mode Control

The diagram of a buck converter with voltage control is provided in Fig. 9.2. When this converter operates in continuous conduction mode, the typical waveforms are illustrated in Fig. 9.3. The control parameter is the conduction duty cycle of the switch d, which is defined as:

$$d = \frac{t_{on}}{T_S}. \tag{9.7}$$

We can consider the case represented in Fig. 9.2, where the switching period T_S is constant, and t_{on} is the adjustable parameter. The other possibility is to keep t_{on} constant and control the duty cycle by changing the switching frequency. Between times t and $t + dT_S$, the switch S is on. The dynamics are given by:

$$\begin{cases} V_{IN} = L\frac{di_L(t')}{dt'} + v_c(t') = v_L(t') + v_c(t') \\ i_L(t') = C\frac{dv_C(t')}{dt'} + \frac{v_c(t')}{R_L} \end{cases}. \tag{9.8}$$

Between $t + dT_S$ and $t + T_S$, the switch S is off, so the dynamics are now given by:

$$\begin{cases} 0 = L\frac{di_L(t')}{dt'} + v_c(t') = v_L(t') + v_c(t') \\ i_L(t') = C\frac{dv_C(t')}{dt'} + \frac{v_c(t')}{R_L} = i_C(t') + \frac{v_c(t')}{R_L} \end{cases}. \tag{9.9}$$

Averaging these two systems over T_S, we get:

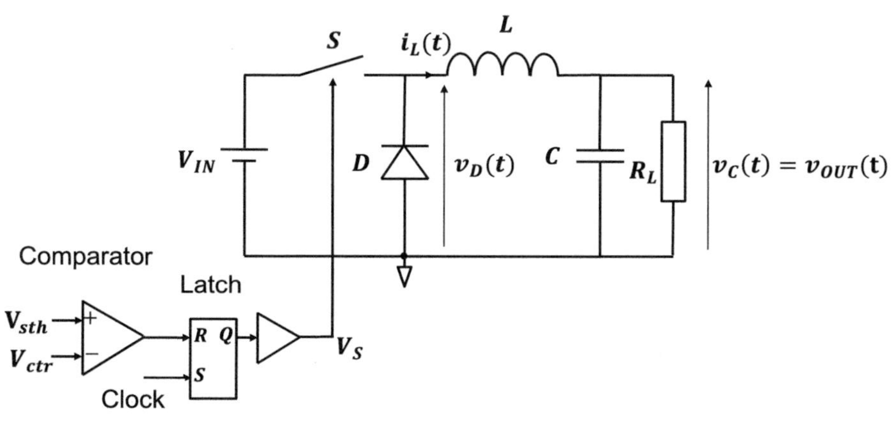

Fig. 9.2 Diagram of a buck converter with voltage-mode control

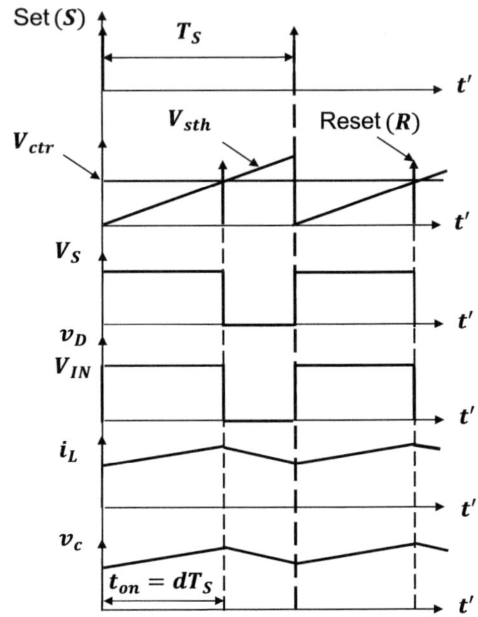

Fig. 9.3 Waveforms of a buck converter operating in CCM

$$\begin{cases} \frac{1}{T_S}\int_t^{t+dT_S} V_{IN}dt' = \frac{1}{T_S}\int_t^{t+T_S} v_L(t')dt' + \frac{1}{T_S}\int_t^{t+dT_S} v_C(t')dt' \\ \frac{1}{T_S}\int_t^{t+T_S} i_L(t')dt' = \frac{1}{T_S}\int_t^{t+T_S} i_C(t')dt' + \frac{1}{T_S}\int_t^{t+dT_S} \frac{v_c(t')}{R_L}dt' \end{cases}. \tag{9.10}$$

Thanks to formulas (9.1), (9.4), (9.6) and (9.10), we get the continuous averaged dynamics:

$$\begin{cases} \frac{dI_L(t)}{dt} = -\frac{V_c(t)}{L} + \frac{V_{IN}}{L}d(t) \\ \frac{dV_C(t)}{dt} = \frac{I_L(t)}{C} - \frac{V_C(t)}{R_L C} \end{cases}. \tag{9.11}$$

In the case where $T_S =$ constant, the control voltage $V_{ctr}(t)$ varies $d(t)$ according to:

$$V_{ctr}(t) = V_P d(t). \tag{9.12}$$

In the resulting system (9.11), the values L, C, R_L and V_{IN} can be considered constant or varying slowly with the time t.

9.1.3 Limitation of the Continuous Averaged Model

To use this model, the ripples of the state variables must be small compared to their averaged values, which is generally the case for a converter operating in continuous mode. The averaged model is very accurate when the frequency under study (f_d) is small compared to the switching frequency f_S. Figure 9.4 shows the current $i_L(t)$ in a buck converter obtained from the averaged model (9.11) and a circuit simulator. The simulations have been performed with the following numerical values:

$$f_S = 100\ \text{kHz},\ L = 400\ \mu\text{H},\ C = 100\ \mu\text{F},\ R_L = 2.5\ \Omega,\ V_{IN} = 50\ \text{V}$$
$$d(t) = 0.5 + 0.05\sin(2\pi f_d t) \quad (9.13)$$

and with the following four values of f_d : 500 Hz, 1 kHz, 5 kHz, 50 kHz. As shown in this figure, the average model can be used for frequencies lower than one twentieth of the switching frequency. This remark is valid for all the averaged models described in the following paragraphs.

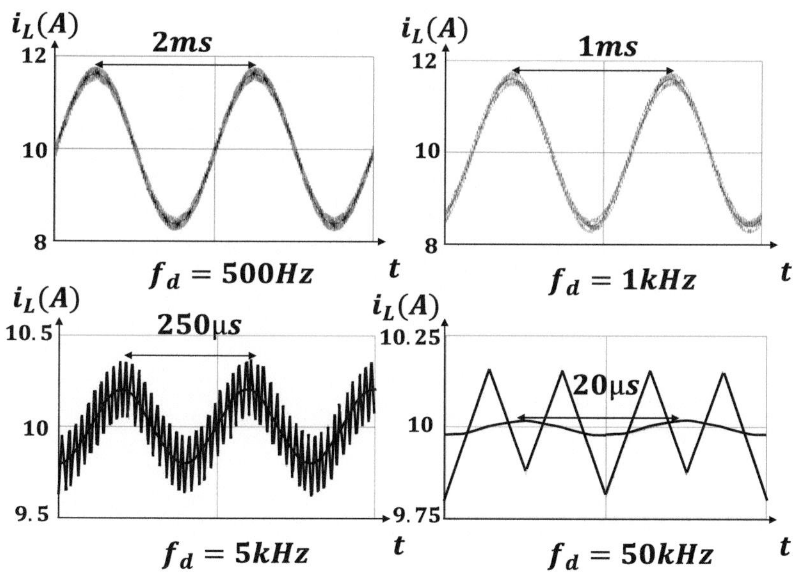

Fig. 9.4 Comparison between the averaged model and circuit simulations

9.2 Averaged Model of CCM Converters with Current-Mode Control

9.2.1 CCM Buck Converter with Current-Mode Control

Current-mode control is a very effective control technique. The advantages of this solution will be discussed in Para. 11. The diagram of a buck converter associated with current-mode control is shown in Fig. 9.5. We only consider here the case where the switching period is constant. The purpose of the artificial ramp is to prevent chaotic behaviour. This point will be detailed in Para. 10.3. It is impossible to analyse these types of behaviour with an averaged model because they occur in a frequency range close to the switching frequency.

The corresponding typical waveforms of this converter are illustrated in Fig. 9.6. In this section, the value of the shunt resistor R_S will be considered small enough to have no effect on the system dynamics.

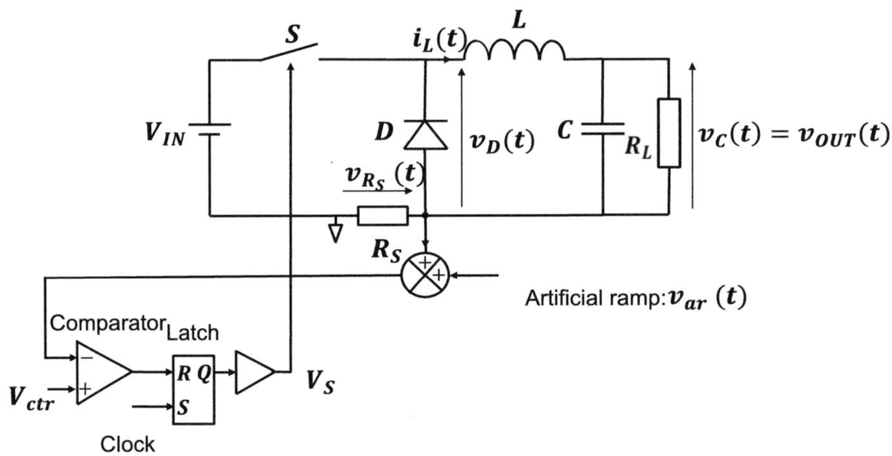

Fig. 9.5 Diagram of a buck converter with current-mode control

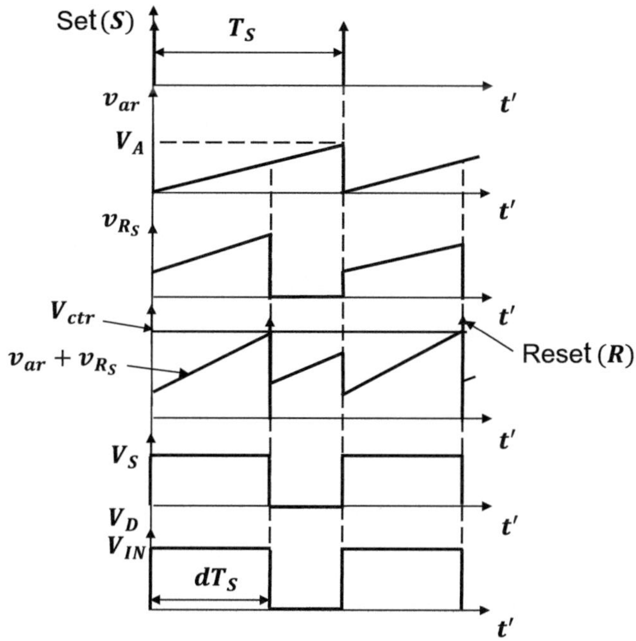

Fig. 9.6 Waveforms of a CCM buck converter with current-mode control

9.2.2 *Simplified Averaged Model*

If the artificial ramp is not used or has a negligible effect $\left(v_{ar} \ll v_{R_S}\right)$, a simplified averaged model can be considered. This model is based on the following approximation:

$$V_{ctr}(t) \simeq \frac{I_L(t)}{R_S}. \tag{9.14}$$

The inductor acts as a current source controlled by $V_{ctr}(t)$. The system becomes a first-order system whose dynamics are given by:

$$\frac{dV_C(t)}{dt} = -\frac{V_C(t)}{R_L C} + \frac{V_{Ctr}(t)}{R_s C}. \tag{9.15}$$

9.2.3 *More Accurate Averaged Model*

A more accurate model that take into account the artificial ramp can be defined from Fig. 9.7. In calculating the average value of $i_L(t)$, we get:

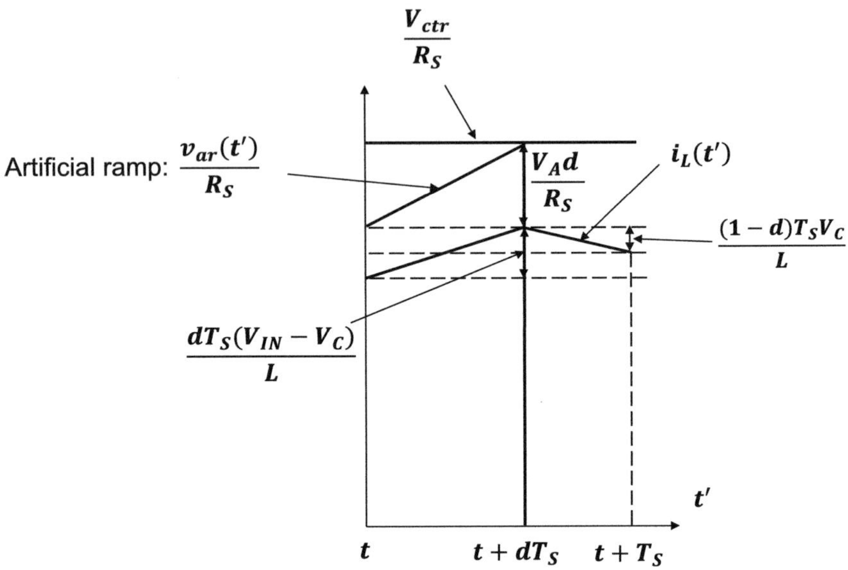

Fig. 9.7 Waveforms of current used to define a more accurate model

$$I_L(t) = \frac{1}{T_S}\int_t^{t+T_S} i_L(t')dt' = \frac{V_{Ctr} - V_A d}{R_s} - \frac{T_S d^2}{2L}(V_{\mathrm{IN}} - V_c) - \frac{(1-d)^2 T_S}{2L} V_c. \tag{9.16}$$

From this equation, we can extract d. We have:

$$d = \frac{V_c}{V_{\mathrm{IN}}} - \frac{V_A L}{V_{\mathrm{IN}} R_S T_S} + \frac{L}{V_{\mathrm{IN}} T_S} \sqrt{\left(\frac{V_A}{R_S} - \frac{T_S V_c}{L}\right)^2 - \frac{2T_S V_{\mathrm{IN}}}{L}\left(\frac{T_S V_c}{2L} - \frac{V_{ctr}}{R_S} + I_L\right)}. \tag{9.17}$$

The dynamics defined by Eq. (9.11) are still valid. Taking into account the new law of the duty cycle given by Eq. (9.17), we obtain the nonlinear system of a CCM buck with current-mode control:

$$\begin{cases} \frac{dI_L(t)}{dt} = -\frac{V_A}{T_S R_S} + \frac{1}{T_S}\sqrt{\left(\frac{V_A}{R_S} - \frac{T_S V_c}{L}\right)^2 - \frac{2T_S V_{\mathrm{IN}}}{L}\left(\frac{T_S V_c}{2L} - \frac{V_{ctr}}{R_S} + I_L\right)} \\ \frac{dV_C(t)}{dt} = \frac{I_L(t)}{C} - \frac{V_C(t)}{R_L C} \end{cases}. \tag{9.18}$$

9.3 Averaged Model of DCM Converters

9.3.1 Averaged Model of a DCM Buck Converter with Voltage-Mode Control

Typical waveforms associated with a buck converter operating in discontinuous conduction mode with voltage control are shown in Fig. 9.8. The control parameter is the duty cycle $d = \frac{t_{\text{on}}}{T_S}$. We can consider the cases with T_S = constant or t_{on} = constant. The average current of $I_L(t)$ over one period T_S is given by:

$$I_L(t) = \frac{1}{T_S}\int_t^{t+T_S} i_L(t')dt' = \frac{V_{\text{IN}} - V_C}{2L}d^2T_S + \frac{V_{\text{IN}} - V_C}{2L}dT_Sd_1. \tag{9.19}$$

The averaged voltage across L being always equal to zero, it follows that:

$$d_1 = \frac{V_{\text{IN}} - V_C}{V_C}d. \tag{9.20}$$

Since $\frac{dI_L(t)}{dt} = \frac{V_L(t)}{L} = 0$, the system becomes a first-order one with the dynamics:

$$\frac{dV_C(t)}{dt} = -\frac{V_C(t)}{R_LC} + \frac{V_{\text{IN}} - V_C(t)}{2LC}\frac{V_{\text{IN}}}{V_C(t)}d^2T_S. \tag{9.21}$$

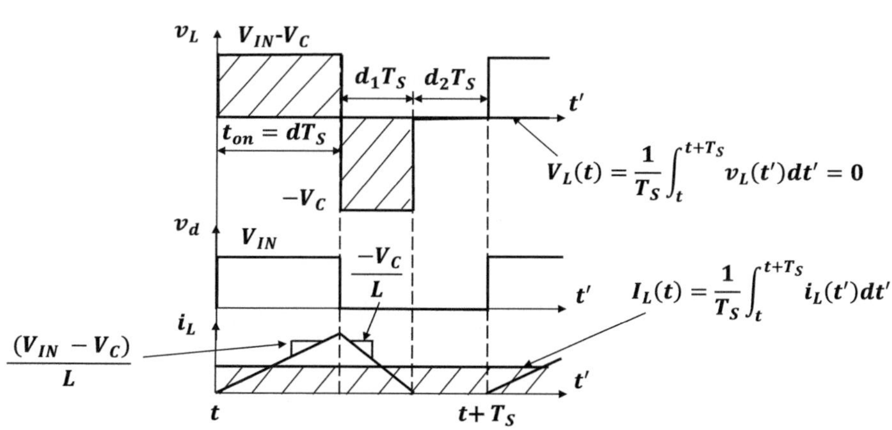

Fig. 9.8 Waveforms of a DCM buck converter with voltage control

9.3.2 Averaged Model of a Critical Conduction Mode Buck Converter with Voltage-Mode Control

Critical conduction mode (CCM) is the boundary between continuous and discontinuous conduction modes. In critical conduction mode, the control parameter is t_{ON} (the switching frequency is variable). The switch S turns on when the current through the inductor L is equal to zero, as shown in Fig. 9.9. This control mode is almost always associated with voltage-mode control. From Fig. 9.10, we can easily deduce the following expression of the average value of i_L:

$$I_L(t) = \frac{1}{T_S}\int_{t}^{t+T_S} i_L(t')dt' = \frac{V_{IN} - V_C}{2L}t_{on}. \tag{9.22}$$

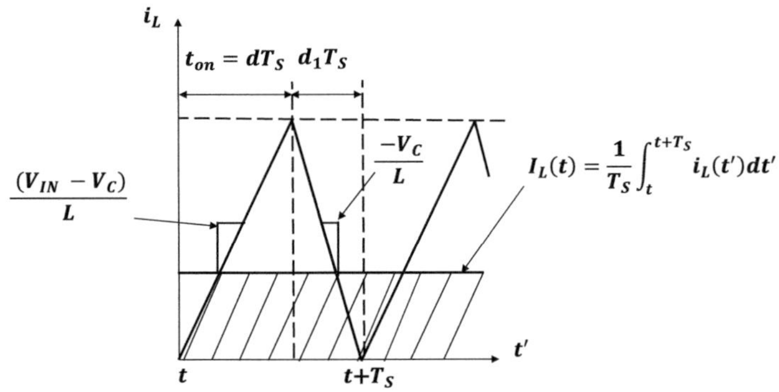

Fig. 9.9 Waveforms of a critical conduction mode buck converter with voltage-mode control

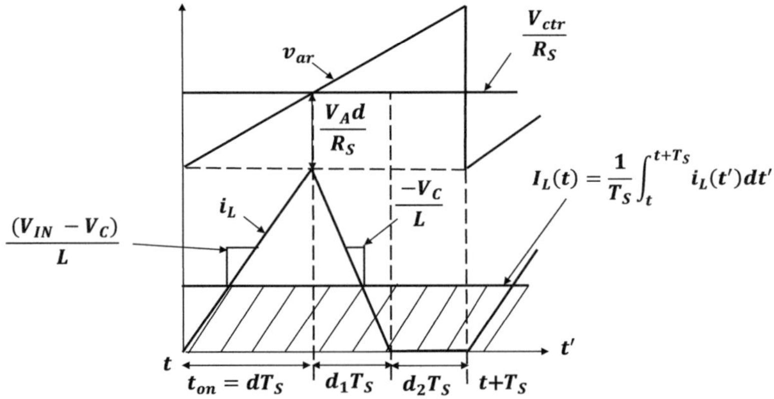

Fig. 9.10 Waveforms of a DCM buck converter with current-mode control

This leads to the system dynamics:

$$\frac{dV_C(t)}{dt} = \frac{V_{\text{IN}} - V_C}{2LC} t_{\text{on}} - \frac{V_C(t)}{R_L C}. \tag{9.23}$$

9.3.3 Averaged Model of a DCM Buck Converter with Current-Mode Control

At low load, a continuous conduction mode converter controlled with current-mode control generally operates in discontinuous continuous mode.

In practice, this is the most common configuration encountered. As discussed in Para. 9.2.1, the switching frequency is generally fixed. The waveforms corresponding to this mode are shown in Fig. 9.10. We can deduce the following equation, which gives a relationship between d and V_{ctr} :

$$d = \frac{V_{ctr}}{R_s\left(\frac{V_{\text{IN}} - V_C}{L}\right)T_s + V_A}. \tag{9.24}$$

From Eqs. (9.21) and (9.24), we obtain the first-order differential equation that describe the dynamics of this system:

$$\frac{dV_C(t)}{dt} = -\frac{V_C(t)}{R_L C} + \frac{V_{\text{IN}} - V_C(t)}{2LC}\frac{V_{\text{IN}}}{V_C(t)}\left(\frac{V_{ctr}}{R_s\left(\frac{V_{\text{IN}} - V_C(t)}{L}\right)T_s + V_A}\right)^2 T_S. \tag{9.25}$$

9.4 Discrete Models of CCM, DCM and Critical Conduction Mode Buck Converter

Discrete models are very useful for analysing high-frequency phenomena close to the switching frequency, such as chaotic behaviours (Erickson & Maksimovic, 1996; Erickson & Maksimovic, 2001; Erickson & Maksimović, 2020; Ferrieux & Forest, 1976; Johns & Sudhoff, 1985).

9.4.1 Discrete Model of a CCM Buck with a Voltage-Mode Control

The nonlinear recursive system is defined from the figure below Fig. 9.11.

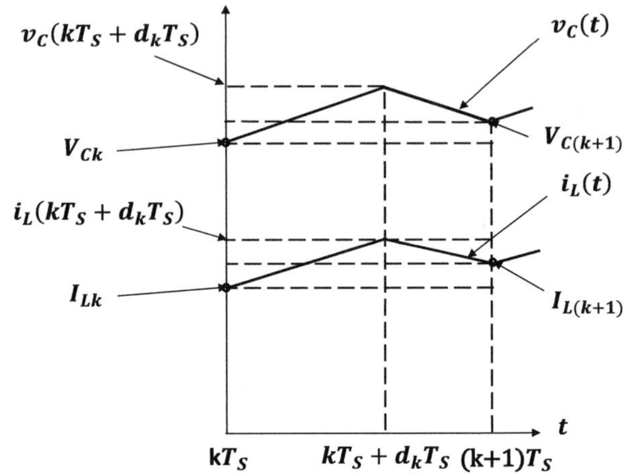

Fig. 9.11 Waveforms of $v_C(t)$ and $i_L(t)$ for a CCM buck converter

To simplify expressions while maintaining accuracy, we consider:

$$v_C(t) = V_{Ck} \quad \text{for} \quad t \in [kT_S, (k+1)T_S]. \tag{9.26}$$

We get:

- for $t \in [kT_S, kT_S + d_kT_S]$:

$$i_L(t) = I_{Lk} + \frac{V_{\text{IN}} - V_{Ck}}{L} t \tag{9.27}$$

- for $t \in [kT_S + d_kT_S, (k+1)T_S]$:

$$i_L(t) = I_{Lk} + \frac{V_{\text{IN}d_kT_S}}{L} - \frac{V_{Ck}}{L} t \tag{9.28}$$

With this approximation, we can determine the nonlinear discrete system corresponding to a CCM buck converter with voltage-mode control:

$$\begin{cases} I_{L(k+1)} = I_{Lk} - \frac{T_S}{L} V_{Ck} + \frac{V_{IN}T_S}{L} d_k \\ V_{C(k+1)} = (((1-A)V_{Ck} - R_L I_{Lk} + AV_{IN})B - AV_{IN})C \\ \qquad +R_L I_{Lk} + (A-D)V_{Ck} + DV_{IN} d_k \end{cases} \tag{9.29}$$

with

$$A = \frac{{R_L}^2 C}{L}, B = e^{-\frac{T_S d_k}{R_L C}}, C = e^{-\frac{T_S(1-d_k)}{R_L C}}, D = \frac{R_L T_S}{L}. \tag{9.30}$$

9.4.2 Discrete Model of a CCM Buck with Current-Mode Control

Consider:

$$V_{ctr}(t) = V_{ctrk} = \text{constant for } t \in [kT_S, (k+1)T_S]. \quad (9.31)$$

From Fig. 9.7, we get:

$$\frac{V_{ctrk}}{R_S} = I_{Lk} + \frac{V_{\mathrm{IN}} - V_{Ck}}{L} d_k T_S + \frac{V_A}{R_S} d_k. \quad (9.32)$$

This results in:

$$d_k = \frac{(V_{ctrk} - R_S I_{Lk})L}{(V_{\mathrm{IN}} - V_{Ck})T_S R_S + V_A L}. \quad (9.33)$$

By substituting formula (9.33) into the system (9.29), we find the nonlinear discrete system corresponding to a CCM buck converter with current-mode control.

9.4.3 Model of a DCM and Critical Conduction Mode Buck Converters

Using the same methodology, we can easily deduce the discrete system for DCM and critical conduction mode buck converters.

Discrete Model of a DCM Buck Converter with Voltage-Mode Control

It is given by:

$$V_{C(k+1)} = V_{Ck} - \frac{V_{Ck}}{R_L C} T_{Sk} + \frac{V_{\mathrm{IN}}(V_{\mathrm{IN}} - V_{Ck})\mathrm{ton}^2}{2V_{Ck} LC} \quad (9.34)$$

with $t_{\mathrm{ON}} = \text{constant}$ and T_{Sk} as control parameter.

Discrete Model of a DCM Buck Converter with Current-Mode Control

We get:

$$V_{C(k+1)} = V_{Ck} - \frac{V_{Ck}}{R_L C} T_S + \frac{V_{\mathrm{IN}}(V_{\mathrm{IN}} - V_{Ck})}{2V_{Ck} LC} \left(\frac{V_{ctr} T_S L}{V_A L + (V_{\mathrm{IN}} - V_{Ck}) R_S T_S} \right)^2 \quad (9.35)$$

with V_{ctr} as control parameter.

Discrete Model of a Critical Conduction Mode Buck Converter with Voltage-Mode Control

It is given by:

$$V_{C(k+1)} = V_{Ck} + \frac{V_{\mathrm{IN}}(V_{\mathrm{IN}} - V_{Ck})tonk^2}{2LCV_{Ck}} - \frac{V_{\mathrm{IN}}tonk}{R_L C} \tag{9.36}$$

with t_{onk} as control parameter.

9.5 Hybrid Model of Converters

Hybrid models are well suited to analyse the dynamics of switching converters (Erickson & Maksimovic, 1996, 2001, 2020; Ferrieux & Forest, 1976; Johns & Sudhoff, 1985; Krein, 1998; Priyadarshi et al., 2024). In this section, we present an example of a hybrid model for a buck converter operating in continuous conduction mode and another one using hysteresis control. The use of hybrid models will be illustrated in Section 15.

9.5.1 Hybrid Model of a CCM Buck Converter

Taking into account the definition of a hybrid system given in Para. 3.1.1, a buck converter operating in continuous mode can be represented as follows.

- $Q = \{q_0, q_1\}$ is a set of 2 discrete states $\left(n_Q = 2\right)$. q_0 and q_1 correspond, respectively, to the switch S being off and on.
- $I_L, V_C : \mathbb{R} \to \mathbb{R}$ are the two continuous state variables ($n = 2$).
- $Init \in Q \times \mathbb{R}^2$ is a set of $1 + n = 3$ initial conditions.
- $Dom = \left\{D_{q_0}, D_{q_1}\right\}$ is a set of 2 continuous domains with $D_{q_0} = D_{q_1} = [0, I_{L\max}] \times [0, V_{C\max}]$.$I_{L\max}$ and $V_{C\max}$ are the current and voltage limitations set by the control circuit.
- $U = \left\{U_{q_0} = 0, U_{q_1} = 1\right\}$ is a set of one-dimensional control vectors ($u = 1$). $U = U_{q_0} = 0$ and $U = U_{q_1} = 1$ correspond, respectively, to the switch S being off and on.
- $\overrightarrow{F_{q_0}}, \overrightarrow{F_{q_1}} : \mathbb{R}^3 \to \mathbb{R}^2$ are 2 dimensional vector fields describing the dynamics of the system. We get:

$$q = q_0 : \begin{cases} \frac{dI_L(t)}{dt} = -\frac{V_c(t)}{L} \\ \frac{dV_C(t)}{dt} = \frac{I_L(t)}{C} - \frac{V_C(t)}{R_L C} \end{cases},$$

$$q = q_1 : \begin{cases} \frac{dI_L(t)}{dt} = -\frac{V_c(t)}{L} + \frac{V_{\mathrm{IN}}}{L} \\ \frac{dV_C(t)}{dt} = \frac{I_L(t)}{C} - \frac{V_C(t)}{R_L C} \end{cases}. \tag{9.37}$$

- $E = \{(q_0, q_1), (q_1, q_0)\}$ is a set of 2 discrete transitions.
- $G = \{G_e, e \in E\}$ is a set of guards that prescribe when a discrete state transition occurs. In this example, (q_0, q_1) and (q_1, q_0) are controlled, respectively, by the Set and Reset signals (see Fig. 9.3).
- In this example, there is no discontinuity in the state parameters, so we do not need to use a reset function R_e.

The overall dynamics of this hybrid model can be expressed in the following matrix form:

$$\begin{pmatrix} \frac{dI_L}{dt} \\ \frac{dV_C}{dt} \end{pmatrix} = \begin{pmatrix} 0 & \frac{-1}{L} \\ \frac{1}{C} & \frac{-1}{R_L C} \end{pmatrix} \begin{pmatrix} I_L \\ V_C \end{pmatrix} + \begin{pmatrix} \frac{V_{IN}}{L} \\ 0 \end{pmatrix} U. \tag{9.38}$$

9.5.2 *Hybrid Model of Buck Converter Using Hysteresis Control*

A buck converter using hysteresis control can be considered an autonomous hybrid system, since there is no control vector.

The switching between the two discrete states is defined solely by the limit values: $I_{L\max}, I_{L\min}, V_{C\max}, V_{C\min}$ as represented in Fig. 9.12. This converter operates in both continuous and discontinuous conduction mode, so we need to add one discrete state. The model is similar to that given in the Para. 9.5.1, with the following differences:

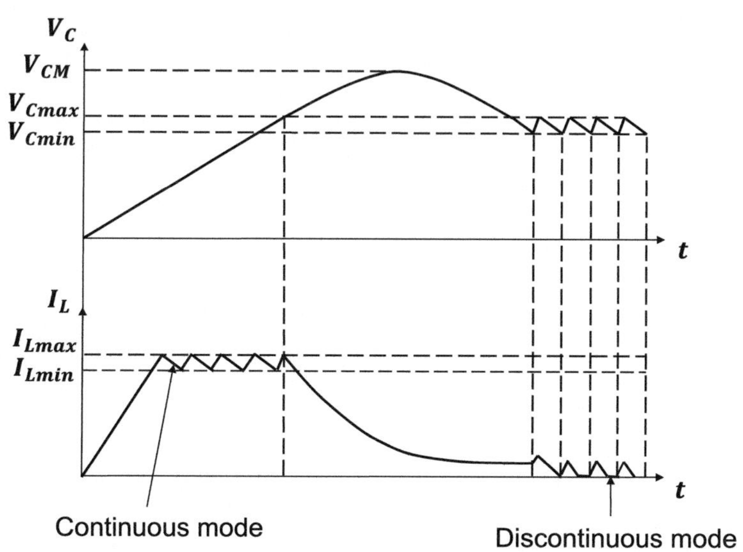

Fig. 9.12 $I_L(t)$ and $V_c(t)$ for a buck converter with hysteresis control

- $Q = \{q_0, q_1, q_2\}$ is a set of 3 discrete states $(n_Q = 3)$. q_1 corresponds to the switch S being on and q_0, q_2 to the switch being off.
- $Dom = \{D_{q_0}, D_{q_1}, D_{q_2}\}$ is a set of 3 continuous domains with $D_{q_0} = [0, I_{L\max}], [0, V_{CM}]$, $D_{q_1} = [0, I_{L\max}], [0, V_{CM}]$, $D_{q_2} = [0, 0], [0, V_{CM}]$.
- $\overrightarrow{F_{q_0}}, \overrightarrow{F_{q_1}}, \overrightarrow{F_{q_2}} : \mathbb{R}^3 \to \mathbb{R}^2$ are 2 dimensional vector fields describing the dynamics of the system. $\overrightarrow{F_{q_0}}$ and $\overrightarrow{F_{q_1}}$ are the same as in the previous paragraph, while $\overrightarrow{F_{q_2}}$ is defined by:

$$q = q_2 : \begin{cases} \frac{dI_L(t)}{dt} = 0 \\ \frac{dV_C(t)}{dt} = -\frac{V_C(t)}{R_L C} \end{cases} . \tag{9.39}$$

- $E = \{(q_0, q_1), (q_1, q_0), (q_2, q_1), (q_0, q_2)\}$ is a set of 4 discrete transitions (see Fig. 9.13).
- $G = \{G_e, e \in E\}$ are defined by inequalities on the state parameters, as shown in Fig. 9.13.

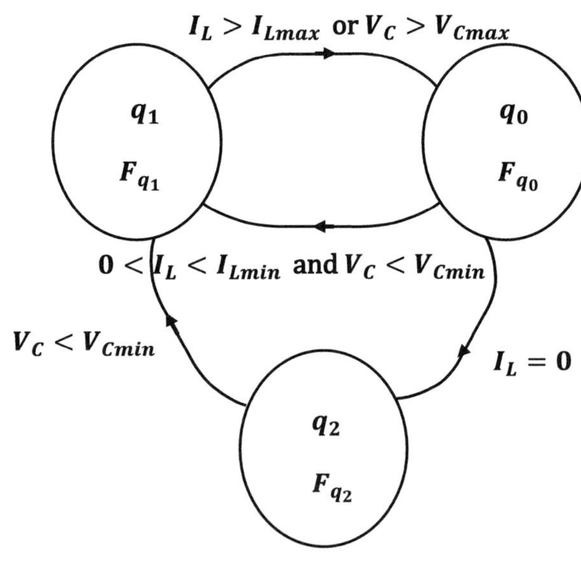

Fig. 9.13 Autonomous hybrid system of a buck converter with hysteresis control

9.6 First Harmonic Model of Resonant Converters

In this paragraph, the very popular LLC resonant half-bridge converter is used as an illustration (Lavigne, 2019). The basic topology and its associated waveforms are shown in Figs. 9.14 and 9.15. The control parameter is the switching frequency $F_S(t)$, fixed by a Voltage Controlled Oscillator (VCO) via $V_{ctr}(t)$.

The frequency control f_S of the LLC converter operates in the vicinity of the series resonance frequency f_o. Thus, the main component of the current in the resonant network L_r, C_r is close to fundamental frequency f_o. This means an approximation using the single fundamental harmonic of V_P can be made. f_o is given by:

$$f_0 = \frac{1}{2\pi\sqrt{LC}}. \tag{9.40}$$

As shown in Fig. 9.16, the input voltage V_P is represented by its fundamental voltage component as follows

$$v_{Peq1}(t) = \frac{2V_{\mathrm{IN}}}{\pi}\sin(2\pi F_S t). \tag{9.41}$$

Now, we need to determine the equivalent resistor R_{eq} connected in parallel with the magnetizing inductor L_m, according to the waveforms shown in Fig. 9.15. We can deduce the fundamental voltage of v_{L_m}:

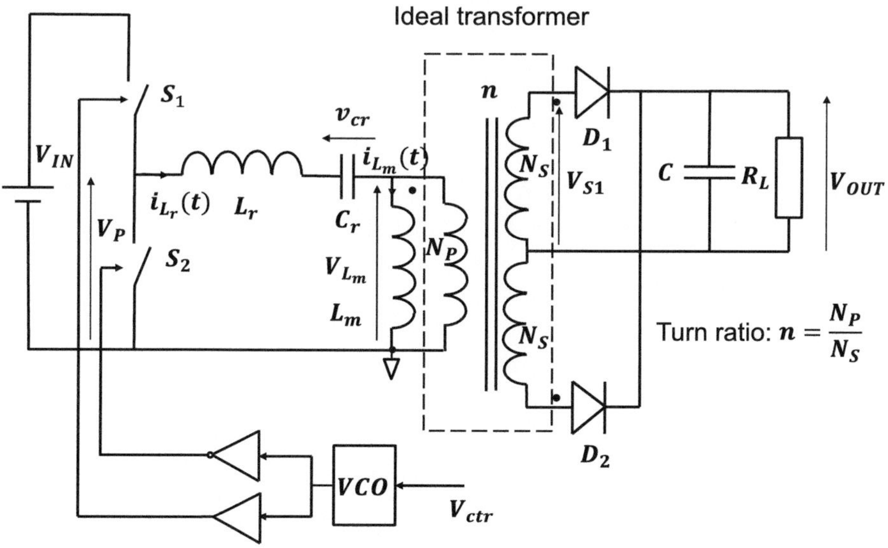

Fig. 9.14 LLC resonant half-bridge converter

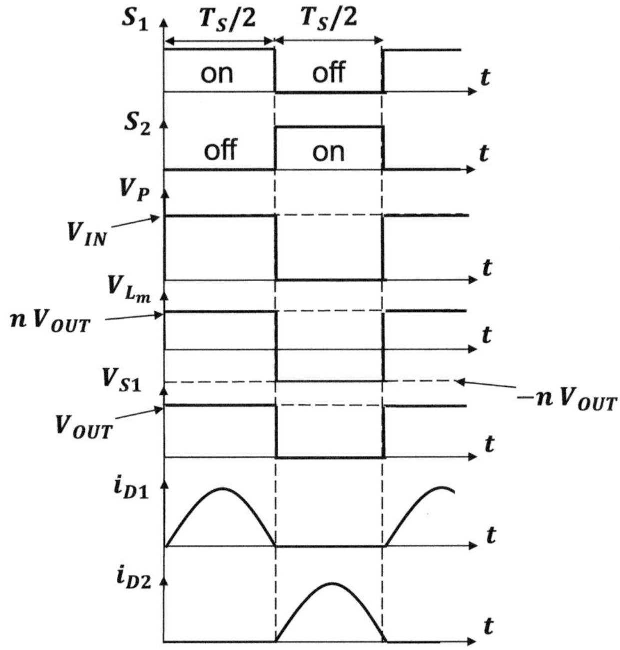

Fig. 9.15 Waveforms of an LLC resonant half-bridge converter

$$v_{L_{m1}} = \frac{4nV_{\text{OUT}}}{\pi}\sin(2\pi F_S t). \tag{9.42}$$

If we consider the transformer and the diodes D_1 and D_2 as ideal components (not generating power losses), we can write:

$$\frac{\left(\frac{4nV_{\text{OUT}}}{\sqrt{2}\pi}\right)^2}{R_{eq}} = \frac{V_{\text{OUT}}^2}{R_L}, \tag{9.43}$$

leading to the expression of R_{eq}

$$R_{eq} = \frac{8n^2}{\pi^2}R_L. \tag{9.44}$$

From the final model given in Fig. 9.16, we can easily determine the dynamics of the LCC converter

$$\begin{cases} \frac{dv_{Cr}(t)}{dt} = \frac{1}{C_r} i_{Lr}(t) \\ \frac{di_{Lr}(t)}{dt} = -\frac{1}{L_r} v_{cr}(t) - \frac{8n^2R_L}{\pi^2L_r} i_{Lr}(t) + \frac{8n^2R_L}{\pi^2L_r} i_{Lm}(t) + \frac{2V_{IN}}{\pi L_r}\sin(2\pi F_S(t)t). \\ \frac{di_{Lm}(t)}{dt} = \frac{8n^2R_L}{\pi^2L_m} i_{Lr}(t) - \frac{8n^2R_L}{\pi^2L_m} i_{Lm}(t) \end{cases} \tag{9.45}$$

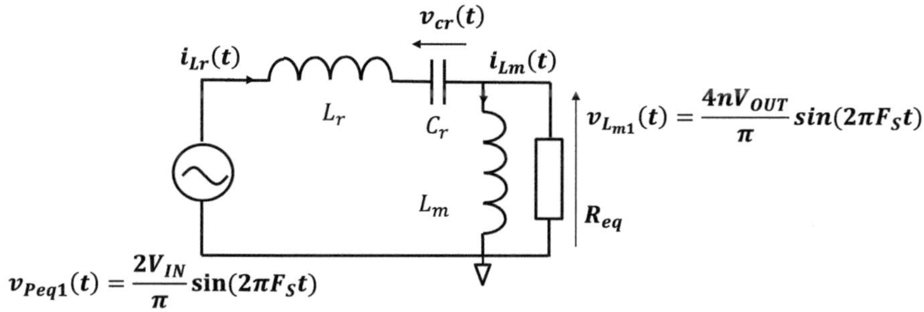

Fig. 9.16 Model of the LLC resonant half-bridge converter

This is a nonautonomous system with the state parameters v_{Cr}, i_{Lr}, i_{Lm} and $f_S(t)$ as control parameters.

9.7 Nonlinear Model Including Parasitic Elements

The models previously described can be easily completed by adding parasitic elements to the various components. Figure 9.17 shows the most common parasitic elements, i.e. the R_{DSON} of a MOSFET transitor, the threshold voltage V_{TO} and series resistor R_D of the diode, the parasitic resistors of the inductor (r_L) and capacitor (R_{ESCR}) and shunt resistor (R_S). As an example, we provide below the averaged model of a CCM buck converter with voltage-mode control, including the parasitic elements R_{DSON}, r_L and R_{ESCR}.

$$\begin{cases} \frac{dI_L(t)}{dt} = -\frac{1}{L}\left(d(t)R_{\mathrm{DSON}} + r_L + \frac{r_{\mathrm{ESCR}}R_L}{r_{\mathrm{ESCR}} + R_L}\right)I_L(t) - \frac{1}{L}\left(\frac{R_L}{R_L + r_{\mathrm{ESCR}}}\right)V_C(t) + \frac{V_{\mathrm{IN}}}{L}d(t) \\ \frac{dV_C(t)}{dt} = \frac{R_L I_L(t)}{C(R_L + r_{\mathrm{ESCR}})} - \frac{V_C(t)}{C(R_L + r_{\mathrm{ESCR}})} \end{cases} \tag{9.46}$$

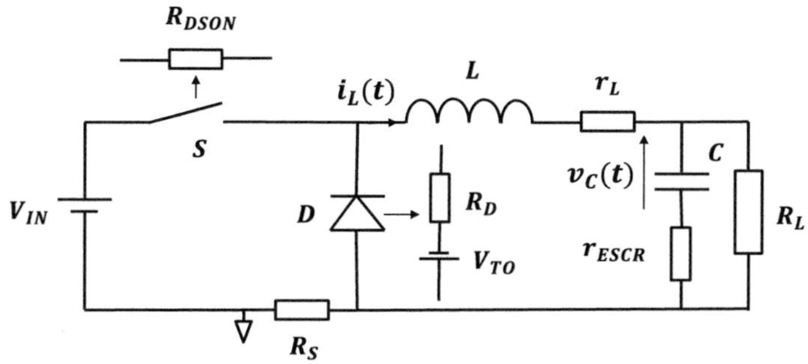

Fig. 9.17 Buck converter with parasitic elements

Chapter 10
A Step-By-Step Controller Design Procedure

In this section, we will examine a step-by-step controller design procedure (Erickson & Maksimovic, 1996; Erickson & Maksimovic, 2001; Erickson & Maksimović, 2020; Middlebrook & Ćuk, 1976). This methodology will be illustrated by the controller design of a CCM buck converter using current-mode control.

10.1 Step 1: Selection of the Topology

The selection of the topology and the values of the various converter components are generally made at a preceding stage. They are generally defined according to the following specifications: input and output voltage ranges, output power range, efficiency requirements and insulation needs. As an example, consider once again a buck converter operating in continuous mode with current mode-control (see Fig. 10.1). All components are considered ideal except for the diode, which is modelled by its forward voltage drop. The study uses the following values.

$$V_{\mathrm{IN}} \in [300\ \mathrm{V}, 400\ \mathrm{V}],\ V_F = 0.8\ \mathrm{V},\ V_{\mathrm{REV}} = 10\ \mathrm{V},\ R_L \in [100\Omega, 200\Omega],\ R_S = 0.1\Omega$$
$$L = 4\ \mathrm{mH},\ C = 100\ \mu\mathrm{F},\ T_S = 10\ \mu\mathrm{s}. \tag{10.1}$$

10.2 Step 2: Regulation Specifications

The objectives of the corrector design are to meet the regulation specifications while accounting for component variations. Below are the specifications for our illustrative example:

B. Rivet, *Theories of Nonlinear Systems for Power Converters*,
https://doi.org/10.1007/978-3-032-25062-9_10

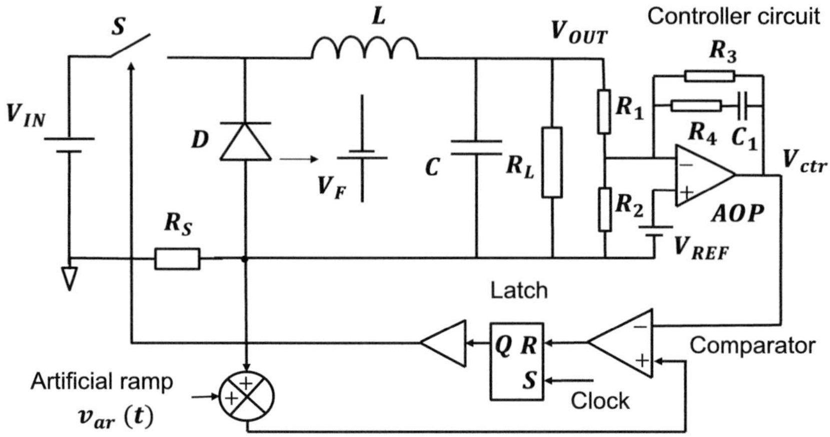

Fig. 10.1 Example of a PID controller using an operational amplifier

- Regulate the output voltage at the targeted level: $V_{OUTtar} = 160$ V with a static error lower than 1 V.
- High-frequency stability for $V_{IN} \in [300\text{ V}, 400\text{ V}]$ and $R_L \in [100\Omega, 200\Omega]$.
- Low-frequency stability for $V_{IN} \in [300\text{ V}, 400\text{ V}]$ and $R_L \in [100\Omega, 200\Omega]$.
- Ripple ratio between V_{IN} and V_{OUT} and at 1 kHz, greater than 100.
- Attenuation of the open-loop gain at the switching frequency higher than 30 dB.
- Response time of V_{OUT} due to a load variation ($R_L = 100\Omega \rightarrow R_L = 200\Omega$) lower than 5 ms and overvoltage lower than 162 V.

10.3 Step 3: Choosing of a Corrector Circuit

This second step consists of choosing the simplest controller circuit that meets all the regulation specifications. Figure 10.2 shows an example of a corrector circuit designed with an operational amplifier (AOP) considered ideal.

In our example, we will first try to meet the regulation specifications with this circuit without the resistor R_5 and the capacitors C_2 and C_3. The full system of the converter with its controller is shown in Fig. 10.1.

10.4 Step 4: Elimination of High-Frequency Instabilities

The third step consists of eliminating instability near to the switching frequency F_S. The best-known example of instability is related to current mode control. The usual solution to this problem is to add an artificial ramp v_{ar} (see Fig. 10.1). This is the technique used in our example. In Chap. 14, we will consider other alternatives using

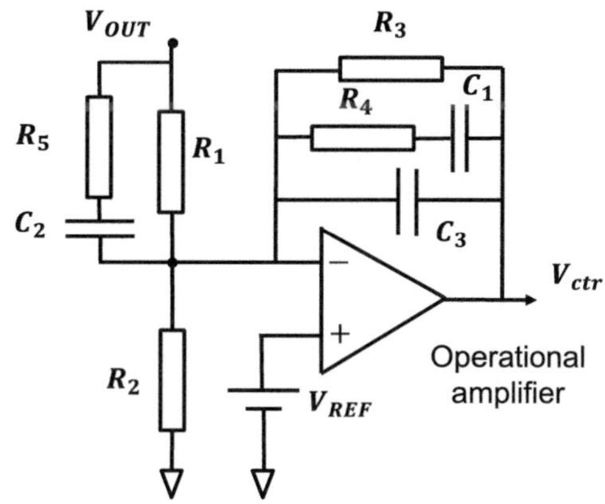

Fig. 10.2 Buck converter with its controller

methods for controlling chaotic systems. As mentioned in the previous section, the best way to study the high-frequency phenomena of a given converter is to determine the corresponding discrete model. To simplify the calculations while retaining a good level of precision, the following assumptions can be made: the input current of the corrector circuit is negligible compared to the load current, and V_{OUT} and V_{ctr} can be considered as constant for several switching periods. The voltage drop across R_S is also negligible compared to V_{IN}. A discrete model can be determined from the waveforms shown in Fig. 10.3.

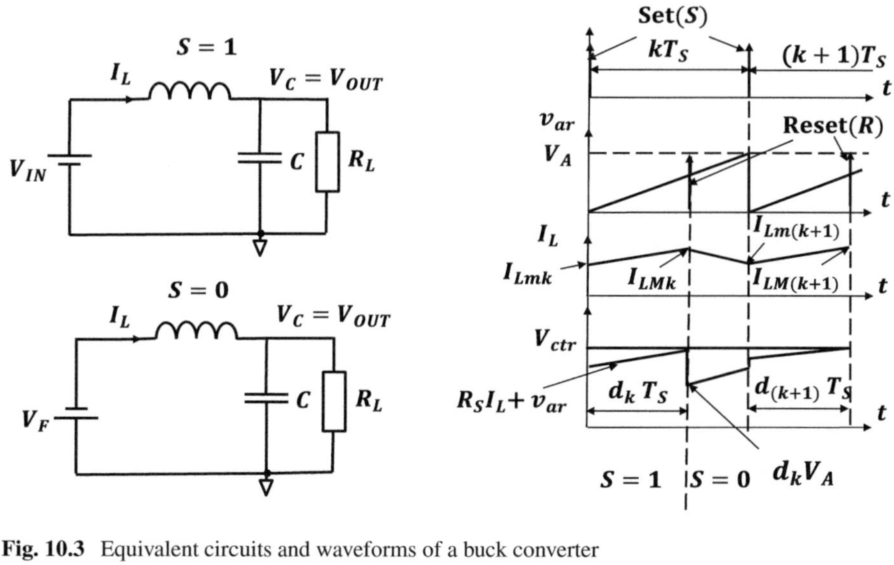

Fig. 10.3 Equivalent circuits and waveforms of a buck converter

Start by calculating the equilibrium point: (d_e, I_{Lme}, I_{LMe}). At the equilibrium, we can write:

$$\begin{cases} V_{\mathrm{IN}} = L\frac{I_{LMe}-I_{Lme}}{d_e T_S} + V_{\mathrm{OUT}} \\ V_F = L\frac{I_{LMe}-I_{Lme}}{(1-d_e)T_S} - V_{\mathrm{OUT}} \\ V_{\mathrm{IN}}\frac{I_{LMe}+I_{Lme}}{2}d_e = \frac{V_{\mathrm{OUT}}^2}{R_L} + V_F\frac{I_{LMe}+I_{Lme}}{2}(1-d_e) \end{cases} . \tag{10.2}$$

The last equation of the system is related to the energy balance during a switching period T_S. It follows that:

$$d_e = \frac{V_F + V_{\mathrm{OUT}}}{V_{\mathrm{IN}} + V_F}, \quad I_{LMe} = \frac{V_{\mathrm{OUT}}^2}{R_L(V_{\mathrm{IN}}d_e - V_F(1-d_e))} + \frac{V_F + V_{\mathrm{OUT}}}{2L}(1-d_e)T_S, \tag{10.3}$$

$$I_{Lme} = I_{LMe} - \frac{(V_F + V_{\mathrm{OUT}})(1-d_e)T_S}{L}. \tag{10.4}$$

From the waveforms in Fig. 10.3, and according to the previously stated assumptions, we can easily calculate the recursive relation between $I_{Lm(k+1)}$ and I_{Lmk} when equilibrium is not reached. We get:

$$\begin{aligned} I_{Lm(k+1)} = I_{Lmk} &+ \frac{V_{\mathrm{IN}} - V_{\mathrm{OUT}}}{L}\frac{V_{ctr} - R_S I_{Lmk}}{V_A + \frac{R_S(V_{\mathrm{IN}}-V_{\mathrm{OUT}})T_S}{L}}T_S \\ &- \frac{(V_F + V_{\mathrm{OUT}})}{L}\left(1 - \frac{V_{ctr} - R_S I_{Lmk}}{V_A + \frac{R_S(V_{\mathrm{IN}}-V_{\mathrm{OUT}})T_S}{L}}\right)T_S. \end{aligned} \tag{10.5}$$

V_{ctr} can be considered constant and can be expressed with the equilibrium point as follows:

$$V_{ctr} = R_S I_{LMe} + V_A d_e. \tag{10.6}$$

Thanks to Eqs. (10.5) and (10.6), we have a relation of the form:

$$I_{Lm(k+1)} = F(I_{Lmk}, V_A). \tag{10.7}$$

It is interesting to note that high-frequency phenomena can be represented with a simple first-order discrete system. Figure 10.4 shows the bifurcation diagram of this discrete system for $V_{\mathrm{IN}} = 300$ V and $V_{\mathrm{IN}} = 400$ V with $R_L = 100\ \Omega$. Without an artificial ramp ($V_A = 0$), the buck converter is instable for $V_{\mathrm{IN}} = 300$ V. For $R_L = 200\ \Omega$, we get the same type of graph with reduced values of I_{Lme}. The system is stable for $V_A > 0.02$ V (see Fig. 10.4).

When I_{Lmk} diverges, the recursive relation (10.5) is no longer valid because the reset function is not always activated at each switching period. Therefore, the bifurcation

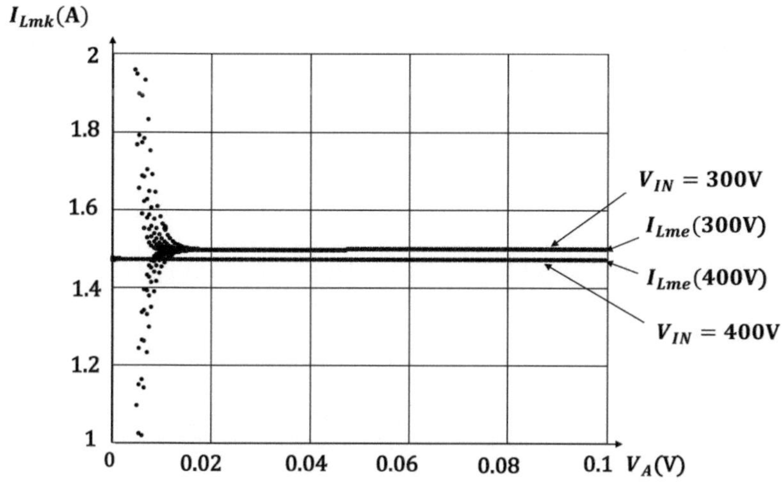

Fig. 10.4 Bifurcation diagram of a buck converter with current-mode control

diagram is inaccurate in this region. This phenomenon is illustrated in Fig. 10.5. This is not important because the key information for the designer is to know value of V_A at which the converter becomes stable.

Generally speaking, the presence of parasitic elements prevents the emergence of high-frequency instabilities. Therefore, the result of the above bifurcation diagram, which takes into account only the parasitic element of the diode, can be considered the worst case. Contrary to a widespread belief, such instabilities can occur for duty cycles lower than 0.5. In order to eliminate these high-frequency instabilities for

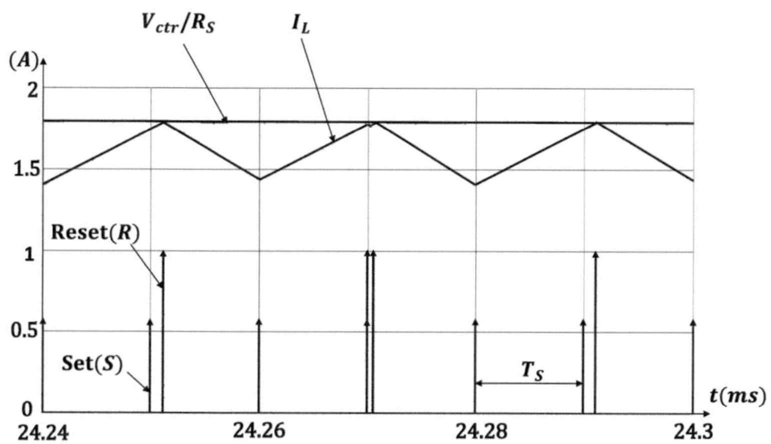

Fig. 10.5 Reset signal with $V_A = 0$ V, $V_{\mathrm{IN}} = 300$ V, $R_L = 100\ \Omega$

$V_{\text{IN}} \in [300\text{ V}, 400\text{ V}]$ and $R_L \in [100\ \Omega, 200\ \Omega]$ with a good safety margin, we choose:

$$V_A = 0.03\text{ V}. \tag{10.8}$$

10.5 Step 5: Equivalent Circuit of the Full Averaged Model

For this step, we attempt to determine an equivalent circuit corresponding to the full averaged model with parasitic elements. The controller circuit will be retained, and the equivalent circuit of the converter will be built using the averaged system. For our example, using the method described in Para. 9.2.3, we have:

$$\begin{cases} dV_{\text{IN}} - (1-d)V_F = L\frac{dI_L}{dt} + V_C \\ I_L = C\frac{dV_C}{dt} + \frac{V_C}{R_L} \\ I_L = \frac{V_{ctr}-dV_A}{R_S} - \frac{d^2T_S}{2L}(V_{\text{IN}} - V_C) - \frac{(1-d)^2T_S}{2L}(V_C + V_F) = G(d, V_{ctr}, V_C) \end{cases}. \tag{10.9}$$

The corresponding equivalent circuit is shown in Fig. 10.6. This type of circuit is very useful. It can be used to:

- Determine the equilibrium point and the static error with respect to one of the circuit parameters.
- Simulate variations in load or input voltage.
- By adding a small AC source to the circuit, simulate the frequency response around the equilibrium point without calculating the transfer functions. The AC source can be placed in different parts of the circuit, for example, in series with C. This method is illustrated in Para. 10.7.

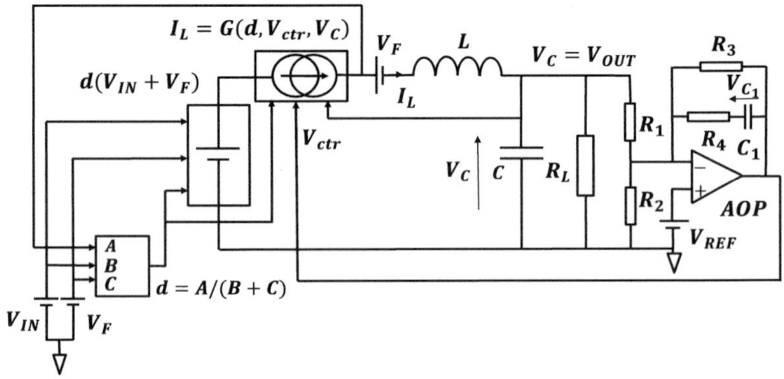

Fig. 10.6 Equivalent circuit of the full averaged model

- Determine the influence of the different parasitic elements on the static error, the frequency response and the load variation. It is easy to add one of the parasitic elements to this equivalent circuit.
- Make the circuit simpler by eliminating parasitic elements that have little influence or by using the simple model of current-mode control. With this simpler model, the transfer function formulas will also be simpler.
- Reduce simulation time compared to simulations performed with a real circuit.

10.6 Step 6: Component Calculation to Achieved an Acceptable Static Error

We begin by calculating the values of R_1 and R_2. The current I_{R1} through R_1 must be low to limit the power losses in it. It is given by:

$$I_{R1} = \frac{V_{OUTtar} - V_{REV}}{R_1}. \tag{10.10}$$

V_{OUTtar} can be adjusted according to:

$$V_{OUTtar} = \frac{R_1 + R_2}{R_2} V_{REV}. \tag{10.11}$$

Using the last two formulas, with $I_{R1} = 1$ mA, we get:

$$R_1 = 150\ \text{k}\Omega, \quad R_2 = 10\ \text{k}\Omega. \tag{10.12}$$

Very often in the literature, the static error is determined using the small-signal system around the equilibrium point. This method gives good results when the static error is small, but leads to inaccuracies otherwise. Here, we propose determining the static error $V_{OUT} - V_{OUTtar}$ using the equivalent circuit of the full averaged model. The advantage of this method is that it maintains a good level of accuracy regardless of the value of the static error. In steady state, the capacitors C and C_1 act as an open circuit; therefore, the only way to reach the static error is by acting on R_3 (with R_1 and R_2 already fixed). We also need to consider the variation of V_{IN} and R_L. Figure 10.7 shows the static error in relation to R_3 for $V_{IN} \in [300\ \text{V}, 400\ \text{V}]$ and $R_L \in [100\ \Omega, 200\ \Omega]$. We observe that the static error decreases as R_3 increases. It is well known that an increase in the static gain of the corrector reduces the static error. We also note that this parameter does not depend on the variation of V_{IN} and R_L (since all the curves are superimposed). This is one of the benefits of current-mode control.

A comparison between voltage-mode and current-mode control will be presented in the next section. To meet the specifications on the static error:

$$V_{OUTe} - V_{OUTtar} < 1\ \text{V} \tag{10.13}$$

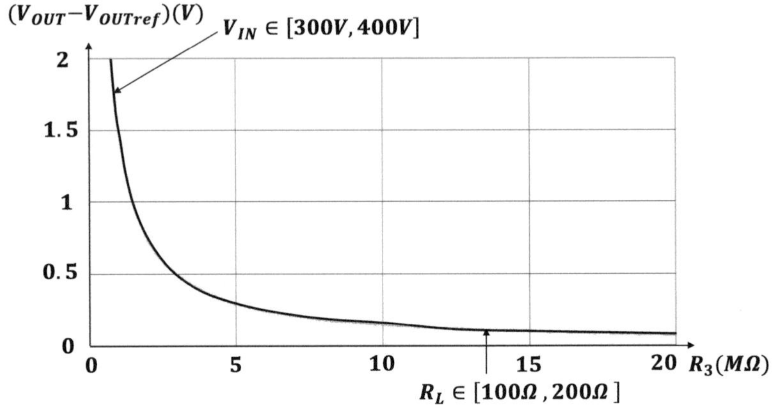

Fig. 10.7 Static error with respect to R_3, V_{IN}, and R_L

with a good safety margin, we take

$$R_3 = 3\ \text{M}\Omega. \tag{10.14}$$

Obviously, when R_3 is fixed, the state parameters corresponding to the equilibrium are also fixed. For $V_{\text{IN}} = 300$ V and $R_L = 100\ \Omega$, the simulation gives:

$$V_{Ce} = V_{\text{OUT}e} \simeq 160.49\ \text{V}, \quad V_{C1e} \simeq 9.81\ \text{V}, \quad I_{Le} \simeq 1.6\ \text{A}. \tag{10.15}$$

10.7 Step 7: Determination of a Simplified Averaged Model

The objective of this step is to obtain a simplified averaged model in order to make the expressions of the transfer functions more manageable while maintaining a good level of accuracy. The influence of each parasitic element on the transfer function will be examined. In the case of current-control, the simplified averaged model will be considered first. As in our example, the current-control mode is employed. We begin by considering the simplified averaged model. Referring to Para. 9.2.2, we obtain:

$$\begin{cases} I_L = C\frac{dV_C}{dt} + \frac{V_C}{R_L} \\ I_L = \frac{V_{ctr}}{R_S} \end{cases}. \tag{10.16}$$

From this system, we can derive the full equivalent circuit of the converter using the simplified averaged model. It is shown in Fig. 10.8.

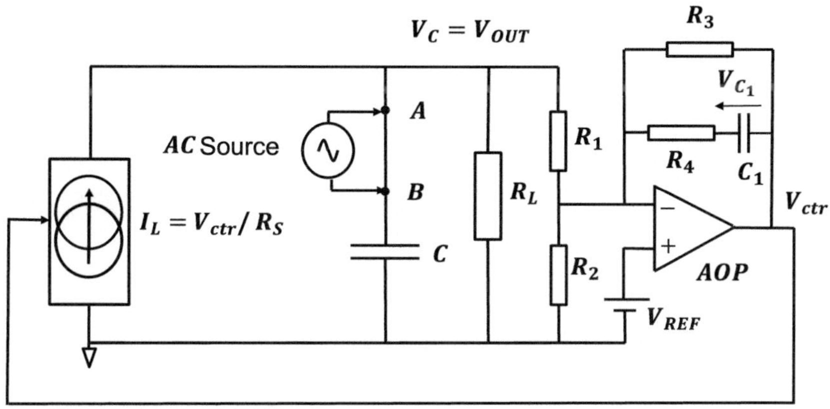

Fig. 10.8 Equivalent circuit of the complete system

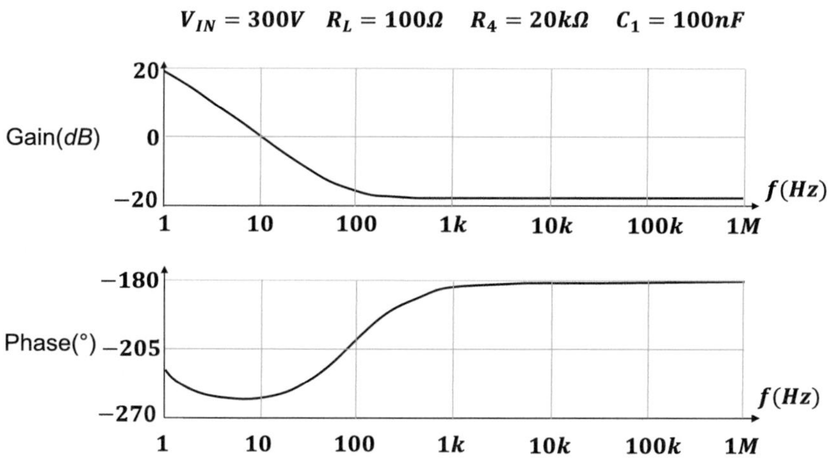

Fig. 10.9 Transfer function of $\frac{V_{ctr}(j\omega)}{V_{OUT}(j\omega)}$ simulated with the two averaged circuits

It is now possible to compare, through simulation, the transfer functions obtained from the two circuits shown in Figs. 10.6 and 10.8. We add an AC source to these two circuits, for example in series with the capacitor C, as illustrated in Fig. 10.8. As mentioned previously, these simulations are useful for obtaining the transfer functions around the equilibrium point without performing any calculation. Simulations show that the transfer functions are identical. An illustrative example is provided in Fig. 10.9. We note that the transfer function $\frac{V_{ctr}(j\omega)}{V_{OUT}(j\omega)}$ obtained with the two averaged circuits are superimposed. Thus, it is possible to use the simplified model to determine all the transfer functions of the complete system. In the simplified model, the parasitic element of the diode is not included. This means that the forward voltage V_F

has no influence on the transfer functions. This represents an additional advantage of current-mode control.

With the simplified model, the inductor L is replaced by the controlled current generator:

$$I_L = \frac{V_{ctr}}{R_S}. \tag{10.17}$$

As a consequence, the system only has two state parameters (V_C and V_{C1}) and becomes a second-order system.

$$\begin{cases} \frac{dV_C}{dt} = -\frac{1}{C}\left(\frac{R_3R_4}{R_SR_1(R_3+R_4)} + \frac{1}{R_L}\right)V_C - \frac{R_3}{CR_S(R_3+R_4)}V_{C1} + \frac{A}{CR_S}V_{REF} = F_1(V_C, V_{C1}) \\ \frac{dV_{C1}}{dt} = \frac{R_3}{C_1R_1(R_3+R_4)}V_C - \frac{1}{C_1(R_3+R_4)}V_{C1} - \frac{R_3(R_1+R_2)}{C_1R_1R_2(R_3+R_4)}V_{REF} = F_2(V_C, V_{C1}) \end{cases} \tag{10.18}$$

with

$$A = 1 + \frac{R_3R_4(R_1+R_2)}{R_1R_2(R_3+R_4)}. \tag{10.19}$$

Note that this system is linear. Thus, this simplification makes it possible to move from a complex third-order nonlinear system to a linear second-order system.

10.8 Step 8: Determination of All the Transfer Functions

At this stage, we use the techniques described in Para. 1.9 to determine all the transfer functions. It is important to remember that these transfer functions generally depend on the equilibrium state parameters and the nominal values of V_{IN} and R_L. The equilibrium state parameters are determined in Para. 10.5. With this simplified model, the state parameter I_{Le} is eliminated. V_{Ce}, V_{C1e} can be found by solving the following system, given by

$$\begin{cases} 0 = F_1(V_{Ce}, V_{C1e}) \\ 0 = F_2(V_{Ce}, V_{C1e}) \end{cases}. \tag{10.20}$$

We obtain:

$$\begin{aligned} V_{Ce} = V_{\mathrm{OUT}e} &= \frac{R_L}{R_2}\frac{R_1R_3 + R_1R_2 + R_2R_3}{R_LR_3 + R_1R_s}V_{\mathrm{REF}} \simeq 160.49\ \mathrm{V}, \\ V_{C1e} &= V_{\mathrm{REF}} - \frac{R_s}{R}V_{\mathrm{OUT}e} \simeq 9.839\ \mathrm{V}. \end{aligned} \tag{10.21}$$

This pedagogical example is simple. As a rule, it is not possible to find a literal expression of the equilibrium state parameters. We note that V_{Ce} has the same value as that determined in Para. 10.6. V_{C1e} is slightly higher. This difference is due to the presence of an artificial ramp in the full model. It is recommended to use the values given in Para. 10.6, even though the two equilibrium points usually yield the same transfer functions. From the system (10.18), we can derive the following linearised small-signal system:

$$\begin{pmatrix} \dot{v}_C \\ \dot{v}_{C1} \end{pmatrix} = \begin{pmatrix} -\frac{1}{R_L C} - \frac{1}{\tau_1} & -\frac{1}{\tau_2} \\ \frac{1}{\tau_3} & -\frac{1}{\tau_4} \end{pmatrix} \begin{pmatrix} v_C \\ v_{C1} \end{pmatrix} + \begin{pmatrix} 0 & \frac{1}{\tau_5} \\ 0 & -\frac{1}{\tau_6} \end{pmatrix} \begin{pmatrix} v_{\mathrm{IN}} \\ v_{\mathrm{REF}} \end{pmatrix} \tag{10.22}$$

with

$$\begin{aligned} \tau_1 &= \frac{CR_S R_1 (R_3 + R_4)}{R_3 R_4}, \tau_2 = \frac{CR_S (R_3 + R_4)}{R_3}, \tau_3 = \frac{C_1 R_1 (R_3 + R_4)}{R_3}, \\ \tau_4 &= C_1 (R_3 + R_4), \tau_5 = \frac{CR_S}{A}, \tau_6 = \frac{C_1 R_1 R_2 (R_3 + R_4)}{R_3 (R_1 + R_2)}, v_{\mathrm{OUT}} = v_C. \end{aligned} \tag{10.23}$$

Since this new system is linear, it is now possible to use the Laplace transform. The linearised full system can be represented as shown in Fig. 10.10. From Eq. (10.22), by eliminating $V_{C1}(s)$ and considering

$$V_{\mathrm{OUT}}(s) = V_C(s), \tag{10.24}$$

it follows that

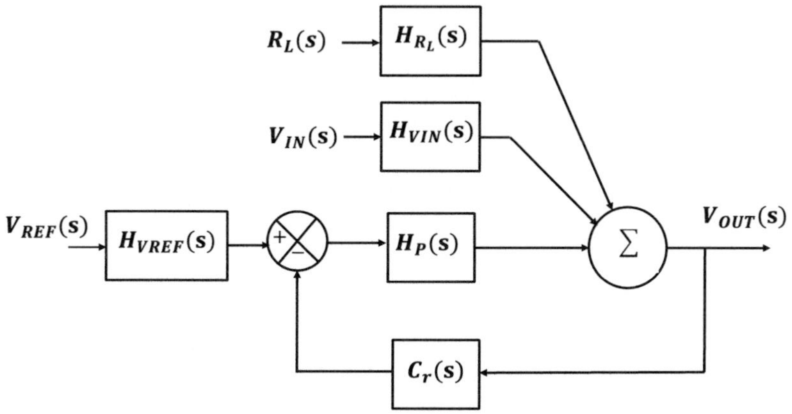

Fig. 10.10 Linearised full system

$$V_{OUT}(s) = \frac{R_L}{R_S(1+R_LCs)}\left(\left(\frac{CR_S}{\tau_5} + \frac{CR_S\tau_4}{\tau_2\tau_6(1+s\tau_4)}\right)V_{REF}(s) - \left(\frac{CR_S}{\tau_1} + \frac{CR_S\tau_4}{\tau_2\tau_3(1+s\tau_4)}\right)V_{OUT}(s)\right). \quad (10.25)$$

From Fig. 10.10, we deduce that

$$V_{OUT}(s) = H_P(s)(H_{VREF}(s)V_{REF}(s) - C_r(s)V_{OUT}(s)) + H_{VIN}(s)V_{IN}(s) + H_{RL}(s)R_L(s). \quad (10.26)$$

Thanks to Eqs. (10.25) and (10.26) (with $R_L(s) = 0$) and after some transformations, we obtain:

$$C_r(s) = \frac{R_3(1+R_4C_1s)}{R_1(1+(R_3+R_4))C_1s)}, \quad H_P(s) = \frac{R_L}{R_S(1+R_LCs)}, \quad H_{VIN}(s) = 0. \quad (10.27)$$

Generally speaking, it is not useful to calculate $H_{VREF}(s)$ unless V_{REF} is not constant. The purpose of this method is to ensure that all the transfer functions have been identified correctly.

In this example the transfer functions do not depend on the state value at the equilibrium point. The only system parameter to consider is the load resistor R_L.

10.9 Step 9: Final Design of the Corrector Circuit

This final step consists of finalising the design of the corrector circuit to satisfy the complete specification. The open-loop transfer function $G_{BO}(jf)$ can be written as follows:

$$G_{BO}(jf) = G_O\frac{\left(1+\frac{jf}{f_z}\right)}{\left(1+\frac{jf}{f_p}\right)\left(1+\frac{jf}{f_{p1}}\right)} \quad (10.28)$$

with

$$G_O = \frac{R_3R_L}{R_1R_S}, \quad f_z = \frac{1}{2\pi R_4C_1}, \quad f_p = \frac{1}{2\pi R_LC}, \quad f_{p1} = \frac{1}{2\pi(R_3+R_4)C_1}. \quad (10.29)$$

With G_O and f_p fixed, we can only act on f_z and f_{p1} by adjusting R_4 and C_1. The designer of the corrector circuit still faces various trade-offs. In our example, the main trade-off to consider is between the phase margin security φ_m and the response of V_{OUT} to a load variation (the cross-over frequency f_C is fixed at 1.6 kHz).

Figures 10.11 and 10.12 respectively show the open-loop transfer function and the response of V_{OUT} to a load variation for $R_4 = 10\ \text{k}\Omega, C_1 = 20$ nF and

$R_4 = 10$ kΩ, $C_1 = 500$ nF. These two figures show that with a high phase margin ($\varphi_m = 90°$), the output reacts slowly to a load variation. With a lower phase margin ($\varphi_m = 53°$), the system reacts faster due to increased gain at low frequencies. If we decrease φ_m further, the system becomes unstable. Another way to reduce the overvoltage is to increase the crossover frequency. We will keep the following for the final design:

$$R_4 = 10\ \text{k}\Omega,\ C_1 = 20\ nF,\ \text{giving}\ fz = 0,795\ \text{Hz}, f_{p1} = 2.6\ \text{Hz}. \tag{10.30}$$

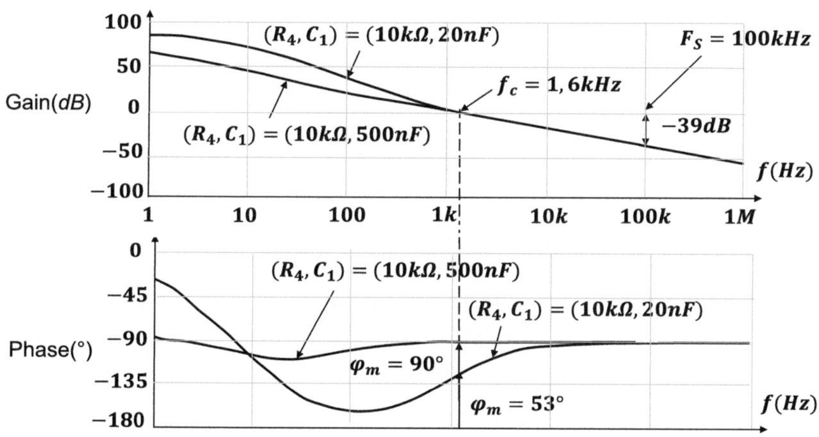

Fig. 10.11 Frequency response of $G_{BO}(jf)$ with $R_L = 100\ \Omega$

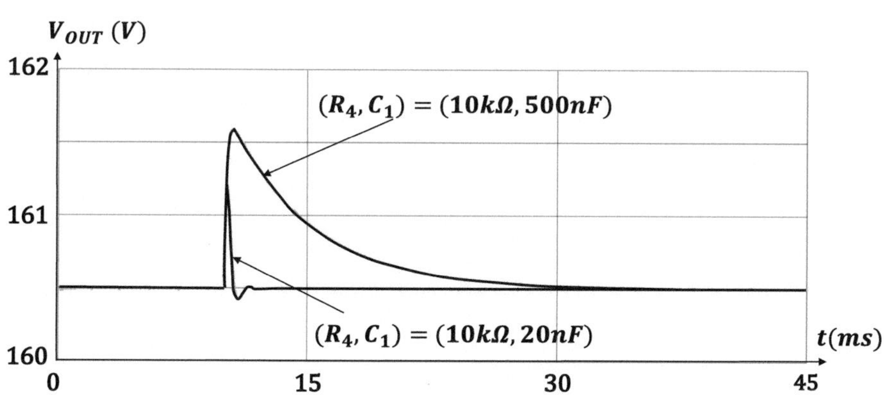

Fig. 10.12 Response of V_{OUT} to a load variation: $R_L = 100\ \Omega \rightarrow R_L = 200\ \Omega$

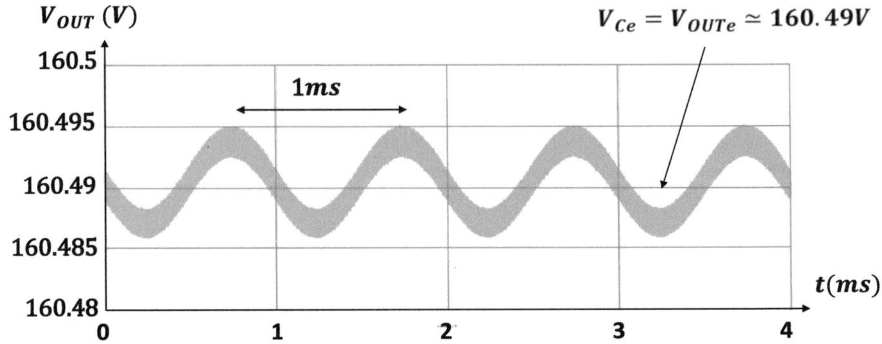

Fig. 10.13 Output response to a sinusoidal input at 10 V at 1 kHz

With these values, the response time is lower than 5 ms, and the over voltage less than 162 V. The attenuation of the open-loop gain is higher than 30 dB (−39 dB) at the switching frequency F_S.

We can easily verify that the phase margin is sufficiently high with $R_L = 200\Omega$. Finally, consider the requirement of V_{IN} ripples divided by V_{OUT} ripples at 1 kHz higher than 100. In theory, since $H_{\text{VIN}}(s) = 0$, there are no V_{OUT} ripples. We only need to verify this point with a simulation of the real circuit. Figure 10.13 shows the actual voltage output when a sinusoidal ripple of 10 V at 1 kHz is applied on V_{IN}.

We note that the output ripple is indeed very low, and the attenuation from V_{IN} to V_{OUT} is higher than 1000, thus well within the specification. All the specification requirements are met, so it is not necessary to consider a more complex corrector circuit.

Chapter 11
Examples of Common Control Issues

In this chapter, several common control issues will be examined.

11.1 Comparison of Current-Mode Control and Voltage-Mode Control

The comparison between current-mode and voltage-mode controls will be analysed using the buck converter presented in Para. 10. First, we will design a controller based on voltage-mode control to meet the specification given in Para. 10.2. Then, we will compare the performances obtained with that of the current-mode control (Mammano, 1994). Let us follow the procedure described in the previous section, step by step, to design the new controller. Since the regulation specification and the choice of controller are the same as in the previous example, the first three steps are identical. In theory, a buck converter operating in continuous mode with voltage-mode control exhibits no instabilities in high-frequencies near the crossover frequency F_S. Thus, we can proceed directly to step five of the design procedure.

Using standard techniques, the equivalent circuit of the complete averaged model (including parasitic elements) for voltage-mode control is obtained (see Fig. 11.1). $V_P = 1$ V represents the magnitude of the sawtooth generator. The calculation of R_1 and R_2 corresponding to step six remains the same. The simulation results show that the steady-state error is practically independent of V_{IN} and R_L. We also keep:

$$R_3 = 3\ \mathrm{M}\Omega \tag{11.1}$$

which yields the following value for the steady-state error:

$$V_{\mathrm{OUT}} - V_{\mathrm{OUT}tar} \simeq 0.47\ \mathrm{V}. \tag{11.2}$$

B. Rivet, *Theories of Nonlinear Systems for Power Converters*,
https://doi.org/10.1007/978-3-032-25062-9_11

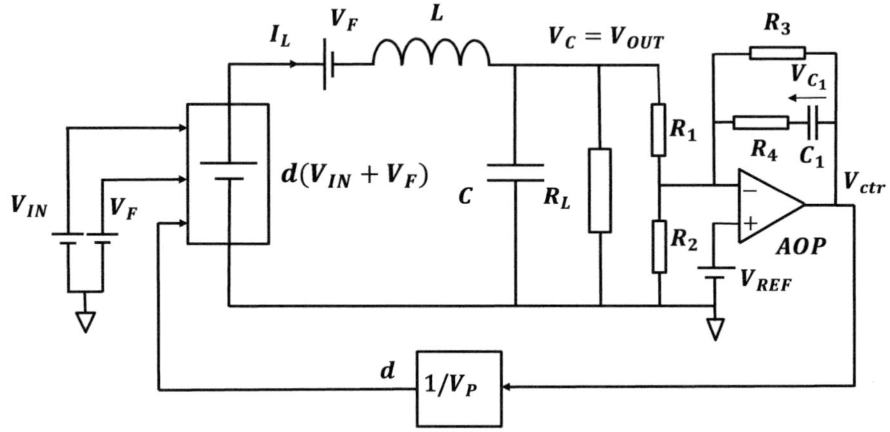

Fig. 11.1 Equivalent circuit for the complete averaged model

In this example, since the only parasitic element to be taken into consideration is V_F, the only way to simplify the circuit is to eliminate it. The simulation results of the transfer functions based on the circuit in Fig. 11.1 show that the effect of the diode's forward voltage drop is negligible. Thus, the simplified circuit is obtained and the transfer functions are calculated by taking:

$$V_F = 0\ \text{V}. \tag{11.3}$$

We begin step height by presenting the differential equations corresponding to the circuit in Fig. 11.1 with $V_F = 0$ V. We obtain:

$$\begin{cases} \frac{dV_C}{dt} = -\frac{1}{R_L C} V_C + \frac{1}{C} I_L = F_1(V_C, I_L) \\ \frac{dV_{C1}}{dt} = \frac{1}{\tau_1} V_C - \frac{1}{\tau_2} V_{C1} - \frac{1}{\tau_3} V_{\text{REF}} = F_2(V_C, V_{C1}, V_{\text{REF}}) \\ \dfrac{dI_L}{dt} = \dfrac{-1 + k_1(V_{\text{IN}})}{L} V_C + \dfrac{k_2(V_{\text{IN}})}{L} V_{C1} + \dfrac{k_3(V_{\text{IN}})}{L} V_{\text{REF}} \\ \quad = F_3\big(V_C, V_{C1}, V_{\text{REF}}, V_{\text{IN}}\big) \end{cases} \tag{11.4}$$

with

$$\tau_1 = \frac{R_1(R_3 + R_4)}{R_3} C_1,\ \tau_2 = (R_3 + R_4) C_1,\ \tau_3 = \frac{R_1 R_2 (R_3 + R_4)}{R_3 (R_1 + R_2)} C_1, \tag{11.5}$$

and

$$\begin{aligned} k_1(V_{\text{IN}}) &= -\frac{R_3 R_4 V_{\text{IN}}}{R_1 (R_3 + R_4) V_P},\ k_2(V_{\text{IN}}) = -\frac{R_3 V_{\text{IN}}}{(R_3 + R_4) V_P}, \\ k_3(V_{\text{IN}}) &= \left(1 + \frac{R_3 R_4 (R_1 + R_2)}{R_1 R_2 (R_3 + R_4)}\right) \frac{V_{\text{IN}}}{V_P}, \end{aligned} \tag{11.6}$$

Note that with voltage mode-control, the system has one order more than under current-mode control. The state parameters are (V_C, V_{C1}, I_L) with $V_{\text{OUT}} = V_C$, and the system parameters to be taken into account are V_{IN} and R_L. The equilibrium point is determined by solving the corresponding S_e system. It follows that:

$$\begin{aligned} &V_{Ce} \simeq 160.47\ \text{V}, V_{C1e} \simeq 9.43\ \text{V}, I_{Le} \simeq 1.6\ \text{A}, \quad \text{with} \\ &V_{\text{IN}} = 300\ \text{V}, R_L = 100\ \Omega. \end{aligned} \tag{11.7}$$

Using the usual techniques, we obtain the $H_P(s)$ and $C_r(s)$ transfer functions (see Para. 10.8)

$$H_P(s) = \frac{V_{\text{IN}}}{V_A\left(1 + \frac{L}{R_L}s(1 + R_L C s)\right)}, \; C_r(s) = \frac{R_3}{R_1}\frac{(1 + R_4 C_1 s)}{(1 + (R_3 + R_4)C_1 s)}. \tag{11.8}$$

The first observation regarding the comparison between current-mode and voltage-mode controls is the difference in transfer functions shown in Fig. 11.2.

For current-mode control, $H_P(j\omega)$ is of the first order, so the phase tends to $-90°$ when $f \to \infty$, in contrast for current-mode control, $H_P(j\omega)$ is of the second order and the phase tends to $-180°$. With the control circuit used, we obtain:

$$f_z = \frac{1}{2\pi R_4 C_1} > f_{p1} = \frac{1}{2\pi (R_3 + R_{4)} C_1}. \tag{11.9}$$

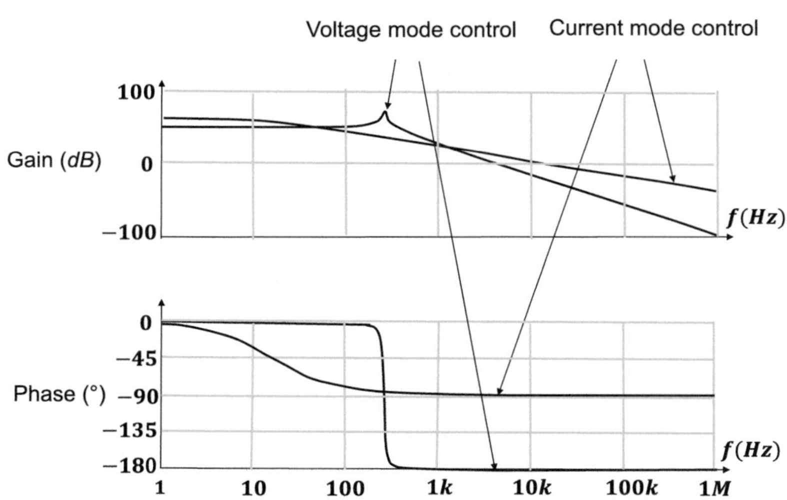

Fig. 11.2 Transfer function of $H_P(j\omega)$ with current- and voltage-mode controls

Thus, this type of compensator cannot increase the phase margin. Consequently, it is not possible to satisfy all the specifications, such as low-frequency stability, with this compensator. Now try to perform the design using the corrector circuit shown in Fig. 10.1 (without the capacitor C_3 and with $R_4 = 0$). The first seven steps of the control design are maintained. In steady-state mode, capacitors C_1 and C_2 act as open circuits, so the equilibrium values of V_{Ce}, V_{C1e} and I_{Le} and the static error are the same as with the previous corrector circuit. V_{C2e} is given by:

$$V_{C2e} = V_{Ce} - V_{\text{REF}} \simeq 150.47 \text{ V with } V_{\text{IN}} = 300 \text{ V}, R_L = 100\ \Omega \tag{11.10}$$

Now, the system containing one inductor and three capacitors becomes a fourth-order system. The dynamics of the averaged model are given by:

$$\begin{cases} \frac{dV_C}{dt} = -\frac{1}{R_L C} V_C + \frac{I_L}{C} = F_1(V_C, I_L) \\ \frac{dV_{C1}}{dt} = \frac{1}{\tau_1} V_C - \frac{1}{\tau_2} V_{C1} - \frac{1}{\tau_3} V_{C2} - \frac{1}{\tau_4} V_{\text{REF}} = F_2(V_C, V_{C1}, V_{C2}, V_{\text{REF}}) \\ \frac{dV_{C2}}{dt} = \frac{1}{\tau_5} V_C - \frac{1}{\tau_5} V_{C2} - \frac{1}{\tau_5} V_{\text{REF}} = F_3\big(V_C, V_{C2}, V_{\text{REF}}\big) \\ \frac{dI_L}{dt} = \frac{-1}{L} V_C - \frac{V_{\text{IN}}}{LV_P} V_{C1} + \frac{V_{\text{IN}}}{LV_P} V_{\text{REF}} = F_4\big(V_C, V_{C1}, V_{\text{REF}}, V_{\text{IN}}\big) \end{cases} \tag{11.11}$$

with

$$\tau_1 = \frac{R_1 R_5}{(R_1 + R_5)} C_1, \quad \tau_2 = R_3 C_1, \tau_3 = R_5 C_1, \tau_4 = \frac{R_1 R_2 R_5}{R_1 R_2 + R_1 R_5 + R_2 R_5} C_1, \tau_5 = R_5 C_2. \tag{11.12}$$

This system is not completely linear due to the expression of $F_4\big(V_C, V_{C1}, V_{\text{REF}}, V_{\text{IN}}\big)$. Using the linearisation technique described in Para. 1.8, we obtain the following linearised system, represented by its Jacobian and associated M_p matrixes:

$$\begin{pmatrix} \dot{v}_C \\ \dot{v}_{C1} \\ \dot{v}_{C2} \\ \dot{i}_L \end{pmatrix} = \begin{pmatrix} -\frac{1}{R_L C} & 0 & 0 & \frac{1}{C} \\ \frac{1}{\tau_1} & -\frac{1}{\tau_2} & -\frac{1}{\tau_3} & 0 \\ \frac{1}{\tau_5} & 0 & -\frac{1}{\tau_5} & 0 \\ -\frac{1}{L} & -\frac{V_{\text{IN}}}{LV_P} & 0 & 0 \end{pmatrix} \begin{pmatrix} v_C \\ v_{C1} \\ v_{C2} \\ i_L \end{pmatrix} + \begin{pmatrix} 0 & 0 \\ -\frac{1}{\tau_4} & 0 \\ -\frac{1}{\tau_5} & 0 \\ M_{41} & M_{42} \end{pmatrix} \begin{pmatrix} v_{\text{REF}} \\ v_{\text{IN}} \end{pmatrix} \tag{11.13}$$

with

$$M_{41} = \left(\frac{\partial F_4\left(V_C, V_{C1}, V_{\text{REF}}, V_{\text{IN}}\right)}{\partial V_{\text{REF}}} \right)_{V_{Ce}, V_{C1e}, V_{\text{REF}}, V_{\text{IN}}} = \frac{V_{\text{IN}}}{LV_P}, \tag{11.14}$$

and

$$M_{42} = \left(\frac{\partial F_4(V_C, V_{C1}, V_{\text{REF}}, V_{\text{IN}})}{\partial V_{\text{IN}}} \right)_{V_{Ce}, V_{C1e}, V_{\text{REF}}, V_{\text{IN}}} = \frac{V_{\text{REF}} - V_{C1e}}{LV_P}. \tag{11.15}$$

Now, by applying the Laplace transform and eliminating $V_{C1}(s), V_{C2}(s), I_L(s)$ from the system (11.13), taking into account that $V_{\text{OUT}}(s) = V_C(s)$ and identifying each term with the following equation

$$\begin{aligned} V_{\text{OUT}}(s) = H_P(s)[H_{\text{REF}}(s)V_{\text{REF}}(s) - C_r(s)V_{\text{OUT}}(s)] \\ + H_{V\text{IN}}(s)V_{\text{IN}}(s) \end{aligned} \tag{11.16}$$

after some calculations, we obtain

$$\begin{aligned} H_P(s) &= \frac{V_{IN}}{V_A\left(1 + \frac{L}{R_L}s(1 + R_L C s)\right)}, \\ C_r(s) &= \frac{R_3}{R_1} \frac{(1 + (R_1 + R_5)C_2 s)}{(1 + R_3 C_1 s)(1 + R_5 C_2 s)}, \end{aligned} \tag{11.17}$$

and

$$H_{V\text{IN}}(s) = \frac{V_{\text{REF}} - V_{C1e}}{V_P\left(1 + \frac{L}{R_L}s(1 + R_L C s)\right)}. \tag{11.18}$$

Here also, it is unnecessary to calculate $H_{V\text{REF}}(s)$. The open-loop transfer function $H_{BO}(jf)$ can be written as follows:

$$H_{BO}(jf) = H_O \frac{\left(1 + \frac{jf}{f_z}\right)}{\left[1 + \frac{jf}{f_1}\left(1 + \frac{jf}{f_p}\right)\right]\left(1 + \frac{jf}{f_{p1}}\right)\left(1 + \frac{jf}{f_{p2}}\right)} \tag{11.19}$$

with

$$\begin{aligned} H_O &= \frac{R_3 V_{IN}}{R_1 V_P}, f_1 = \frac{R_L}{2\pi L}, f_p = \frac{1}{2\pi R_L C}, f_{p1} = \frac{1}{2\pi R_5 C_2}, \\ f_{p2} &= \frac{1}{2\pi R_3 C_1}, f_z = \frac{1}{2\pi (R_1 + R_5) C_2}. \end{aligned} \tag{11.20}$$

Now we can act on f_z, f_{p1} and f_{p2} by adjusting R_5, C_1 and C_2. The Proportional-Derivative compensator, consisting of f_z and f_{p1}, will be used to increase the phase margin φ_m, whereas f_{p2} will decrease the gain of $G_{BO}(jf)$ at the switching frequency f_s. We chose:

$$R_5 = 5\ \text{k}\Omega,\ C_1 = 0.5\ \text{pF},\ C_2 = 0.15\ \text{nF}, \tag{11.21}$$

given

$$f_{p1} \simeq 212\ \text{kHz}, f_{p2} \simeq 106\ \text{kHz}, f_z \simeq 6.8\ \text{kHz}. \tag{11.22}$$

The Fig. 11.3 shows the transfer functions of the PD compensator and the attenuator.

The open-loop transfer functions of the current-mode and voltage-mode controls are shown in Fig. 11.4.

They are simulated with $V_{\text{IN}} = 300$ V and $R_L = 100\ \Omega$. This figure shows that the phase margin is sufficient. Note that the phase margin remains satisfactory for $V_{\text{IN}} \in [300\ \text{V}, 400\ \text{V}]$ and $R_L \in [100\ \Omega, 200\ \Omega]$. If the designer has doubts about the validity of the Bode criteria, the stability of the converter can be confirmed by plotting the Nyquist diagram of G_{BO}.

As shown in Fig. 11.5, the Nyquist diagram of G_{BO} for the voltage-mode control confirms the stability of the system. Unlike in current mode, the transfer function $H_{\text{VIN}}(jf)$ is not equal to zero. The transfer function between V_{IN} and V_{OUT} is given by:

$$H_{VINBC}(jf) = \frac{V_{\text{OUT}}(jf)}{V_{\text{IN}}(jf)} = \frac{H_{\text{VIN}}(jf)}{1 + H_{BO}(jf)}. \tag{11.23}$$

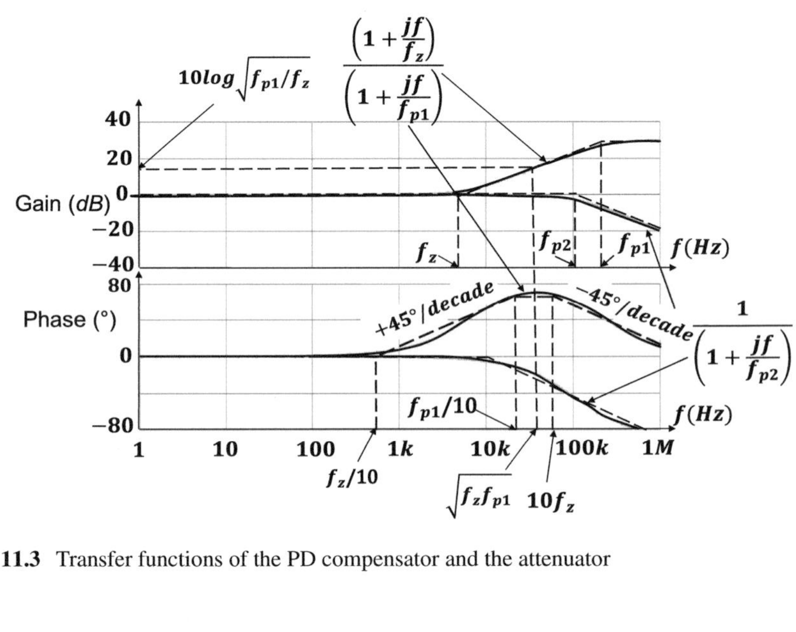

Fig. 11.3 Transfer functions of the PD compensator and the attenuator

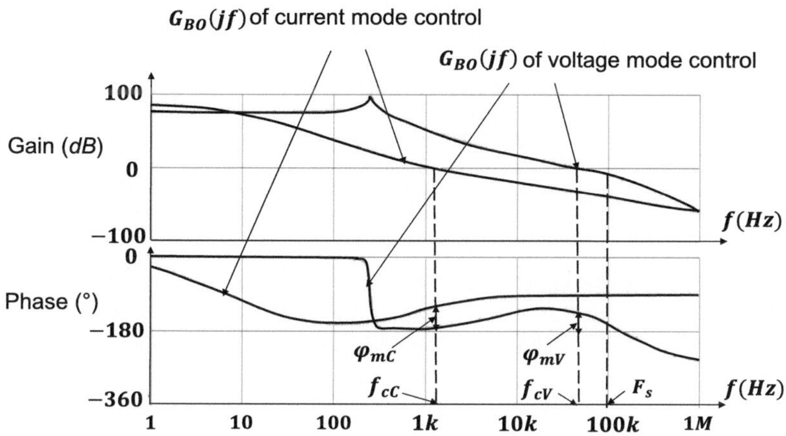

Fig. 11.4 Comparison of voltage-mode and current-mode controls

As this formula shows, the well-known method to reduce the effect of the input on the output at a given frequency is to increase the gain of $G_{BO}(jf)$ at that frequency. The attenuation of the ripple at 1 kHz is:

$$|H_{VINBC}(1\text{ kHz})| \simeq 9.4.10^{-5}. \tag{11.24}$$

So, the gain of $G_{BO}(1\text{ kHz})$ is sufficiently large to attenuate the disturbances within the limits specified. Figure 11.4 shows that the attenuation at the switching frequency F_S is not sufficient. To achieve the attenuation required by the specification, a more complex control circuit would be necessary (this will not be implemented).

Fig. 11.5 Nyquist diagram of $G_{BO}(jf)$ for the voltage-mode control

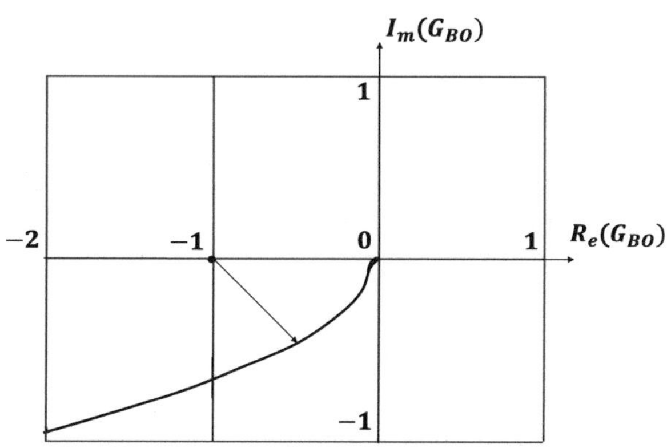

Through this example, we see some of the differences between the voltage-mode and the current-mode control. The following paragraph provides more information on the comparison between these two control modes.

Comparison of Voltage-Mode and Current-Mode Control

From an historical perspective, the voltage-mode control was developed before the current-mode control. It is a very intuitive control method, which use the error signal (the difference between the actual and desired output voltage) to adjust the circuit's average input voltage through the PWM duty cycle. Thus, it is a two-pole feedback loop. With current-mode control, the peak inductor current is directly controlled by the error signal, therefore, it is a one-pole feedback loop. Consequently, current-mode control provides a better response to input voltage and load variations.

As shown in the illustrative example, it is easier to design a compensator circuit that meets the full specifications with current-mode control. However, voltage-mode control is easier to implement than current-mode control because voltage sensing is simpler than current sensing, and under certain conditions slope compensation is required to eliminate oscillations. Current-mode control provides better current limiting thereby improving reliability under transient current conditions. In isolated converters, current-mode control facilitates the balancing of the hysteresis loop.

11.2 Controller Design for a Converter Operating in Both CCM and DCM

Very often, at low load, a converter designed primarily to operate in continuous mode can also operate in discontinuous mode. In this case, the designer generally needs verify that the compensator circuit optimised for continuous mode ensures the converter's stability at low load (Anonyme, s.d.; Yan, Shu, Sharkh, Wu, & Chen, n.d.).

Consider again the example of the buck converter with current-mode control and its optimised controller studied in Para. 10. As shown in Fig. 11.6, for $R_L = 1.7\ \text{k}\Omega$, the converter operates in critical mode, which is the boundary between continuous and discontinuous conduction modes. Using the differential equation given in Para. 9.3, we obtain the following nonlinear system corresponding to the averaged model of a DCM converter with current-mode control and its controller:

$$\begin{cases} \frac{dV_C}{dt} = -\frac{V_C}{R_L C} + \frac{V_{IN}-V_C}{2LC}\frac{V_{IN}}{V_C}\left(\frac{V_{ctr}}{R_s\left(\frac{V_{IN}-V_C}{L}\right)T_s+V_A}\right)^2 T_S = F(V_C, V_{C1}, V_{REF}) \\ \frac{dV_{C1}}{dt} = \frac{R_3}{R_1(R_3+R_4)C_1}V_C - \frac{1}{(R_3+R_4)C_1}V_{C1} - \frac{R_3(R_1+R_2)}{R_1R_2(R_3+R_4)C_1}V_{REF} \end{cases} \tag{11.25}$$

with

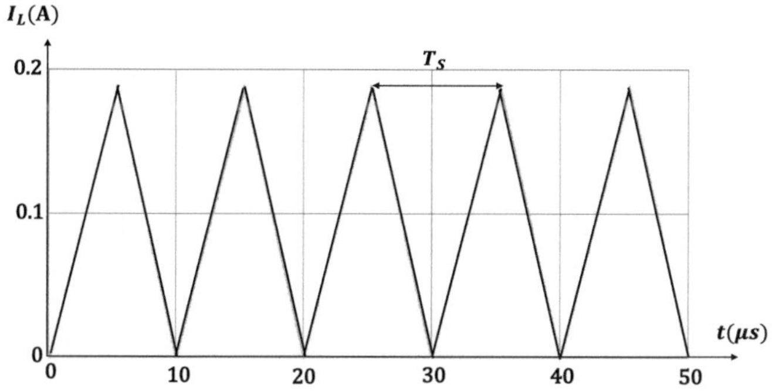

Fig. 11.6 Buck converter operating in critical mode for $R_L = 1.7$ kΩ

$$V_{ctr} = -\frac{R_3R_4}{R_1(R_3+R_4)}V_C - \frac{R_3}{R_3+R_4}V_{C1} + \left[1 + \frac{R_3R_4(R_1+R_2)}{R_1R_2(R_3+R_4)}\right]V_{\text{REF}}. \quad (11.26)$$

In this example, the converter and the compensator are already dimensioned. Thus, it is not necessary to calculate the open-loop transfer function. First, we draw the full averaged model from system (11.25). From this model we determine the equilibrium value of V_C. We find:

$$V_{Ce} \simeq 160.5 \text{ V}. \quad (11.27)$$

Then, we open the circuit between the load and the corrector circuit and add a DC voltage equal to V_{Ce} in series with an AC source at the input of the compensator. This circuit, shown in Fig. 11.7 allows simulation of the open-loop transfer function. The open-loop transfer is determined by

$$H_{BO}(jf) = H_P(jf)C_r(jf) = \frac{-V_{\text{OUTin}}(jf)}{V_{\text{OUT}}(jf)}. \quad (11.28)$$

In order to obtain the correct phase of $H_{BO}(jf)$, the simulation must be performed by considering the sign-in front of $V_{\text{OUTin}}(jf)$. The simulation results ($R_L = 1.7$ kΩ and $V_{\text{IN}} = 300$ V) shown in Fig. 11.8 confirm that the converter remains stable in discontinuous mode. Other simulations with lower values of R_L and different values of V_{IN} show the low impact of these parameters on stability. Note that the attenuation at the switching frequency also remains sufficient.

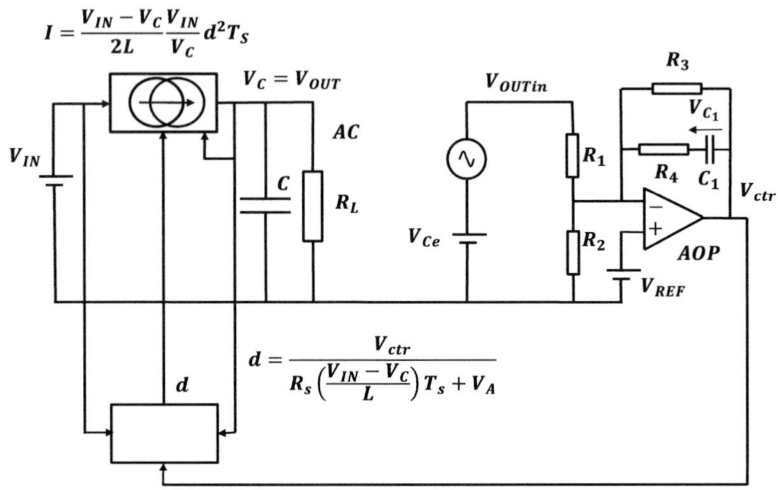

Fig. 11.7 Full averaged model used to simulate the open-loop transfer function

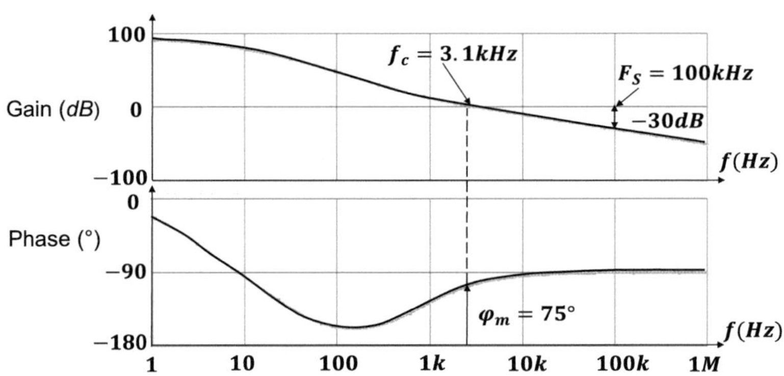

Fig. 11.8 Simulation results of the open-loop transfer function

11.3 Controller Design for a Controller with a RHPZ

When operating in continuous mode, some converters such as boost, buck-boost, and flyback are more difficult to control. This difficulty is linked to the fact that these converters initially respond to a circuit disturbance in the opposite direction to that commanded by the control signal (Goudarzian & Heidar, 2022). Consider a boost converter with its control circuit as shown in Fig. 11.9.

If a sudden increase in load occurs, the control signal V_{ctr} will increase the duty cycle of the switch. The result will be that the capacitor C will be disconnected from the inductor L for a longer time, and the output voltage V_{OUT} will initially decrease. This phenomenon is inherent to the converter and cannot be eliminated by the control

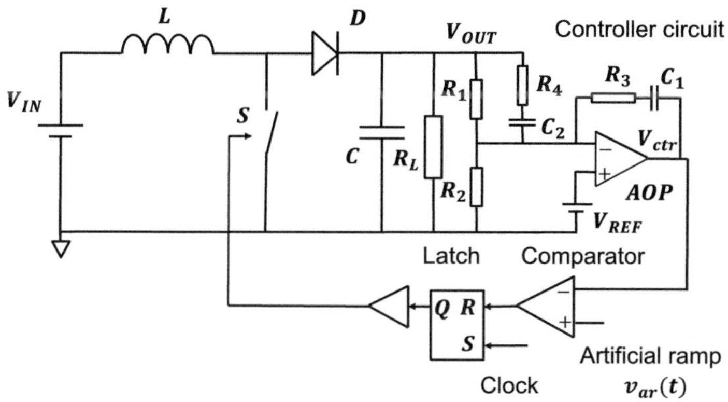

Fig. 11.9 Boost converter with its controller circuit

circuit. The values of the different parameters are:

$$
\begin{aligned}
V_{\mathrm{IN}} &= 10\ \mathrm{V},\, V_A = 1\ \mathrm{V}\ V_{\mathrm{REF}} = 10\ \mathrm{V},\, T_s = 10\ \mu\mathrm{s},\, L = 300\ \mu\mathrm{H}, \\
C &= 100\ \mu\mathrm{F},\, R_L = 100\ \Omega\ R_1 = 15\ \mathrm{k}\Omega,\, R_2 = 60\ \mathrm{k}\Omega, \\
R_3 &= 30\ \mathrm{k}\Omega,\, R_4 = 1\ \mathrm{k}\Omega,\, C_1 = 100\ \mathrm{nF},\, C_2 = 5\ \mathrm{nF}.
\end{aligned}
\tag{11.29}
$$

The simulation in Fig. 11.10 shows the output response $V_{\mathrm{OUT}}(t)$ when the load suddenly changes from $R_L = 200\ \Omega$ to $R_L = 100\ \Omega$. Using the usual linearisation techniques and assuming no parasitic components, we obtain the following transfer function of the boost converter operating in continuous mode:

$$
H_P(jf) = \frac{V_{\mathrm{OUT}}(jf)}{V_{ctr}(jf)} = \frac{V_{\mathrm{IN}}}{V_P} \frac{\left(1 - \frac{f}{f_{\mathrm{RHP}}} j\right)}{\left(\frac{L}{R_L} 2\pi f j - LC(2\pi f)^2 + \frac{V_{\mathrm{IN}}^2}{V_{\mathrm{OUT}e}^2}\right)} \tag{11.30}
$$

with

$$
f_{\mathrm{RHP}} = \frac{R_L V_{\mathrm{IN}}^2}{2\pi L V_{\mathrm{OUT}e}^2}. \tag{11.31}
$$

The previously described effect results in the presence of a Right Half Plane Zero (corresponding to the frequency f_{RHP}) in the transfer function of the boost converter. Due to the RHPZ, the phase of $H_P(jf)$ tends to $-270°$ (see Fig. 11.11). The lower the value of f_{RHP}, the more difficult the controller design will be. As shown by Eq. (11.31), the worst-case situation occurs for the lower values of R_L and V_{IN} and the higher values of L. In this example, the static gain of the controller tends to infinity and $V_{\mathrm{OUT}e}$ tends to:

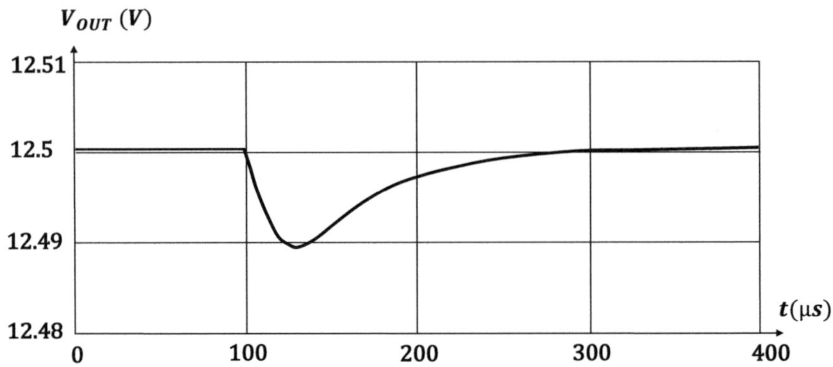

Fig. 11.10 Output response to a load variation from $R_L = 200\ \Omega$ to $R_L = 100\ \Omega$

$$V_{\text{OUT}e} \rightarrow \frac{R_1 + R_2}{R_2} V_{\text{REF}} = 12.5\ \text{V}. \tag{11.32}$$

Thus, we obtain:

$$f_{\text{RHP}} \simeq 34\ \text{kHz}. \tag{11.33}$$

The transfer function of $H_P(jf)$ is plotted in Fig. 11.11. The transfer function of the corrector is given by:

$$C_r(jf) = \frac{(1 + 2\pi jfR_3C_1)(1 + 2\pi jf(R_1 + R_4)C_2)}{2\pi jfR_1C_1(1 + 2\pi jfR_4C_2)}. \tag{11.34}$$

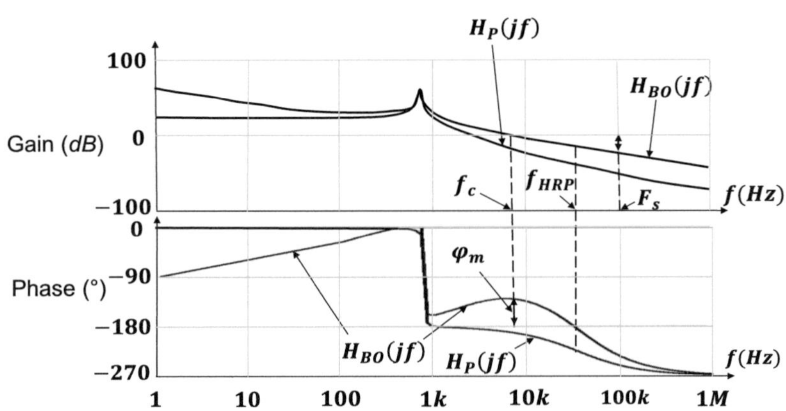

Fig. 11.11 Transfer function of $H_P(jf)$ and $H_{BO}(jf)$

Generally, the crossover frequency f_C is chosen to be much smaller than f_{RHP}. In this example

$$f_C \simeq 8.7 \text{ kHz}. \tag{11.35}$$

The purpose of C_1 is to increase the gain at low frequencies in order to reduce the steady-state error. This relatively high value (100 nF) makes the capacitor C_1 behave like a short circuit at higher frequencies. R_3, R_4 and C_2 are determined so as to increase the phase margin φ_m. With the given values, we obtain:

$$\varphi_m \simeq 47^\circ. \tag{11.36}$$

The open-loop transfer function $H_{B0}(jf)$ is shown in Fig. 11.11.

11.4 Controller Design of a Power Factor Correction Circuit

The Power Factor Correction circuit, using a boost converter placed between the bridge and the storage capacitor C, as shown in Fig. 11.12, is a widely used solution today (Infineon Technologies, 2016).

The primary advantage of this circuit is that it transforms the load seen from the AC line into a purely resistive load, ensuring compliance with standards such as IEC 61000-3-2. Another benefit of this solution is that it provides a constant output voltage V_{OUT} regardless of the magnitude of the AC line V_{mains}. In this section, we

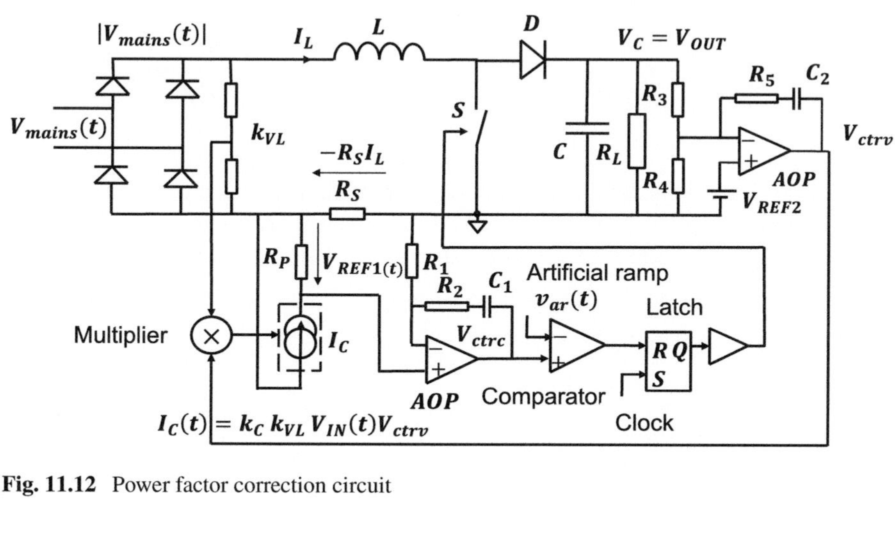

Fig. 11.12 Power factor correction circuit

consider a PFC operating in continuous conduction mode and at a constant switching frequency F_s, designed with the following parameters:

$$\begin{aligned} L &= 500\ \mu\text{H},\ C = 500\ \mu\text{F},\ R_L = 400\ \Omega,\ k_{VL} = 0.01, \\ k_C &= 0.001\ \text{AV}^{-2},\ R_S = 0.01\ \Omega \\ R_P &= 10\ \Omega,\ F_s = 100\ \text{kHz},\ V_A = 1\ \text{V}. \end{aligned} \tag{11.37}$$

In our example, the AC line voltage is defined by:

$$V_{\text{mains}}(t) = V_M \sin(2\pi f_{\text{mains}} t) \text{ with } V_M = 200\ \text{V and } f_{\text{mains}} = 50\ \text{Hz}. \tag{11.38}$$

The targeted output voltage is:

$$V_{\text{OUT}tar} = 400\ \text{V}. \tag{11.39}$$

PFC control is performed using current regulation, with a fast response time, combined with voltage regulation that operates with a much slower response time. The current regulation shapes the inductor current waveform into a sinusoidal waveform that matches the shape of the input voltage $V_{\text{mains}}(t)$.

Due to the faster response time of the current regulation, the control signal V_{ctrv} and the voltage V_{OUT} can be considered constant during a line period T_{mains}. The averaged circuit of the current regulation is shown in Fig. 11.13. If the current regulation is well designed, we obtain the following expression for the averaged value of I_L :

$$I_L(t) = \frac{R_P}{R_S} k_{VL} k_C V_M |\sin(2\pi f_{\text{mains}} t)| V_{ctrv}. \tag{11.40}$$

$I_L(t)$ is a function with two time periods f_{mains}. The above expression is valid for

$$t \in \left[0, \frac{T_{\text{mains}}}{2}\right]. \tag{11.41}$$

To achieve our objective, the response time of the current regulation must be faster than the variation of the line voltage, so the open-loop transfer function can be determined by considering $V_{\text{mains}}(t)$ and the reference signal $V_{\text{REF1}}(t)$ as constant.

Using the usual techniques, we can deduce the open-loop transfer function (with I_L as input and output parameter) of the current regulation when:

$$V_{\text{OUT}} \simeq V_{\text{OUT}tar} = 400\ \text{V}. \tag{11.42}$$

We obtain:

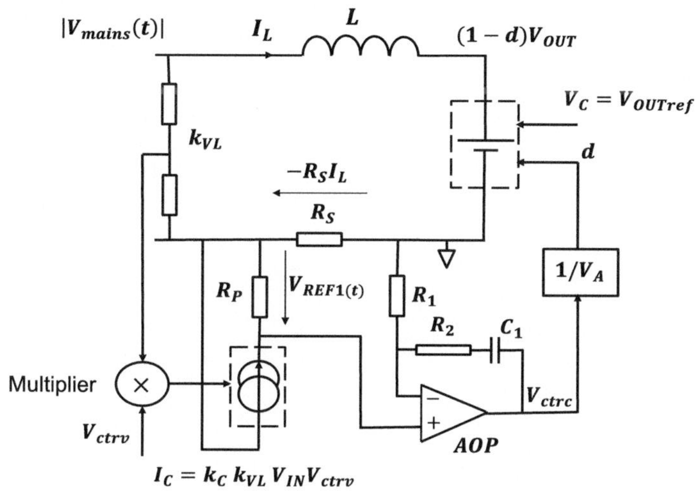

Fig. 11.13 Averaged circuit of the current regulation loop

$$H_{BOC}(jf) = \frac{I_{\text{LOUT}}(jf)}{I_{\text{LIN}}(jf)} = \frac{V_{\text{OUT}tar} R_S \left[1 + (R_1 + R_2) j2\pi f C_1\right]}{V_A L R_1 C_1 (j2\pi)^2}. \tag{11.43}$$

We choose:

$$R_1 = 3\ \text{k}\Omega, R_2 = 10\ \text{k}\Omega, \tag{11.44}$$

The corresponding transfer function is presented in Fig. 11.14. The gain:

$$G_{BOC}(2f_L) \simeq 70.6\ \text{dB} \tag{11.45}$$

is sufficiently high at the frequency $2f_{\text{mains}}$ to ensure that $I_L(t)$ follows the waveform of $V_{\text{mains}}(t)$, as shown in Fig. 11.8. Moreover, we have a good stability:

$$\varphi_m \simeq 50° \tag{11.46}$$

and a good attenuation at the switching frequency:

$$G_{BOc}(f_s) \simeq -25.2\ \text{dB}. \tag{11.47}$$

The averaged circuit of the complete PFC system is shown in Fig. 11.15. In order to obtain the targeted output voltage: $V_{\text{OUT}tar} = 400$ V with $V_{\text{REF2}} = 10$ V, we take:

$$R_3 = 390\ \text{k}\Omega, \text{and}\ R_4 = 10\ \text{k}\Omega. \tag{11.48}$$

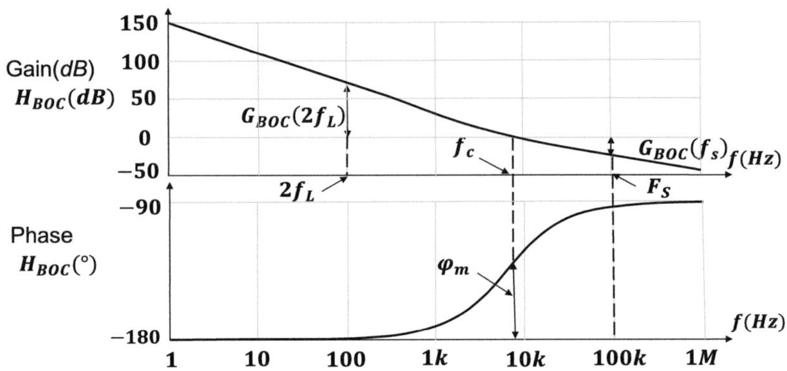

Fig. 11.14 Transfer function of $H_{BOC}(jf)$

The next step consists of defining the open-loop transfer function of the voltage regulation. To this end, we begin by calculating the average value of the current generator:

$$(1 - d(t))I_L(t) \tag{11.49}$$

over half a period of $V_{\text{mains}}(t)$ $\left(\frac{T_{\text{mains}}}{2}\right)$ in steady state mode. Note that this is also the average value of the current in the diode I_{DAV}. We can write:

$$I_{DAV} = \frac{2}{T_L} \int_0^{\frac{T_L}{2}} (1 - d(t))I_L(t)dt. \tag{11.50}$$

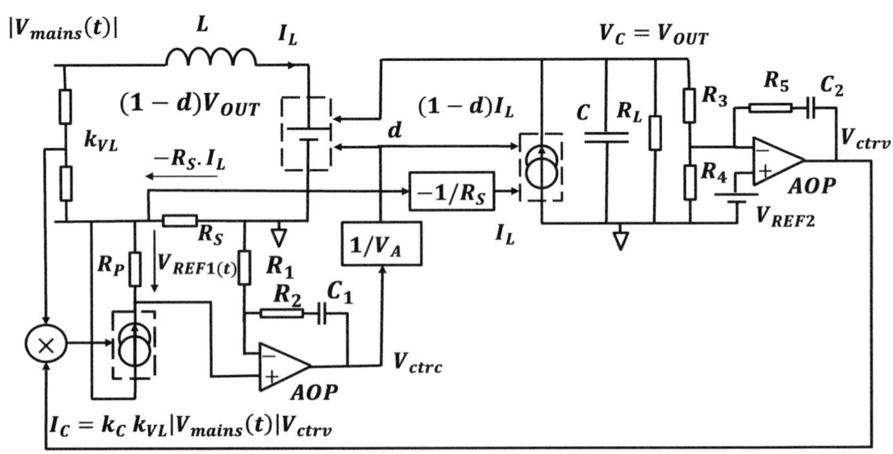

Fig. 11.15 Averaged circuit of the complete PFC system

The expression of $I_L(t)$ is given by Eq. (11.40). In a PFC application, the expression $d(t)$ is not exactly the same as that of a boost converter with a constant input voltage V_{IN}, which is given by:

$$d(t) = 1 - \frac{V_{\mathrm{IN}}}{V_{\mathrm{OUT}}}. \tag{11.51}$$

As shown in the Fig. 11.16, for $t \in \left[0, \frac{T_{\mathrm{mains}}}{4}\right]$, the current through the inductor increases, and the duty cycle becomes higher than $d(t)$, from a value of Δd.

On the contrary, for $t \in \left[\frac{T_{\mathrm{mains}}}{4}, \frac{T_{\mathrm{mains}}}{2}\right]$, the current through the inductor decreases and the duty cycle becomes lower than $d(t)$, from a value of Δd. What is important to note is the symmetry of $(1 - d(t))I_L(t)$ with respect to $\frac{T_{\mathrm{mains}}}{4}$. Thanks to this symmetry, Δd can be eliminated from the calculation, and we obtain:

$$I_{\mathrm{DAV}} = \frac{2}{T_L} \int_0^{\frac{T_{\mathrm{mains}}}{2}} \left(\frac{V_M \sin(2\pi f_L t)}{V_{\mathrm{OUT}}} \right) \frac{R_P}{R_S} k_{VL} k_C V_M \sin(2\pi f_{\mathrm{mains}} t) V_{ctrv} dt. \tag{11.52}$$

V_{OUT} and V_{ctrv} can be considered constant over $\frac{T_L}{2}$. The calculation gives:

$$I_{\mathrm{DAV}} = \frac{R_P k_{VL} k_C V_M^2}{2 R_S V_{\mathrm{OUT}}} V_{ctrv} = K \frac{V_{ctrv}}{V_{\mathrm{OUT}}}. \tag{11.53}$$

The equivalent circuit for small-signal analysis becomes simple. It consists of the controlled current generator $K \frac{V_{ctrv}}{V_{\mathrm{OUT}}}$ replacing the current generator $(1 - d)I_L$ in the schematic of Fig. 11.15. The following open-loop transfer function (with V_{OUT} as

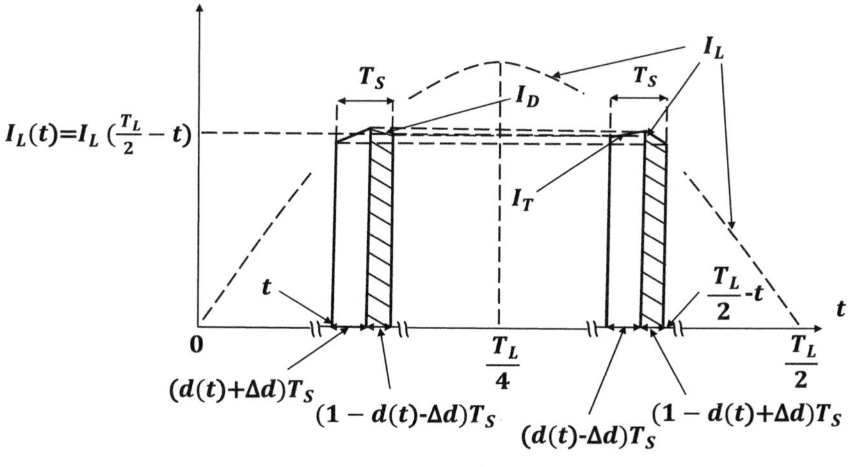

Fig. 11.16 Symmetry of $(1 - d(t))I_L(t)$ with respect to $\frac{T_{\mathrm{mains}}}{4}$

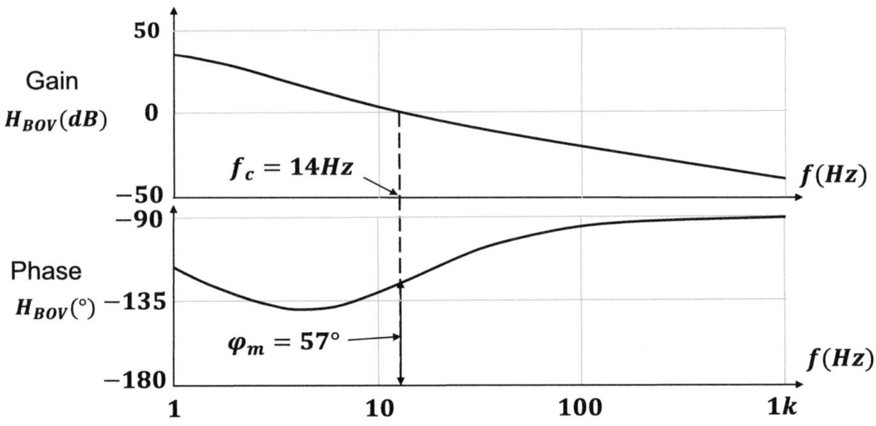

Fig. 11.17 Transfer function of $H_{\mathrm{BOV}}(jf)$

input and output parameter) can be easily deduced from this circuit:

$$H_{\mathrm{BOV}}(jf) = \frac{KR_L(1 + j2\pi f R_5 C_2)}{V_{\mathrm{OUT}ref}(2 + j2\pi f R_L C)j2\pi f R_3 C_2} \tag{11.54}$$

Simulation of $H_{\mathrm{BOV}}(jf)$ performed with $R_5 = 30$ kΩ and $C_2 = 500$ nF (see Fig. 11.17) shows satisfactory voltage regulation. For a PFC application, f_{RHP} is given by:

$$f_{\mathrm{RHP}} = \frac{R_{AC}}{2\pi L} \tag{11.55}$$

where R_{AC} is the equivalent input resistor $\left(\frac{V_{\mathrm{mains}}}{I_{\mathrm{mains}}}\right)$. In our example, we obtain $R_{AC} = 50\ \Omega$ and $f_{\mathrm{RHP}} = 15.9$ kHz. f_{RHP} being much lower than the crossover frequency of H_{BOV} ($f_c = 14$ Hz) there is no issue with the RHPZ. In a PFC, this is practically always the case.

Figure 11.18 displays $V_{\mathrm{OUT}}(t)$, $|V_{\mathrm{mains}}(t)|$, $I_L(t)$ and $d(t)$, simulated with the circuit of the Fig. 11.15 when steady state is reached. We can observe a small 100 Hz ripple on $V_{\mathrm{OUT}}(t)$.

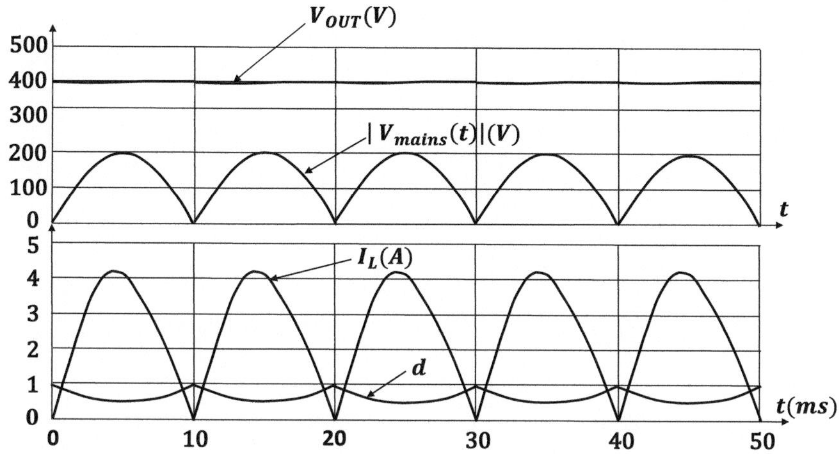

Fig. 11.18 Steady state waveforms of $V_{\mathrm{OUT}}(t)$, $|V_{\mathrm{mains}}(t)|$, $I_L(t)$ and $d(t)$

11.5 Controller Design for a Charge Pump Converter

In this example, we consider the two-stage charge pump associated with the voltage regulator shown in Fig. 11.19 (Shin et al. 2005).

The regulation is carried out using the operational transconductance amplifier (OTA) and two transistors. The dynamic behaviours of the three components is neglected, so that we can write:

$$I_{ctr} = g_m(V_{\mathrm{REF}} - \frac{R_2}{R_1 + R_2} V_{\mathrm{OUT}}), \text{ and } I_C = \beta g_T V_{C1}. \tag{11.56}$$

The voltage generator $V_{\mathrm{SWP}}(t)$ is defined by its magnitude V_{SWP} and duty cycle d, generated by a buck or a boost converter. The diodes D_1 and D_2 are modelled

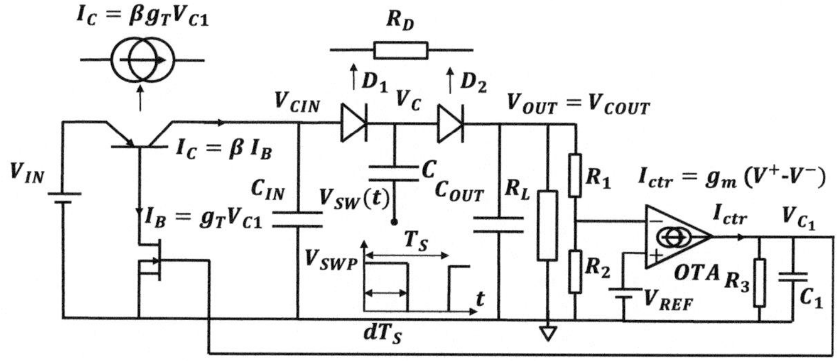

Fig. 11.19 Two-stage charge pump converter with voltage regulation

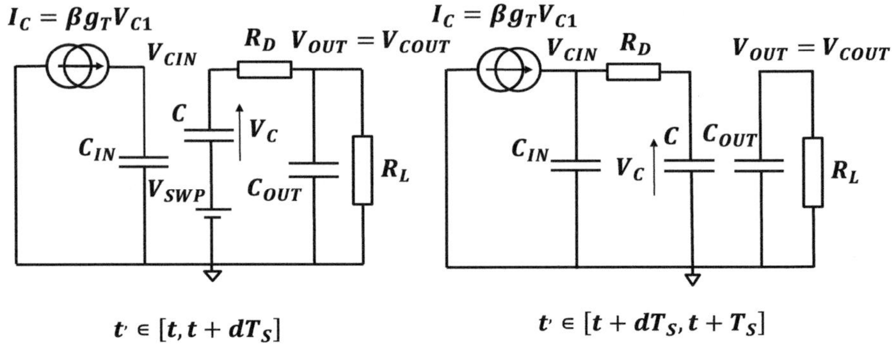

Fig. 11.20 Equivalent circuits for $V_{\mathrm{SW}}(t\prime) = V_{\mathrm{SWP}}$ and $V_{\mathrm{SW}}(t') = 0$

in forward mode by a resistor R_D (in fact, a resistor is connected in series with each diode), while the other parasitic components are neglected. The targeted output voltage is $V_{\mathrm{OUT}tar} = 15$ V. The numerical values of the parameters are:

$$
\begin{aligned}
&V_{\mathrm{IN}} = 10\ \mathrm{V},\ V_{\mathrm{REF}} = 5\ \mathrm{V},\ R_L = 100\ \Omega,\ R_1 = 600\ \mathrm{k}\Omega,\ R_2 = 300\ \mathrm{k}\Omega\\
&R_3 = 1\ \mathrm{k}\Omega,\ R_D = 5\ \Omega,\ C_{\mathrm{IN}} = 2\ \mu\mathrm{F},\ C = 2\ \mu\mathrm{F},\ C_{\mathrm{OUT}} = 5\ \mu\mathrm{F},\\
&C_1 = 5\ \mu\mathrm{F},\ V_{\mathrm{SWP}} = 10\ \mathrm{V},\ d = 0.5,\ \beta g_T = 8\ \mathrm{mS},\ g_m = 80\ \mathrm{mS},\\
&F_S = 100\ \mathrm{kHz}(T_S = 10\ \mu\mathrm{s}).
\end{aligned}
\tag{11.57}
$$

The averaged model of the charge pump converter is defined by the two equivalent circuits shown in Fig. 11.20, corresponding to:

$$
V_{\mathrm{SW}}(t) = V_{\mathrm{SWP}}(D_1\mathrm{off},\ D_2\mathrm{on})\mathrm{and}\ V_{SW}(t) = 0(D_1\mathrm{on},\ D_2\mathrm{off}) \tag{11.58}
$$

We obtain:

$$
\text{for } t' \in [t, t + dT_S],\ V_{\mathrm{SW}}(t') = V_{\mathrm{SWP}}
$$

$$
\begin{cases}
I_C = C_{\mathrm{IN}}\frac{dV_{\mathrm{CIN}}}{dt}\\
V_C + V_{\mathrm{SWP}} = -R_D C\frac{dV_C}{dt} + V_{\mathrm{COUT}}\\
C_{\mathrm{OUT}}\frac{dV_{\mathrm{COUT}}}{dt} + \frac{V_{\mathrm{COUT}}}{R_L} = \frac{V_C + V_{\mathrm{SWP}} - V_{\mathrm{COUT}}}{R_D}
\end{cases}. \tag{11.59}
$$

$$
\text{for } t' \in [t + dT_S, t + T_S],\ V_{\mathrm{SW}}(t') = 0
$$

$$
\begin{cases}
I_C = C_{\mathrm{IN}}\frac{dV_{\mathrm{CIN}}}{dt} + \frac{V_{\mathrm{CIN}} - V_C}{R_D}\\
V_{\mathrm{CIN}} = R_D C\frac{dV_C}{dt} + V_C\\
C_{\mathrm{OUT}}\frac{dV_{\mathrm{COUT}}}{dt} + \frac{V_{\mathrm{COUT}}}{R_L} = 0
\end{cases}. \tag{11.60}
$$

By averaging these systems over a switching period T_S and including the differential equation due to the corrector circuit, we obtain the averaged model of the complete system:

$$\begin{cases} \frac{dV_{CIN}}{dt} = -\frac{(1-d)}{R_D C_{IN}} V_{CIN} + \frac{(1-d)}{R_D C_{IN}} V_C + \frac{\beta g_T}{C_{IN}} V_{C1} \\ \frac{dV_C}{dt} = \frac{(1-d)}{R_D C} V_{CIN} - \frac{1}{R_D C} V_C + \frac{d}{R_D C} V_{OUT} - \frac{dV_{SWP}}{R_D C} \\ \frac{dV_{OUT}}{dt} = \frac{d}{R_D C_{OUT}} V_C - \left(\frac{d}{R_D C_{OUT}} + \frac{1}{R_L C_{OUT}}\right) V_{OUT} + \frac{dV_{SWP}}{R_D C_{OUT}} \\ \frac{dV_{C1}}{dt} = -\frac{g_m R_2}{(R_1+R_2)C_1} V_{OUT} - \frac{1}{R_3 C_1} V_{C1} + \frac{g_m V_{REF}}{C_1} \end{cases} \quad (11.61)$$

Since the charge pump converter contains four capacitors, the complete system is naturally defined by four state parameters ($V_{CIN}, V_C, V_{COUT}, V_{C1}$) and four differential equations. Solving the corresponding system S_e yields:

$$V_{OUTe} = 14.3 \text{ V}. \quad (11.62)$$

Using the standard techniques, we obtain the following transfer functions of the corrector:

$$C_r(s) = \frac{V_{C1}(s)}{V_{OUT}(s)} = \frac{g_m R_2 R_3}{(R_1 + R_2)(1 + R_3 C_1 s)}. \quad (11.63)$$

The transfer function of the charge pump converter $H_P(s) = \frac{V_{OUT}(s)}{V_{C1}(s)}$ is given by:

$$H_P(s) = \frac{d(1-d)\beta g_T R_D R_L}{(R_L R_D C_{OUT} s + dR_L + R_D)\left[(1-d+R_D C_{IN} s)(1+R_D C s) - (1-d)^2\right] - d^2 R_L (1-d+R_D C_{IN})}. \quad (11.64)$$

R_3 and C_1 were chosen to achieve a good regulation according to the usual criterion (see the open-loop transfer function in Fig. 11.21).

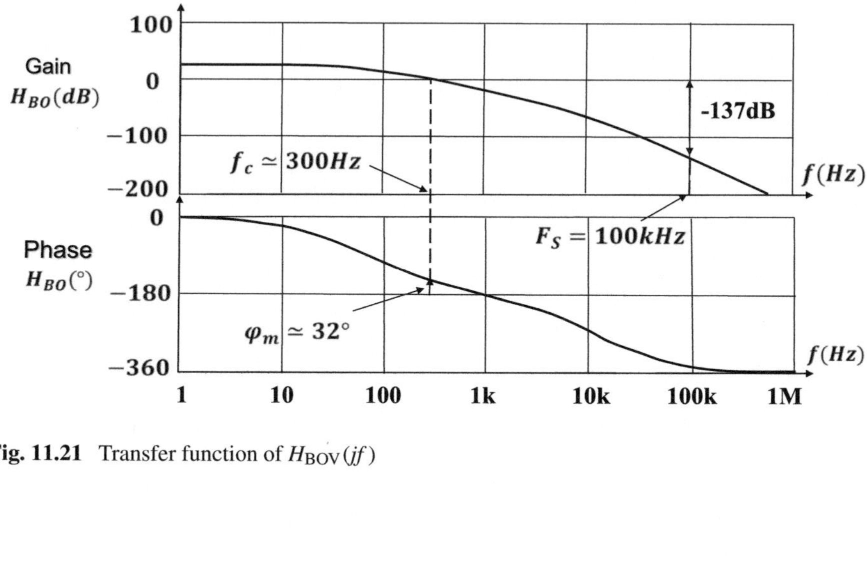

Fig. 11.21 Transfer function of $H_{BOV}(jf)$

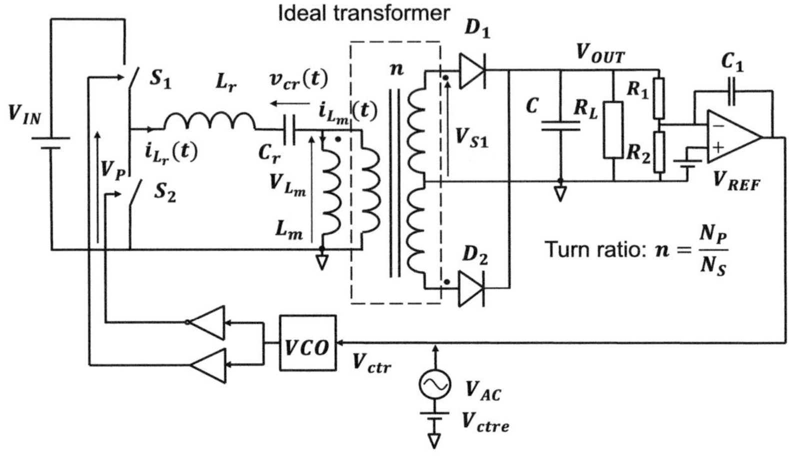

Fig. 11.22 LLC converter with its controller

11.6 Controller Design of an *LLC* Converter

In Fig. 11.22, we consider an *LLC* converter associated with its controller (Texas Instruments, 2010). In this example the input voltage $V_{\rm IN}$ and the load R_L are constant. The numerical values of the parameters are:

$$V_{\rm IN} = 400\ \text{V},\ V_{{\rm OUT}tar} = 20\ \text{V},\ V_{\rm REF} = 10\ \text{V},\ R_L = 2\ \Omega,\ L_r = 50\ \mu\text{H},$$
$$L_m = 250\ \mu\text{H}, C_r = 25\ \text{nF},\ C = 200\ \mu\text{F},\ C_1 = 400\ \text{nF},\ R_1 = 50\ \text{k}\Omega,$$
$$R_2 = 50\ \text{k}\Omega,\ n = 10. \tag{11.65}$$

Using the first harmonic approximation reviewed in Para. 9.6, we can deduce the following static voltage gain $A(F_n)$ of an *LLC* converter:

$$A(F_n) = \frac{V_{\rm OUT}}{V_{\rm IN}} = \frac{L_n F_n^2}{2n\sqrt{\left[(1+L_n)F_n^2 - 1\right]^2 + \left[\left(F_n^2 - 1\right)F_n L_n Q_e\right]^2}} \tag{11.66}$$

with

$$R_e = 8\frac{n^2}{\pi^2}R_L,\ Q_e = \frac{\sqrt{L_r/C_r}}{R_e}, f_0 = \frac{1}{2\pi\sqrt{L_r C_r}},\ F_n = \frac{f_S}{f_0},\ L_n = \frac{L_m}{L_r}. \tag{11.67}$$

The switching frequency f_S is controlled by V_{ctr} via the voltage controlled oscillator (*VCO*). The relationship given F_S in relation to V_{ctr} is:

$$F_S(t) = f_0 + 250000(1 - V_{ctr}(t)). \tag{11.68}$$

The various components were selected so that the converter operates in steady state-mode in the vicinity of f_0. Thus, in this mode, we obtain:

$$V_{\mathrm{OUT}} \simeq V_{\mathrm{OUT}e} \simeq 20\ \mathrm{V},\ V_{ctr} \simeq V_{ctre} \simeq 1\ \mathrm{V}\ F_S = f_0 \simeq 142\ \mathrm{kHz}. \tag{11.69}$$

The curve of Fig. 11.23 shows the dependency of V_{OUT} in relation to V_{ctr}, calculated using formulas (11.66) and (11.68) and simulated with the circuit of Fig. 11.22.

These curves show the good accuracy of the first-harmonic model near the operating mode. To design the compensator circuit, we need to know the small-signal model of the converter. Unfortunately, the state-space averaging model cannot be applied for an *LLC* resonant converter because the switching frequency is close to the natural frequency of the resonant tank network. No simple formulas exist to obtain the frequency transfer function of a *LLC* converter. However, it is possible to simulate the output voltage response V_{OUT} to a sinusoidal control signal V_{ctr} at a given frequency. To do this, we open the circuit at the output of the compensator and inject a DC voltage equal to V_{ctre} in series with a small AC sinusoidal source (see Fig. 11.22). Figure 11.24 shows the simulation results when the frequency of the AC source is equal to 10 kHz and its amplitude is 1 mV.

We can perform the same procedure for different frequencies and obtain the transfer function of the LLC converter. For all the reasons mentioned previously, designing the compensator of an *LLC* converter is more difficult than designing that of a conventional converter. It is generally useful to determine the transfer functions and to adjust the components to meet the regulation specification. In our illustrative example, the input voltage and the load are constant and we only require the system to be stable. The simple compensator circuit of Fig. 11.22 is sufficient to achieve

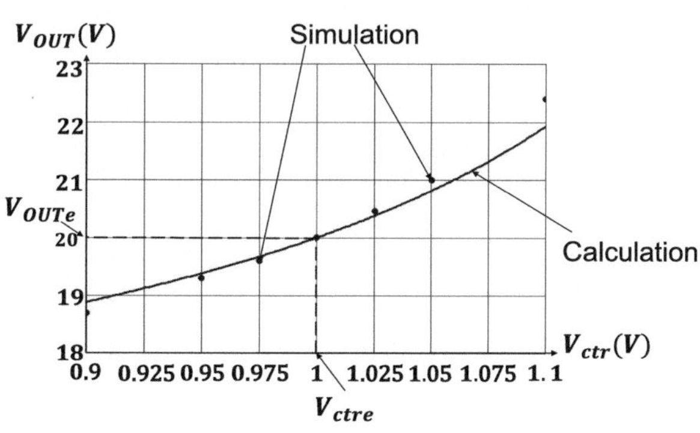

Fig. 11.23 V_{OUT} versus V_{ctr} obtained by calculation and simulation

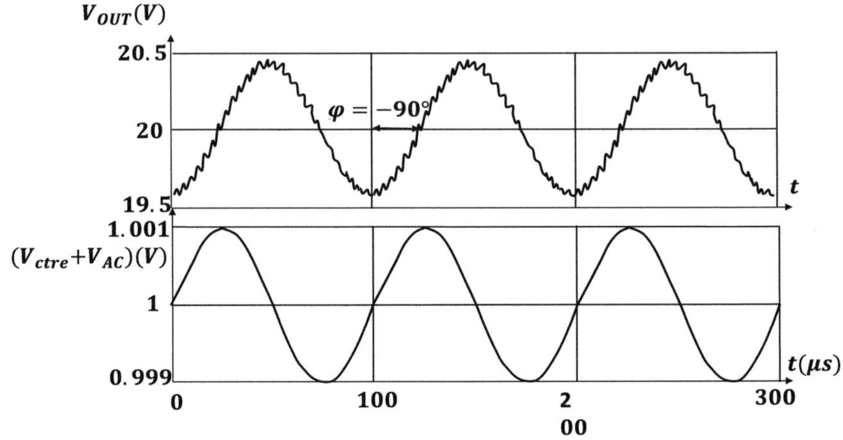

Fig. 11.24 Temporal response of V_{OUT} at 10 kHz

stability. Figure 11.25 presents the steady-state waveforms of V_{OUT}, $I_{D1} + I_{D2}$, V_{ctr} obtained using the compensator circuit shown in Fig. 11.22.

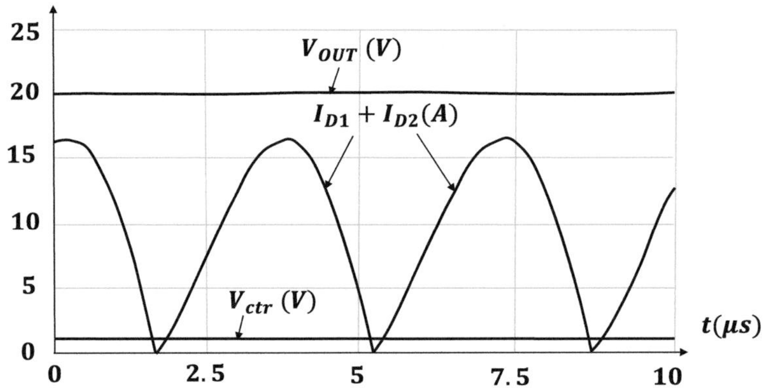

Fig. 11.25 Steady-state waveforms of V_{OUT}, $I_{D1} + I_{D2}$, V_{ctr}

Chapter 12
Examples of Stabilities Studies in the Power Converter Field

In this chapter, several examples of stabilities studies in the power field will be examined.

12.1 Stability of a First-Order Thermal System

12.1.1 System Description

The first example to be studied is a thermal system consisting of a Schottky diode mounted individually on its own heatsink (see Fig. 12.1).

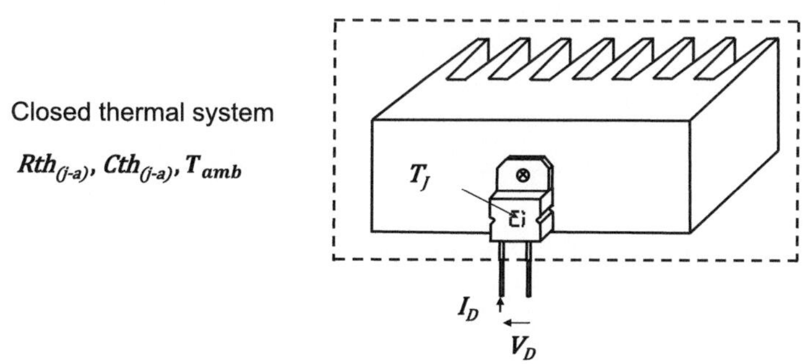

Fig. 12.1 Thermal system of a Schottky diode with its heatsink

B. Rivet, *Theories of Nonlinear Systems for Power Converters*,
https://doi.org/10.1007/978-3-032-25062-9_12

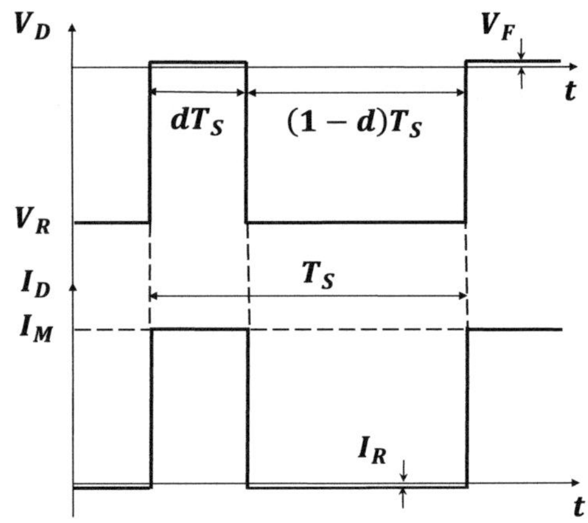

Fig. 12.2 Ideal voltage and current waveforms for a Schottky diode

The Schottky diode is diode is used as a rectifier diode in a switching power supply. The ideal voltage and current waveforms (V_D, I_D) during one switching period T_S are shown in Fig. 12.2. With a pure Schottky diode, the switching power losses are negligible. We need only to consider the conduction and reverse power losses, which are respectively generated by the forward voltage drop V_F and the leakage current I_R. These two static parameters depend on the junction temperature of the diode T_J according to the following formulas:

$$\begin{aligned} V_F(I_M, T_J) &= V_F(I_M, T_{ref})(1 + \alpha_{VF}(T_J - T_{ref})), \\ I_R(V_R, T_J) &= I_R(V_R, T_{ref})e^{c(T_J - T_{rev})}. \end{aligned} \tag{12.1}$$

I_M and V_R are defined in Fig. 12.2, α_{VF} and c are temperature coefficients, and T_{ref} is the reference junction temperature. The leakage current increases very rapidly with temperature following an exponential law, while the forward voltage drop decreases slowly following a linear law.

The associated conduction power losses (P_{COND}) are given by:

$$P_{COND}(T_J) = V_F(I_M, T_{ref})(1 + \alpha_{VF}(T_J - T_{ref}))dI_M, \tag{12.2}$$

and the reverse power losses (P_{REV}) by:

$$P_{REV}(T_J) = I_R(V_R, T_{ref})e^{c(T_J - T_{ref})}(1 - d)V_R. \tag{12.3}$$

d is the duty cycle (see Fig. 12.2). The switching period T_S is very short compared to the thermal time constant. So, it can be considered that on the time scale of the thermal system, P_{COND} and P_{REV} occur simultaneously. Of course, this is not true at the time scale of the switching period, where P_{COND} and P_{REV} are respectively generated during dT_S and $(1 - d)T_S$. Consequently, at a given time (at the thermal system's time scale), the total power losses P_T is the sum of P_{COND} and P_{REV}:

$$P_T(T_J) = P_{COND}(T_J) + P_{REV}(T_J). \tag{12.4}$$

The thermal model is defined by the thermal resistor $Rth_{(j-a)}$ and the thermal capacitor $Cth_{(j-a)}$ between the junction (T_J) and the ambient temperature (T_{amb}). The basic thermal law is given by:

$$P_T(T_J) = \frac{T_J - T_{amb}}{Rth_{(j-a)}} + Cth_{(j-a)} \frac{dT_J}{dt}. \tag{12.5}$$

The expression of the system S becomes:

$$\frac{dT_J}{dt} = \frac{1}{Cth_{(j-a)}} \left(P_T(T_J) - \frac{T_J - T_{amb}}{Rth_{(j-a)}} \right). \tag{12.6}$$

This is a first-order system with T_J as the state parameter ($n = 1$) and 11 system parameters ($p = 11$): $I_M, V_R, V_F(I_M, T_{ref}), d, I_R(V_R, T_{rev}), \alpha_{VF}, c, Rth_{(j-a)}, Cth_{(j-a)}, T_{amb}$ and T_{ref}.

12.1.2 Determination of the Equilibrium Points

The equilibrium points (T_{Jei}) are calculated by solving the corresponding system S_e. We obtain:

$$P_T(T_{Jei}) = \frac{T_{Jei} - T_{amb}}{Rth_{(j-a)}}. \tag{12.7}$$

There is no analytical solution to this equation. Take the following numerical values:

$$I_M = 50\,\text{A},\ V_R = 30,\ d = 0.3,\ V_F(I_M, T_{REF}) = 0.5\,\text{V},\ I_R(V_R, T_{REF}) = 20\,\text{mA},$$
$$\alpha_{VF} = -1.5\,\text{mV}\,°\text{C}^{-1},\ c = 0.05\,°\text{C}^{-1},\ Rth_{(j-a)} = 8°\,\text{CW}^{-1},\ Cth_{(j-a)} = 250\,\text{sW}\,°\text{C}^{-1},$$
$$T_{amb} = 40\,°\text{C},\ T_{ref} = 100\,°\text{C}. \tag{12.8}$$

Figure 12.3 shows the variation of P_T in relation to T_J. For low values of T_J, P_T decreases slowly due to the effect of the negative forward voltage temperature coefficient (α_{VF}). For higher T_J, the strong dependency of the leakage current on T_J becomes dominant. The crossing points of the two curves (see Fig. 12.3) indicate that there are two equilibrium points:

$$T_{Je1} = 103.7\,°\text{C},\ T_{Je2} = 159.1\,°\text{C}. \tag{12.9}$$

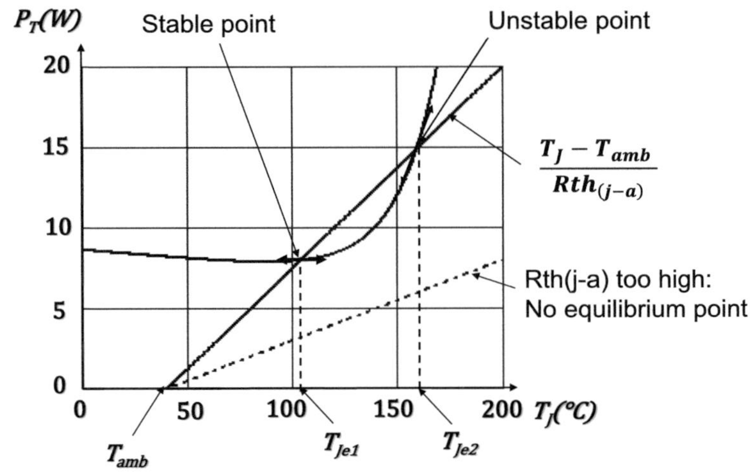

Fig. 12.3 Static thermal model of the system

Under different application conditions (for example, higher $Rth_{(j-a)}$), there may be only one or even no equilibrium point. Obviously, without an equilibrium point, the thermal system is unstable.

12.1.3 Linearisation Around the Equilibrium Points T_{Jei}

Consider small variations of t_J around the equilibrium points T_{Jei}, with $i = 1, 2$, we have:

$$T_J = T_{Jei} + t_J. \tag{12.10}$$

The system S becomes:

$$\frac{d(T_{Jei} + t_J)}{dt} = \frac{1}{Cth_{(j-a)}}\left(P_T(T_{Jei} + t_J) - \frac{T_{Jei} + t_J - T_{amb}}{Rth_{(j-a)}}\right). \tag{12.11}$$

By using the definition of the equilibrium point, we obtain:

$$\frac{d(T_{Jei})}{dt} = \frac{1}{Cth_{(j-a)}}\left(P_T(T_{Jei}) - \frac{T_{Jei} - T_{amb}}{Rth_{(j-a)}}\right) = 0. \tag{12.12}$$

Moreover, the Taylor series expansion gives:

$$P_T(T_{Jei} + t_J) = P_T(T_{Jei}) + \left(\frac{dP_T}{dT_J}\right)_{T_{Jei}} t_J. \quad (12.13)$$

Thanks to Eqs. (12.11), (12.12) and (12.13) we can define the linear system around the equilibrium pointT_{Jei} :

$$\frac{dt_J}{dt} = \frac{1}{Cth_{(j-a)}}\left(\left(\frac{dP_T}{dT_J}\right)_{T_{Jei}} - \frac{1}{Rth_{(j-a)}}\right)t_J\,. \quad (12.14)$$

12.1.4 *Local Stability Around Each Equilibrium Point T_{Jei}*

Consider that at time $t = 0$, when the system is stabilised at T_{Jei}, a small disturbance Δ_{Tj0} occurs. Solving the linear differential equation (12.14), we obtain:

$$\Delta_{Tj}(t) = \Delta_{Tj0} e^{\frac{t}{Cth_{(j-a)}}\left(\left(\left(\frac{dP_T}{dT_J}\right)_{T_{Jei}} - \frac{1}{Rth_{(j-a)}}\right)\right)t}. \quad (12.15)$$

It is easy to conclude that the system is locally stable when:

$$\left(\frac{dP_T}{dT_J}\right)_{T_{Jei}} < \frac{1}{Rth_{(j-a)}}. \quad (12.16)$$

Figure 12.3 shows this condition is met for T_{Je1} but not for T_{Je2}. Consequently, T_{Je1} is a locally stable equilibrium point, and T_{Je2} is an unstable equilibrium point.

12.1.5 *Finding a Stability Limit Cycle Defined by the Adjoint System*

As mentioned in Para. 2.3.3 (Pelegrin et al., 1960; Naslin, 1962; Llibre & Zhang, 2019), the first step is to calculate:

$$H = \frac{d\left(\frac{T_J}{dt}\right)}{dT_J}. \quad (12.17)$$

The result is:

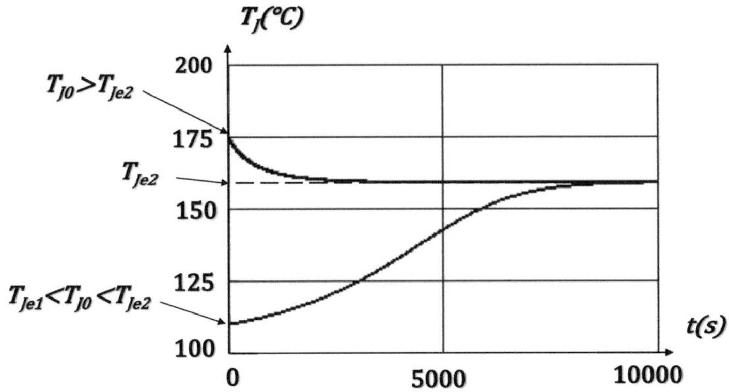

Fig. 12.4 Stability region defined by the adjoint system

$$H = \frac{1}{Cth_{(j-a)}}\left(\left(\frac{\mathrm{d}P_{\mathrm{T}}}{\mathrm{d}T_{\mathrm{J}}}\right) - \frac{1}{Rth_{(j-a)}}\right). \tag{12.18}$$

The nonexistence theorem for limit cycles does not permit concluding the absence of a limit cycle in the region of $T_{\mathrm{J}} < T_{\mathrm{Je2}}$ since the sign of H varies within this region. Let's try to use the following adjoint system to determine the possible limit cycle:

$$\frac{\mathrm{d}T_{\mathrm{J}}}{\mathrm{d}t} = \frac{-1}{Cth_{(j-a)}}\left(P_{\mathrm{T}}(T_{\mathrm{J}}) - \frac{T_{\mathrm{J}} - T_{\mathrm{amb}}}{Rth_{(j-a)}}\right). \tag{12.19}$$

Figure 12.4 shows the temporal response of the adjoint system simulated with an initial condition (T_{J0}) between T_{Je1} and T_{Je2}, and another one higher than T_{Je2}.

This simulation shows that the system converges towards T_{Je2}. In conclusion, the stability area of the stable point T_{Je1} is defined by:

$$T_{\mathrm{J}} < T_{\mathrm{Je2}}. \tag{12.20}$$

It means that if the initial temperature T_{J0} is greater than T_{Je2}, the diode will be damaged due to the thermal runaway phenomenon.

12.1.6 Finding a Stability Region Using Lyapunov Theory

In this paragraph, we will try to find a stability region using the Lyapunov function (Lyapunov, 1992):

$$E(T_J) = \frac{Cth_{(j-a)}}{2}(T_J - T_{Je1})^2. \quad (12.21)$$

This function is appropriate because

$$E(T_{Je1}) = 0,$$

- for a given $E_i \geq 0$, the condition $E(T_J) \leq E_i$ defines a closed region, which in this case, is a closed segment,
- for $E_1 > E_2$, the segment corresponding to E_2 is included in the segment corresponding to E_1.

The calculation of $\dot{E}(T_J)$ gives:

$$\dot{E}(T_J) = \frac{dE(T_J)}{dt} = \frac{dE(T_J)}{dT_J}\frac{dT_J}{dt}$$
$$= (T_J - T_{Je1})\left(P_T(T_J) - \frac{T_J - T_{amb}}{Rth_{(j-a)}}\right). \quad (12.22)$$

We have:

- $\dot{E}(T_{Je1}) = 0$,
- $\dot{E}(T_J) < 0$ for $T_J < T_{Je2}$ and $T_J \neq T_{Je1}$ (see Fig. 12.5).

Thanks to Lyapunov theory, we once again find that the stability region of the equilibrium T_{Je1}point is defined by:

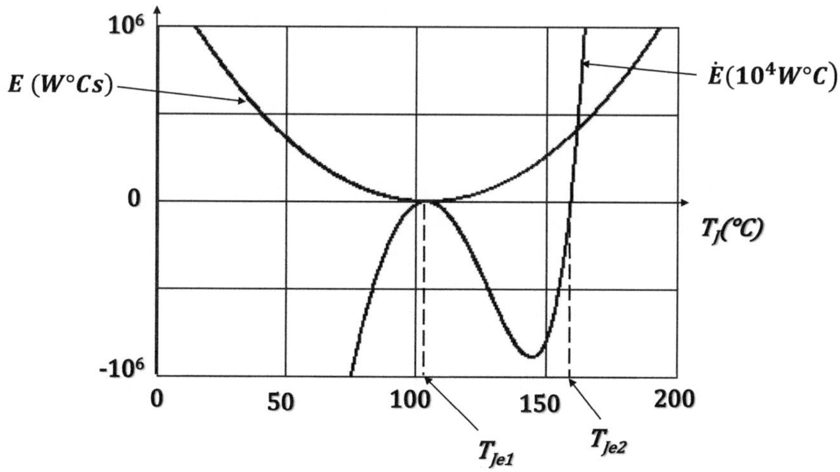

Fig. 12.5 Sign of E and $\dot{E}$ with respect to T_J

$$T_J < T_{Je2}. \tag{12.23}$$

12.1.7 Particular Case of a Schottky Diode Operating as an Or Function

It is interesting to consider the cases where the Schottky diodes operate separately in two distinct modes: the on state and off state. Typical applications include bypass diodes in solar panels and Schottky diodes performing an OR function. This last example is illustrated in Fig. 12.6. This system is used in applications where a high level of safety is required, such as in power supplies for servers. n power supplies are connected in parallel via n Schottky diodes performing an OR function. Under normal conditions, a DC current flows through each of the Schottky diodes and is given by:

$$I_{OUT}/n. \tag{12.24}$$

Assume that, at the time $t = 0$, the SMPSi fails, its output becomes a short circuit. The corresponding diode transitions D_i from the on state to the off state at a DC blocking voltage equal to V_{out}. Each diode is also mounted individually on its own heatsink. In the on state, the total losses in the diode D_i are equal to the conduction losses:

$$P_T(T_J) = P_{COND}(T_J) = V_F(I_{OUT}/n, T_{ref})$$
$$(1 + \alpha_{VF}(T_J - T_{ref})I_{OUT}/n). \tag{12.25}$$

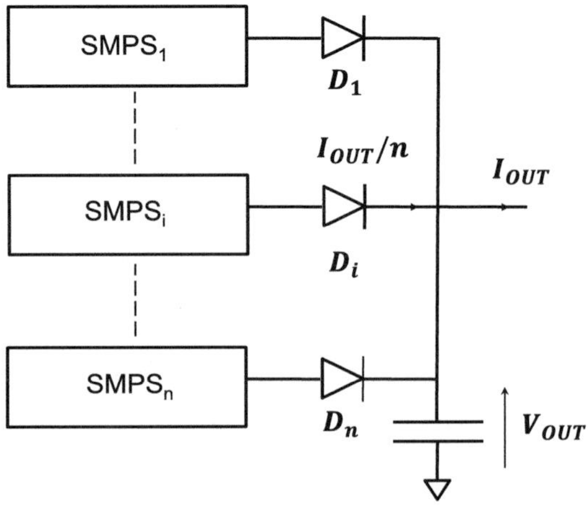

Fig. 12.6 Schottky diodes in an OR application

In the off-state mode, the total power losses consist solely of the reverse power losses:

$$P_{\mathrm{REV}}(T_{\mathrm{J}}) = I_{\mathrm{R}}(V_{\mathrm{OUT}}, T_{\mathrm{ref}})\mathrm{e}^{c(T_{\mathrm{J}}-T_{\mathrm{ref}})}V_{\mathrm{OUT}}. \tag{12.26}$$

After time $t = 0$, the system is described by Eq. (12.26) with two equilibrium points T_{Je1}, T_{Je2}, with an initial junction temperature T_{J0} set by the system operating in the first mode and determined by Eq. (12.25).

If T_{J0}, in the first mode is lower than T_{Je2}, the system will be stable. If $T_{\mathrm{J0}} > T_{\mathrm{Je2}}$, the diode will fail due to the thermal runaway. Figure 12.7 illustrates this specific thermal runaway mechanism. Two levels of conduction losses are shown. With the lower conduction loss level, when the diode D_i switches to reverse mode, the initial junction temperature T_{J01} is lower than T_{Je2}, thereby ensuring that the thermal system evolves towards the stability point T_{Je1}. Conversely, with the higher level of conduction losses, the initial junction temperature T_{J02} exceeds the instability point, causing the system to become unstable. In these applications, the stability criterion is defined as:

$$P_{\mathrm{COND}}(T_{\mathrm{J0}}) > P_{\mathrm{REV}}(T_{\mathrm{J0}}). \tag{12.27}$$

Another case where a thermal system can exceed the instability zone is when a component is subjected to a high current stress. The junction temperature can temporarily enter the instability zone. If the nominal conditions are immediately restored after this stress, the thermal system will become unstable.

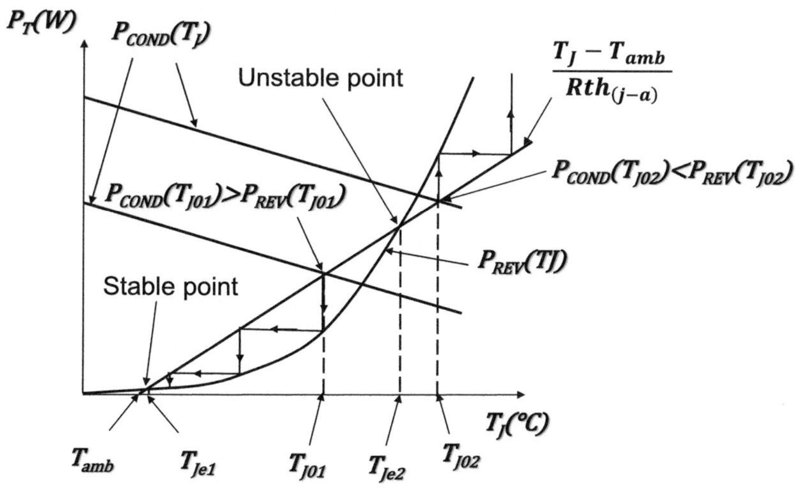

Fig. 12.7 Thermal runaway mechanism in OR-ing diodes

12.2 Thermal System of a Schottky Diode and a MOSFET Transistor

12.2.1 System Description

Consider now a thermal system consisting of a Schottky diode (D) and a MOSFET transitor (T_r) mounted on the same heat sink. These two components operate in a boost converter. The ideal current and voltage waveforms of D and T_r are shown in Fig. 12.8.

In this example, we consider only the predominant power losses, namely the conduction losses generated by T_r (P_{CONDTR}), the conduction losses (P_{CONDD}) and the reverse losses (P_{REVD}) of D. The total power losses in the Schottky diode

$$P_{\text{TD}} = P_{\text{CONDD}} + P_{\text{REVD}} \tag{12.28}$$

depend on its junction temperature (T_{JD}) according to the following relationship:

$$\begin{aligned} P_{\text{TD}}(T_{\text{JD}}) = {} & (V_{\text{F}}(I_{\text{M}}, T_{\text{ref}}) + \alpha_{\text{VF}}(T_{\text{JD}} - T_{\text{ref}})) \\ & (1 - d)I_{\text{M}} + I_{\text{R}}(V_{\text{R}}, T_{\text{ref}})\mathrm{e}^{c(T_{\text{JD}} - T_{\text{ref}})}\mathrm{d}V_{\text{R}} \end{aligned} \tag{12.29}$$

The conduction power losses in the MOSFET transistor depend on its on-state resistance (R_{DON}) at T_{ref} and its junction temperature (T_{JT}) according to

$$P_{\text{CONDT}}(T_{\text{JT}}) = (R_{\text{DON}} + \alpha_{\text{RD}}(T_{\text{JT}} - T_{\text{ref}}))\mathrm{d}I_{\text{M}}^2 \tag{12.30}$$

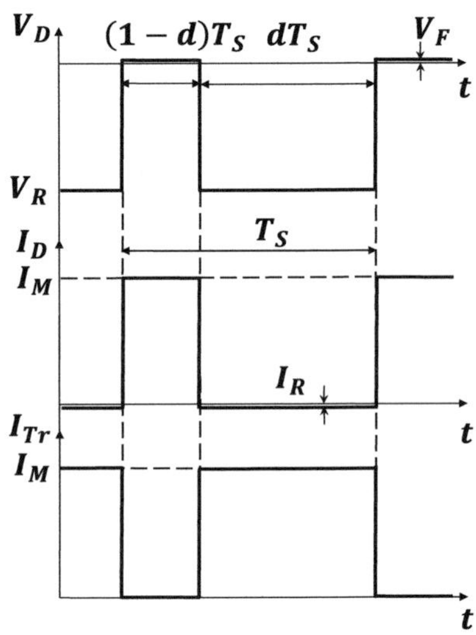

Fig. 12.8 Ideal voltage and current waveforms for D and T_r

This thermal system is represented by the thermal model shown in Fig. 12.9. The dynamics of this third-order system can be deduced from the model below:

$$\begin{cases} \dfrac{dT_{JD}}{dt} = \dfrac{P_{TD}(T_{JD})}{Cth_D} - \dfrac{T_{JD}}{Rth_D Cth_D} + \dfrac{T_C}{Rth_D Cth_D} \\ \quad + \dfrac{P_{TD}(T_{JD}) + P_{CONDT}(T_{JT})}{Cth_H} - \dfrac{T_C}{Rth_H Cth_H} + \dfrac{T_{amb}}{Rth_H Cth_H} \\ \dfrac{dT_{JT}}{dt} = \dfrac{P_T(T_{JT})}{Cth_T} - \dfrac{T_{JT}}{Rth_T Cth_T} + \dfrac{T_C}{Rth_T Cth_T} \\ \quad + \dfrac{P_{TD}(T_{JD}) + P_{CONDT}(T_{JT})}{Cth_H} - \dfrac{T_C}{Rth_H Cth_H} + \dfrac{T_{amb}}{Rth_H Cth_H} \\ \dfrac{dT_C}{dt} = \dfrac{P_{TD}(T_{JD}) + P_{CONDT}(T_{JT})}{Cth_H} - \dfrac{T_C}{Rth_H Cth_H} + \dfrac{T_{amb}}{Rth_H Cth_H} \end{cases} \tag{12.31}$$

We will use the following numerical values:

$$\begin{aligned} I_M &= 20\,\text{A},\, V_R = 20\,\text{V},\, d = 0.3, \\ V_F(I_M, T_{REF}) &= 0.5\,\text{V},\, I_R(V_R, T_{REF}) = 100\,\text{mA}, \\ \alpha_{VF} &= -0.5\,\text{mV}\,{}^\circ\text{C}^{-1},\, c = 0.06\,{}^\circ\text{C}^{-1}, \\ R_{DON} &= 50\,\text{m}\,\Omega,\, \alpha_{RD} = 0.1\,\text{m}\,\Omega\,{}^\circ\text{C}^{-1}, \\ Rth_D &= 1\,{}^\circ\text{C}\,\text{W}^{-1},\, Cth_D = 1\,\text{s}\,\text{W}\,{}^\circ\text{C}^{-1},\, Rth_T = 2\,{}^\circ\text{C}\,\text{W}^{-1}, \\ Cth_T &= 1\,\text{s}\,\text{W}\,{}^\circ\text{C}^{-1},\, Rth_H = 5\,{}^\circ\text{C}\,\text{W}^{-1} \\ Cth_H &= 150\,\text{s}\,\text{W}\,{}^\circ\text{C}^{-1},\, T_{amb} = 40\,{}^\circ\text{C},\, T_{ref} = 125\,{}^\circ\text{C}. \end{aligned} \tag{12.32}$$

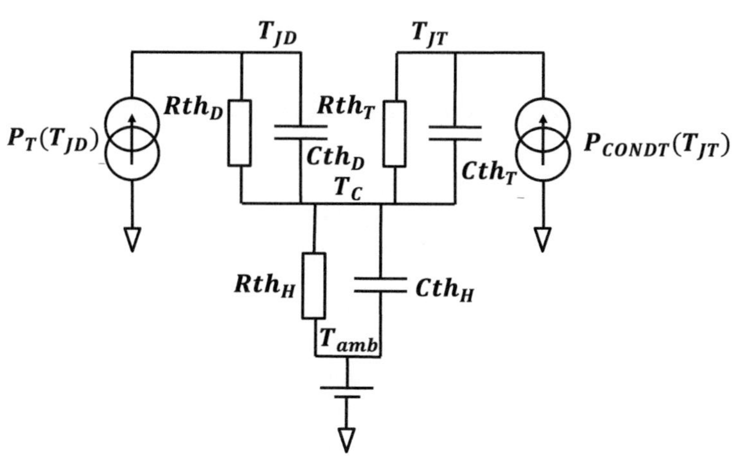

Fig. 12.9 Thermal model of a Schottky diode and a transistor on a heatsink

12.2.2 Equilibrium Points

The equilibrium points $(T_{\mathrm{JDei}}, T_{\mathrm{JTei}}, T_{\mathrm{Cei}})$ are calculated by solving the corresponding system S_e $\left(\frac{\mathrm{d}T_{\mathrm{JD}}}{\mathrm{d}t} = \frac{\mathrm{d}T_{\mathrm{JT}}}{\mathrm{d}t} = \frac{\mathrm{d}T_{\mathrm{C}}}{\mathrm{d}t} = 0\right)$. We obtain the following two equilibrium points (in °C)

$$\begin{aligned} T_{\mathrm{JDe1}}, T_{\mathrm{JTe1}}, T_{\mathrm{Ce1}}) &\simeq (113.9, 118.4, 106.5), \\ (T_{\mathrm{JDe2}}, T_{\mathrm{JTe2}}, T_{\mathrm{Ce2}}) &\simeq (171.6, 168.1, 155.1) \end{aligned} \tag{12.33}$$

12.2.3 Linearisation Around the Two Equilibrium Points

The corresponding Jacobian matrix is:

$$J = \begin{pmatrix} \frac{A}{Cth_{\mathrm{D}}Cth_{\mathrm{T}}} - \frac{1}{Rth_{\mathrm{D}}Cth_{\mathrm{D}}} & \frac{B}{Cth_{\mathrm{H}}} & \frac{1}{Rth_{\mathrm{D}}Cth_{\mathrm{D}}} - \frac{1}{Rth_{\mathrm{H}}Cth_{\mathrm{H}}} \\ \frac{(1-d)I_{\mathrm{M}}\alpha_{\mathrm{VF}}+C}{Cth_{\mathrm{H}}} & \frac{B(Cth_{\mathrm{H}}+Cth_{\mathrm{T}})}{Cth_{\mathrm{H}}Cth_{\mathrm{T}}} - \frac{1}{Rth_{\mathrm{T}}Cth_{\mathrm{T}}} & \frac{1}{Rth_{\mathrm{T}}Cth_{\mathrm{T}}} - \frac{1}{Rth_{\mathrm{H}}Cth_{\mathrm{H}}} \\ \frac{(1-d)I_{\mathrm{M}}\alpha_{\mathrm{VF}}+C}{Cth_{\mathrm{H}}} & \frac{B}{Cth_{\mathrm{H}}} & -\frac{1}{Rth_{\mathrm{H}}Cth_{\mathrm{H}}} \end{pmatrix} \tag{12.34}$$

with

$$\begin{aligned} A &= \left((1-d)I_{\mathrm{M}}\alpha_{\mathrm{VF}} + dcV_{\mathrm{R}}I_{\mathrm{R}}(V_{\mathrm{R}}, T_{\mathrm{REF}})\mathrm{e}^{c(T_{\mathrm{JDei}}-T_{\mathrm{ref}})}\right)(Cth_{\mathrm{D}} + Cth_{\mathrm{T}}), \\ B &= d\alpha_{\mathrm{RD}}I_{\mathrm{M}}^2,\ C = cdV_{\mathrm{R}}I_{\mathrm{R}}(V_{\mathrm{R}}, T_{\mathrm{REF}})\mathrm{e}^{c(T_{\mathrm{JDei}}-T_{\mathrm{ref}})} \end{aligned} \tag{12.35}$$

Note that the Jacobian matrix depends on T_{JDei}.

12.2.4 Local Stability Around Each Equilibrium Point

For $(T_{\mathrm{JDe1}}, T_{\mathrm{JTe1}}, T_{\mathrm{Ce1}}) = (113.9, 118.4, 106.5)$, the three real parts of the root of $\det(\mathrm{Is} - \mathrm{J})$ are strictly negative, so the equilibrium point is locally asymptotically stable. For $(T_{\mathrm{JDe2}}, T_{\mathrm{JTe2}}, T_{\mathrm{Ce2}}) = (171.6, 168.1, 155.1)$, the characteristic equation

$$\det(\mathrm{Is} - \mathrm{J}) = 0 \tag{12.36}$$

gives one positive real root, so the equilibrium point is locally unstable.

12.2.5 Region of Stability

We now consider the widely used configuration of a temperature sensor mounted on the heatsink. When the heatsink temperature (T_C) reaches the temperature T_{CMAX}, the control system stops the converter. Thus, we have:

$$T_C < T_{CMAX}. \tag{12.37}$$

Obviously T_{CMAX} has to be higher than T_{Ce1} and lower than T_{Ce2}. To determine the stability region, we consider the worst-case scenario where:

$$T_C = T_{CMAX} = 115\,°C. \tag{12.38}$$

With this assumption, the thermal systems of the diode and the transistor become thermally independent. The thermal system of the diode is similar to that given in the first example. This system also has one stable equilibrium point and one unstable equilibrium point. The unstable equilibrium point is

$$T_{JDMAX} \simeq 208.4\,°C. \tag{12.39}$$

Consequently, a stable region can be defined by:

$$T_C < T_{CMAX} = 115\,°C \text{ and } T_J < T_{JDMAX}. \tag{12.40}$$

Note that there is no limitation on T_{JTMAX}. The stability region is shown in Fig. 12.10.

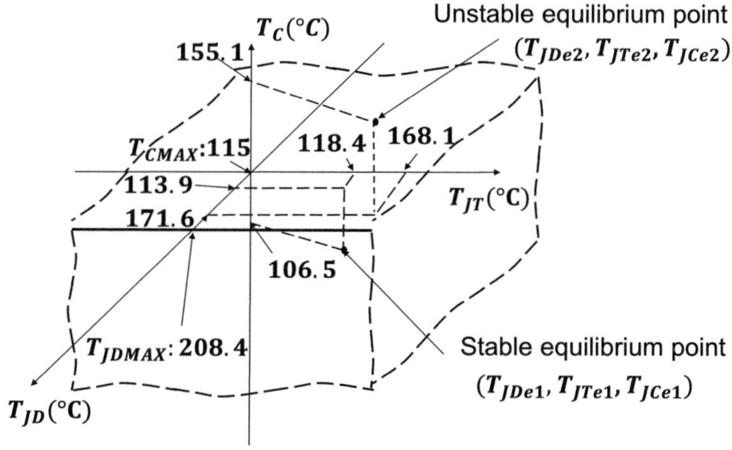

Fig. 12.10 Stability region of the thermal system for D and T_r

12.3 Thermal System of Two Schottky Diodes Connected in Parallel

12.3.1 System Description

In this example, we consider the thermal stability of diodes connected in parallel. Figure 12.11 shows the data for this problem. These two diodes have the following identical electrical characteristics:

$$\begin{aligned} R_{\mathrm{D}}(T_{\mathrm{ref}}) &= 50\,\mathrm{m}\,\Omega,\, \alpha_{\mathrm{RD}} = 0.1\,\mathrm{m}\,\Omega\,{}^{\circ}\mathrm{C}^{-1}, \\ \alpha_{\mathrm{VTO}} &= -1\,\mathrm{mV}\,{}^{\circ}\mathrm{C}^{-1},\, I_{\mathrm{R}}(T_{\mathrm{REF}}) = 50\,\mathrm{mA}, \\ c &= 0.06\,{}^{\circ}\mathrm{C}^{-1},\, T_{\mathrm{ref}} = 125\,{}^{\circ}\mathrm{C} \end{aligned} \tag{12.41}$$

except for the threshold voltage at T_{ref}:

$$\begin{aligned} &V_{\mathrm{TO}}(T_{\mathrm{REF}}) = 0.5\,\mathrm{V}\ \text{for}\ D_2\ \text{and} \\ &V_{\mathrm{TO}}(T_{\mathrm{ref}}) + \Delta V_{\mathrm{TO}}(T_{\mathrm{ref}})\text{for}\ D_1. \end{aligned} \tag{12.42}$$

These two diodes are mounted on two separate heatsinks as shown in Fig. 12.12. They have the following identical thermal characteristics:

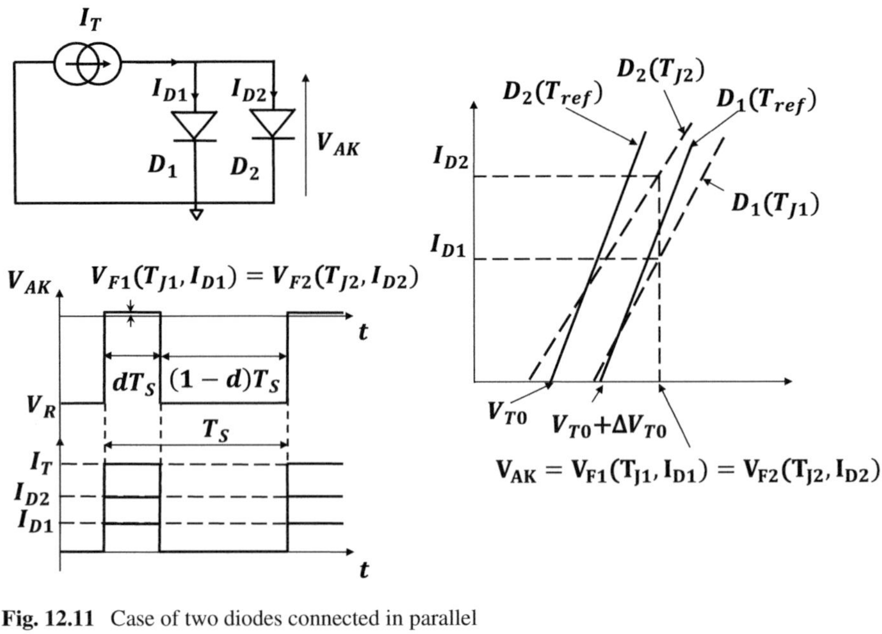

Fig. 12.11 Case of two diodes connected in parallel

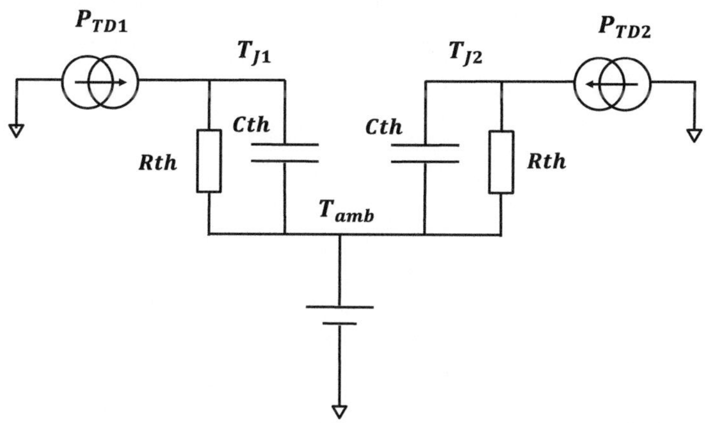

Fig. 12.12 Thermal model of two diodes connected in parallel

$$\begin{aligned} Rth &= 12\,°\mathrm{C\,W}^{-1}, \\ Cth &= 30\,\mathrm{s\,W},\, T_{\mathrm{amb}} = 40\,°\mathrm{C}. \end{aligned} \tag{12.43}$$

The objective of this study is to determine the thermal stability of this system with respect to dispersion ΔV_{TO}. With the two diodes connected in parallel, this induces:

$$I_{\mathrm{T}} = I_{\mathrm{D1}} + I_{\mathrm{D2}} \text{ and } V_{\mathrm{F1}}(T_{\mathrm{J1}}, I_{\mathrm{D1}}) = V_{\mathrm{F2}}(T_{\mathrm{J2}}, I_{\mathrm{D2}}) \tag{12.44}$$

with

$$\begin{aligned} V_{\mathrm{F1}}(T_{\mathrm{J1}}, I_{\mathrm{D1}}) = {} & V_{\mathrm{TO}} + \Delta V_{\mathrm{TO}} + \alpha_{\mathrm{VTO}}(T_{\mathrm{J1}} - T_{\mathrm{ref}}) \\ & + [R_{\mathrm{D}} + \alpha_{\mathrm{RD}}(T_{\mathrm{J1}} - T_{\mathrm{ref}})]I_{\mathrm{D1}} \end{aligned} \tag{12.45}$$

and

$$\begin{aligned} V_{\mathrm{F2}}(T_{\mathrm{J2}}, I_{\mathrm{D2}}) = {} & V_{\mathrm{TO}} + \alpha_{\mathrm{VTO}}(T_{\mathrm{J2}} - T_{\mathrm{ref}}) \\ & + [R_{\mathrm{D}} + \alpha_{\mathrm{RD}}(T_{\mathrm{J2}} - T_{\mathrm{ref}})]I_{\mathrm{D2}}. \end{aligned} \tag{12.46}$$

From Eqs. (12.44), (12.45) and (12.46) we can deduce:

$$I_{\mathrm{D1}}(T_{\mathrm{J1}}, T_{\mathrm{J2}}) = \frac{\alpha_{\mathrm{VTO}}\left(T_{\mathrm{J2}} - T_{\mathrm{j1}}\right) + [R_{\mathrm{D}} + \alpha_{\mathrm{RD}}(T_{\mathrm{J2}} - T_{\mathrm{ref}})]I_{\mathrm{T}} - \Delta V_{\mathrm{TO}}}{2R_{\mathrm{D}} - 2\alpha_{\mathrm{RD}}T_{\mathrm{ref}} + \alpha_{\mathrm{RD}}\left(T_{\mathrm{j1}} + T_{\mathrm{j2}}\right)} \tag{12.47}$$

and

$$I_{\mathrm{D2}}(T_{\mathrm{J1}}, T_{\mathrm{J2}}) = I_{\mathrm{T}} - I_{\mathrm{D1}}(T_{\mathrm{J1}}, T_{\mathrm{J2}}). \tag{12.48}$$

The total power losses of D_1 and D_2 are given by:

$$\begin{aligned} P_{D1}(T_{J1}, T_{J2}) &= [V_{TO} + \Delta V_{TO} + \alpha_{VTO}(T_{J1} - T_{ref})]dI_{D1}(T_{J1}, T_{J2}) \\ &+ [R_D + \alpha_{RD}(T_{J1} - T_{ref})]dI_{D1}^2(T_{J1}, T_{J2}) \\ &+ I_R e^{c(T_{J1} - T_{ref})}(1 - d)V_R \end{aligned} \tag{12.49}$$

$$\begin{aligned} P_{D2}(T_{J1}, T_{J2}) &= [V_{TO} + \alpha_{VTO}(T_{J2} - T_{ref})]dI_{D2}(T_{J1}, T_{J2}) \\ &+ [R_D + \alpha_{RD}(T_{J2} - T_{ref})]dI_{D2}^2(T_{J1}, T_{J2}) \\ &+ I_R e^{c(T_{J2} - T_{ref})}(1 - d)V_R. \end{aligned} \tag{12.50}$$

This system is the second order with the state parameters T_{J1}, T_{J2}. From the thermal model, we obtain the following dynamics

$$\begin{cases} \frac{dT_{J1}}{dt} = \frac{P_{D1}(T_{J1}, T_{J2})}{Cth} - \frac{T_{J1}}{RthCth} + \frac{T_{amb}}{RthCth} \\ \frac{dT_{J2}}{dt} = \frac{P_{D2}(T_{J1}, T_{J2})}{Cth} - \frac{T_{J2}}{RthCth} + \frac{T_{amb}}{RthCth} \end{cases}. \tag{12.51}$$

12.3.2 Stable Equilibrium Points

For each value of ΔV_{TO} approximately less than 97 mV, we can find two equilibrium points. Using the techniques given in Para. 2.1, we can show that one equilibrium point is stable while the other is unstable. The graph in Fig. 12.13 shows the variation of the stable equilibrium (T_{J1e}, T_{J2e}) with the corresponding equilibrium currents (I_{D1e}, I_{D2e}) in relation to ΔV_{TO}.

For ΔV_{TO} approximately higher than 97 mV, the system becomes unstable.

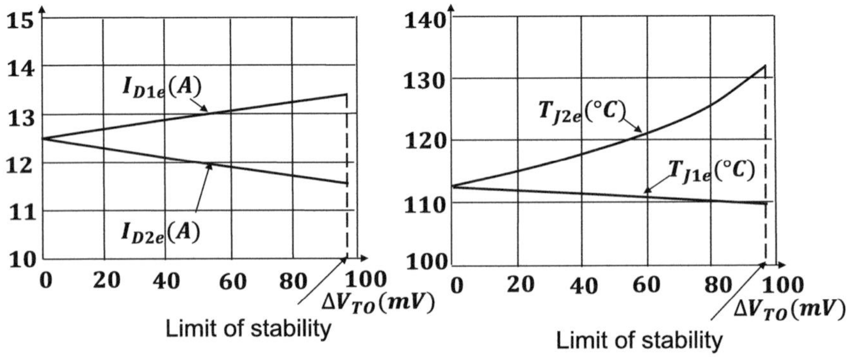

Fig. 12.13 Variation of the stable equilibrium point with respect to ΔV_{TO}

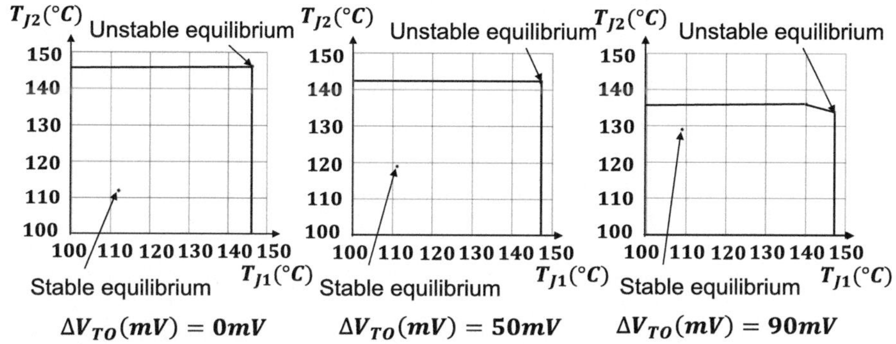

Fig. 12.14 Stability regions for $\Delta V_{\mathrm{TO}} = 0\,\mathrm{mV}$, $\Delta V_{\mathrm{TO}} = 50\,\mathrm{mV}$, $\Delta V_{\mathrm{TO}} = 90\,\mathrm{mV}$

12.3.3 Region of Stability

The graphs presented in Fig. 12.14 illustrate the stability regions for three different values of ΔV_{TO}, along with the positioning of the stable and unstable equilibrium points. We note that the unstable point lies on the boundary between the stable and unstable regions.

12.3.4 Physical Interpretation

The diode D_2 has a lower threshold voltage than D_1. Therefore, the corresponding forward current, power losses and junction temperature of D_2 will be higher than that those of D_1. The negative value of α_{VTO} increases this tendency, while the positive value of α_{RD} has the opposite effect. As ΔV_{TO} increases, I_{D2e} and T_{J2e} increase and consequently, the strong dependence of the reverse losses on the junction temperature (following an exponential law) reduces the stability region. It is always diode D_2 that will be destroyed due to the thermal runaway phenomenon. If this diode becomes a short circuit, D_1 will remain safe. If D_2 becomes an open circuit, D_1 will susceptible to thermal runaway. In this example, we consider only the dispersion of V_{TO}. It is possible to study the influence of the dispersion of other parameters in the same way.

12.3.5 Case of n Diodes Connected in Parallel

Generally, when studying *n* diodes connected in parallel, we consider the worst-case scenario. We also consider only dispersion of $\Delta V_{\mathrm{TO}}(T_{\mathrm{REF}})$. The critical case occurs when one diode, for example D_1 has the highest threshold voltage: $V_{\mathrm{TO}}(T_{\mathrm{REF}}) + \Delta V_{\mathrm{TO}}(T_{\mathrm{REF}})$, while the $n-1$ others diodes $(D_2 \ldots D_n)$ have the lowest: $V_{\mathrm{TO}}(T_{\mathrm{REF}})$.

With these hypotheses, we obtain:

$$I_{\mathrm{D2}} = I_{\mathrm{D3}} = \cdots = I_{\mathrm{D}n}, I_{\mathrm{T}} = I_{\mathrm{D1}} + (n-1)I_{\mathrm{D2}} \tag{12.52}$$

and

$$\begin{aligned} V_{\mathrm{F1}}(T_{\mathrm{J1}}, I_{\mathrm{M1}}) &= V_{\mathrm{F2}}(T_{\mathrm{J2}}, I_{\mathrm{M2}}) = \cdots V_{\mathrm{F}n}(T_{\mathrm{J}n}, I_{\mathrm{M}n}) \\ T_{\mathrm{J2}} &= \cdots T_{\mathrm{J}n}. \end{aligned} \tag{12.53}$$

This system becomes a second order system, and the methodology described in the previous paragraph can be applied to study the thermal stability of the entire system.

12.4 Achieving Power Tracking Control with an Unstable System

12.4.1 Description of the Problem

Consider an electrical generator with the static characteristic:

$$U(I) = U_0 - \frac{U_0}{I_{CC}^2} I^2 \text{ with } U_0 \in [U_{0\,\mathrm{min}}, U_{0\,\mathrm{max}}]. \tag{12.54}$$

The provided power can be expressed as:

$$P(I) = U(I)I = U_0 I - \frac{U_0}{I_{CC}^2} I^3. \tag{12.55}$$

Figure 12.15 shows that this generator provides the maximum power P_{max} for

$$I = \frac{I_{CC}}{\sqrt{3}}. \tag{12.56}$$

The purpose of this study is to define a control law that allows the generator to automatically provide its maximum power regardless the dispersions of U_0. To achieve this, we will utilise the instability inherent in a nonlinear system.

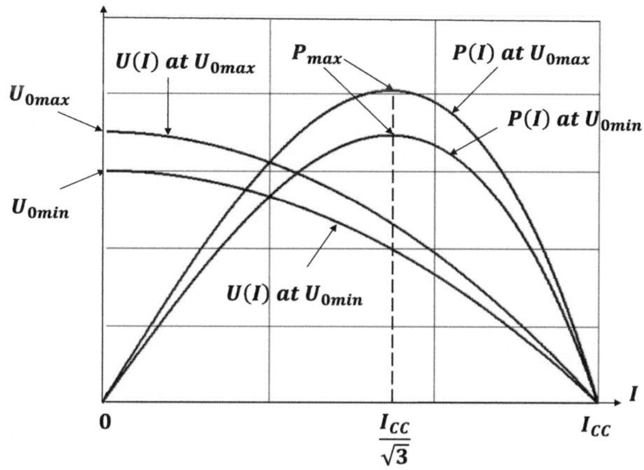

Fig. 12.15 Characteristics of the electrical generator $U(I)$

12.4.2 Finding the Control Law

Consider a control law $I(t)$ whose dynamics are given by:

$$\frac{\mathrm{d}I}{\mathrm{d}t} = k(P_0 - P(I) - U_C I). \tag{12.57}$$

P_0 and U_C are respectively a constant power and a constant voltage. Moreover, as shown in Fig. 12.16, regardless of the dispersions of U_0, we have:

$$P_C(I) - U_C I = P_0 - P(I) - U_C I > 0. \tag{12.58}$$

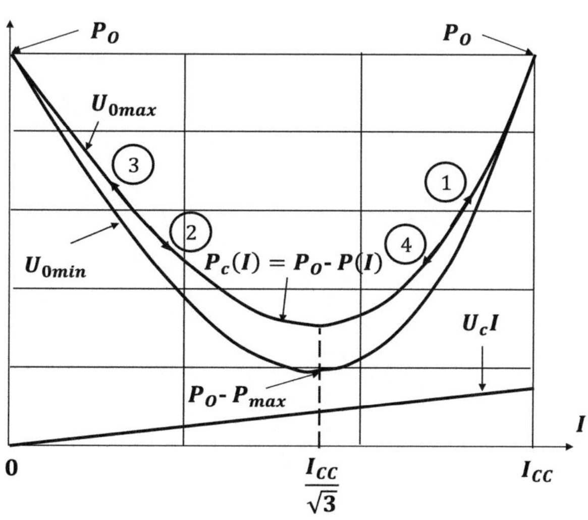

Fig. 12.16 $P_C(I)$ and $U(I)$ with respect to I

Fig. 12.17 k values for the four configurations

$\frac{dI}{dt}$ \ $\frac{dP_C}{dt}$	>0	<0
>0	① ⟹ $\frac{dI}{dt}<0$, $k=-k_o$	② ⟹ $\frac{dI}{dt}>0$, $k=+k_o$
<0	③ ⟹ $\frac{dI}{dt}>0$, $k=+k_o$	④ ⟹ $\frac{dI}{dt}<0$, $k=-k_o$

The above-mentioned formula indicates that for a given k value, the nonlinear system defined by Eq. (12.57) is unstable since there is no equilibrium point. It also shows that the sign of $\frac{dI}{dt}$ is the same as that of k. k is defined by:

$$k = \pm k_0 \text{ with } k_0 = \text{constant} \tag{12.59}$$

The basic idea of this control is to change the sign of k to modify the variation of I in order to guide the generator to supply its maximum power. According to the signs of $\frac{dP_C}{dt}$ and $\frac{dI}{dt}$, four different configurations can be distinguished (see Fig. 12.17). The corresponding value of k for each configuration is shown in the table of Fig. 12.17. For example, if $\frac{dP_C}{dt} > 0$ and $\frac{dI}{dt} > 0$ (case 1), the operating point moves away from P_{max}, so the control must decrease I and set

$$k = -k_0. \tag{12.60}$$

The block diagram of this control is presented in Fig. 12.18.

To sum up, the control law can be defined by:

- If $\frac{dP_C}{dt}\frac{dI}{dt} > 0$ then $k = -k_0$

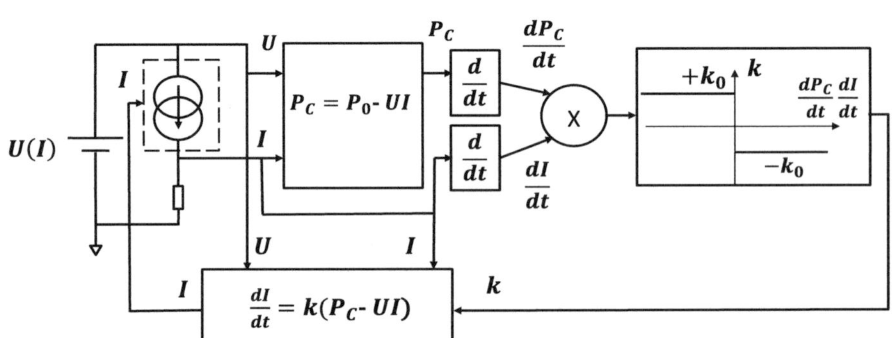

Fig. 12.18 Block diagram of the control system

- If $\frac{dP_C}{dt}\frac{dI}{dt} < 0$ then $k = +k_0$

12.5 Example 5: Stability of a $r, L, C, I(V_C)$ Nonlinear Circuit

12.5.1 System Description

We consider now, the circuit shown in Fig. 12.19 (Barrade, 1997).

A converter (here a buck converter) can be modelled as the following current generator:

$$I = \frac{\eta P_{OUT}}{V_C}. \tag{12.61}$$

P_{OUT} is the output power and η the efficiency of the converter, which is assumed to be constant. With this simplification, the dynamics of the system become:

$$\begin{cases} \frac{dI_L}{dt} = -\frac{rI_L}{L} - \frac{V_C}{L} + \frac{V_{IN}}{L} \\ \frac{dV_C}{dt} = \frac{I_L}{C} - \frac{\eta P_{OUT}}{CV_C} \end{cases}. \tag{12.62}$$

The two state variables ($n = 2$) are I_L and V_C and the six system parameters ($p = 6$) are V_{IN}, r, L, C, P_{OUT} and η.

12.5.1.1 Equilibrium Points

The equilibrium points are calculated by solving the associated system S_e. There are two equilibrium points defined by:

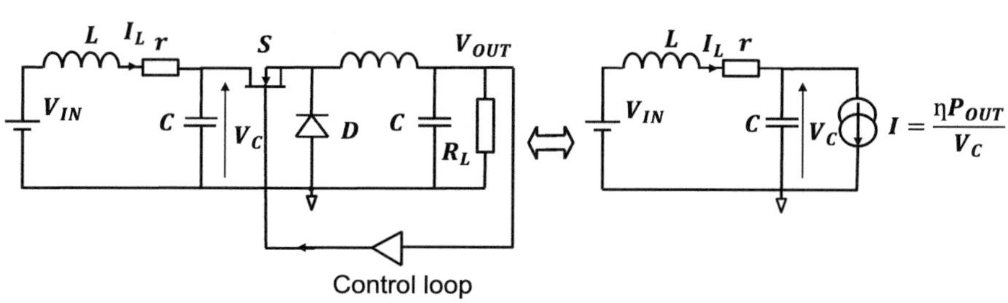

Fig. 12.19 $r, L, C, I(V_C)$ nonlinear circuit

$$(I_{\text{Le1}}, V_{\text{Ce1}}) = \left(\frac{2\eta P_{\text{OUT}}}{V_{\text{IN}} + \sqrt{V_{\text{IN}}^2 - 4\eta P_{\text{OUT}} r}}, \frac{V_{\text{IN}} + \sqrt{V_{\text{IN}}^2 - 4\eta P_{\text{OUT}} r}}{2} \right) \tag{12.63}$$

and

$$(I_{\text{Le2}}, V_{\text{Ce2}}) = \left(\frac{2\eta . P_{\text{OUT}}}{V_{\text{IN}} - \sqrt{V_{\text{IN}}^2 - 4\eta P_{\text{OUT}} . r}}, \frac{V_{\text{IN}} - \sqrt{V_{\text{IN}}^2 - 4\eta . P_{\text{OUT}} r}}{2} \right). \tag{12.64}$$

12.5.2 Linearisation Around Equilibrium Points

The Jacobian matrix J at each equilibrium point $(I_{\text{Lei}}, V_{\text{Cei}})$ is

$$J = \begin{pmatrix} -\frac{r}{L} & -\frac{1}{L} \\ \frac{1}{C} & \frac{P}{CV_{\text{Cei}}^2} \end{pmatrix}. \tag{12.65}$$

12.5.3 Local Stability at Equilibrium Points

As usual, the first step consists of estimating the sign of the root of:

$$\det(\text{Is} - \text{J}) = s^2 + \left(\frac{r}{L} - \frac{\eta P_{\text{OUT}}}{CV_{\text{Cei}}^2} \right) s + \frac{1}{LC} - \frac{\eta P_{\text{OUT}} r}{LCV_{\text{Cei}}^2} = 0. \tag{12.66}$$

The simplest way here is to use Routh's criterion (Routh, 1877). We have:

$$a_2 = 1, a_1 = \frac{r}{L} - \frac{\eta P_{\text{OUT}}}{CV_{\text{Cei}}^2}, a_0 = \frac{1}{LC} - \frac{\eta P_{\text{OUT}} r}{LCV_{\text{Cei}}^2}. \tag{12.67}$$

The stability conditions are

$$\begin{aligned} &a_1 > 0 \\ &\Rightarrow \frac{r}{L} - \frac{\eta P_{\text{OUT}}}{CV_{\text{Cei}}^2} > 0, \end{aligned} \tag{12.68}$$

$$\begin{aligned}
&a_0 > 0 \\
&\Rightarrow \frac{1}{L.C} - \frac{\eta \,.\, P_{\text{OUT}} r}{LCV_{\text{Cei}}^2} > 0.
\end{aligned} \tag{12.69}$$

Consequently, the stability criterion is given by:

$$\frac{r}{L} > \frac{\eta P_{\text{OUT}}}{CV_{\text{Cei}}^2} \text{ and } \frac{V_{\text{Cei}}^2}{r} > \eta P_{\text{OUT}}. \tag{12.70}$$

The last inequality shows that the system cannot be stable without the resistor r. To conclude this study, we use the following numerical values:

$$\begin{aligned}
V_{\text{IN}} &= 50\,\text{V}, L = 100\,\mu\text{H}, C = 100\,\mu\text{F}, \\
r &= 1\,\Omega, \eta = 100\%, P_{\text{OUT}} = 200\,\text{W}.
\end{aligned} \tag{12.71}$$

The corresponding equilibrium points are

$$(I_{\text{Le1}}, V_{\text{Ce1}}) \simeq (4.38, 45.62 \text{ and } (I_{\text{Le2}}, V_{\text{Ce2}}) \simeq (45.62, 4.38) \tag{12.72}$$

The stability criterion indicates that $(I_{\text{Le1}}, V_{\text{Ce1}})$ is locally stable, whereas $(I_{\text{Le2}}, V_{\text{Ce2}})$ is unstable.

12.5.4 Stability Region of (I_{Le1}, V_{Ce1})

In this example, we will determine the stability region around $(I_{\text{Le1}}, V_{\text{Ce1}})$ through simulations. This stability region is shown in the Fig. 12.20. Note that the unstable equilibrium point lies on the boundary of the stability region of the stable equilibrium point. The control of the converter must be adapted to ensure that the state parameters always remain inside the stability region. A better way to stabilise this circuit is to put the resistor r in series with the capacitor, reducing the dissipated power. The same study can be performed with this configuration.

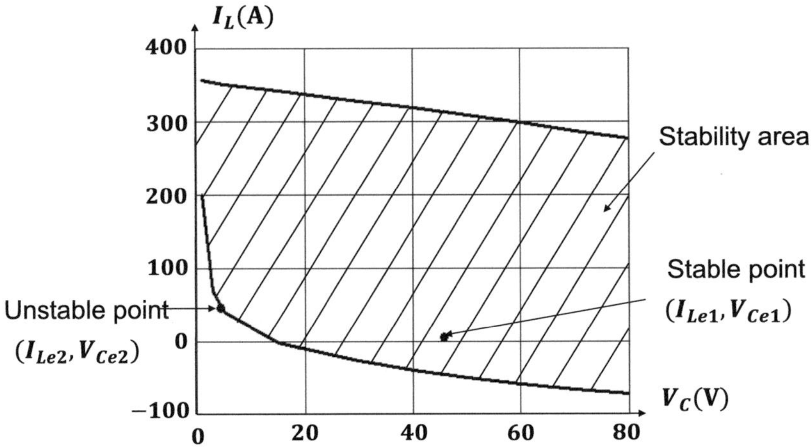

Fig. 12.20 Stability region around $(I_{\mathrm{Le1}}, V_{\mathrm{Ce1}})$

12.6 Example 6: Stability of a $R, L, C, I(V_C)$ Nonlinear Circuit

12.6.1 System Description

Consider again the circuit of Fig. 12.19 without the resistor r ($r = 0$), but with a resistor R in parallel with the inductor L (Barrade, 1997). The dynamics of this system are given by:

$$\begin{cases} \frac{\mathrm{d}I_{\mathrm{L}}}{\mathrm{d}t} = -\frac{V_{\mathrm{C}}}{L} + \frac{V_{\mathrm{IN}}}{L} \\ \frac{\mathrm{d}V_{\mathrm{C}}}{\mathrm{d}t} = \frac{I_{\mathrm{L}}}{C} - \frac{V_{\mathrm{C}}}{RC} - \frac{\eta P_{\mathrm{OUT}}}{CV_{\mathrm{C}}} + \frac{V_{\mathrm{IN}}}{RC} \end{cases}. \tag{12.73}$$

12.6.2 Equilibrium Point

The equilibrium point $(I_{\mathrm{Le}}, V_{\mathrm{Ce}})$is calculated by solving the corresponding system S_{e}. We obtain:

$$(I_{\mathrm{Le}}, V_{\mathrm{Ce}}) = \left(\frac{\eta P_{\mathrm{OUT}}}{V_{\mathrm{IN}}}, V_{\mathrm{IN}}\right). \tag{12.74}$$

12.6.3 Linearisation Around the Equilibrium Point (I_{Le}, V_{Ce})

The Jacobian matrix J at the equilibrium point (I_{Le}, V_{Ce}) is

$$J = \begin{pmatrix} 0 & -\frac{1}{L} \\ \frac{1}{C} & -\frac{1}{RC} + \frac{\eta P_{\mathrm{OUT}}}{C V_{\mathrm{IN}}^2} \end{pmatrix}. \tag{12.75}$$

12.6.4 Local Stability Around the Equilibrium Point (I_{Le}, V_{Ce})

The Routh criterion (Routh, 1877) provides the following stability condition:

$$\frac{1}{R} > \frac{\eta P_{\mathrm{OUT}}}{V_{\mathrm{IN}}^2}. \tag{12.76}$$

This inequality shows that the system cannot be stable without the resistor R. To conclude, let us now use the following numerical values:

$$\begin{aligned} &V_{\mathrm{IN}} = 500\,\mathrm{V}, L = 400\,\mathrm{mH}, C = 100\,\mu\mathrm{F}, \\ &R = 100\,\Omega, \eta = 100\%, P_{\mathrm{OUT}} = 2\,\mathrm{kW}. \end{aligned} \tag{12.77}$$

12.6.5 Determination of a Stability Limit Cycle Using the Adjoint System

As mentioned in Para. 2.3.3 (Pelegrin et al., 1960; Naslin, 1962; Llibre & Zhang, 2019), the first step is to calculate:

$$H = \frac{\mathrm{d}\left(\frac{I_{\mathrm{L}}}{\mathrm{d}t}\right)}{\mathrm{d}I_{\mathrm{L}}} + \frac{\mathrm{d}\left(\frac{V_{\mathrm{C}}}{\mathrm{d}t}\right)}{\mathrm{d}V_{\mathrm{C}}}. \tag{12.78}$$

The result is as follows:

$$H = -\frac{1}{RC} + \frac{\eta P_{\mathrm{OUT}}}{C V_{\mathrm{C}}^2}. \tag{12.79}$$

The theorem on the nonexistence of limit cycles allows us to conclude that without $R(R \to \infty)$, no limit cycle exists because the H quantity is always positive. With a

Fig. 12.21 Limit cycle and stability regions defined by the Lyapunov function

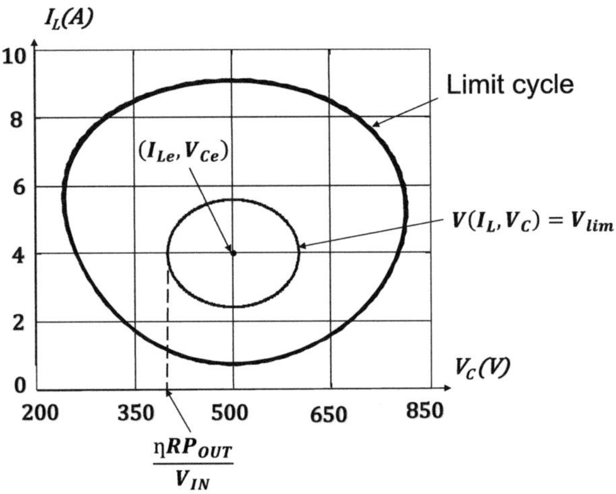

resistor R, the sign of H changes, so we cannot conclude anything about the existence of a limit cycle. The adjoint system is:

$$\begin{cases} \frac{dI_L}{dt} = +\frac{V_C}{L} - \frac{V_{IN}}{L} \\ \frac{dV_C}{dt} = -\frac{I_L}{C} + \frac{V_C}{RC} + \frac{\eta P_{OUT}}{CV_C} - \frac{V_{IN}}{RC} \end{cases}. \tag{12.80}$$

Figure 12.21 shows the limit cycle obtained thanks to the adjoint system.

12.6.6 Determination of a Stability Region Using Lyapunov Theory

In this paragraph, we try to find a stability region using the following Lyapunov function (Lyapunov, 1992):

$$V(I_L, V_C) = \frac{L}{2}(I_L - I_{Le})^2 + \frac{C}{2}(V_C - V_{IN})^2. \tag{12.81}$$

This function is suitable because it satisfies the usual criteria. Its time derivative is:

$$\frac{dV(I_L, V_C)}{dt} = -(V_C - V_{IN})^2\left(\frac{1}{R} - \frac{\eta P_{OUT}}{V_C V_{IN}}\right) \tag{12.82}$$

which is negative if:

$$V_C > \frac{\eta R P_{OUT}}{V_{IN}}. \tag{12.83}$$

confirming that this circuit is always unstable without the resistor R. So, we have:

- $\frac{dV(I_L, V_C)}{dt} = 0$,
- $\frac{dV(I_L, V_C)}{dt} < 0$ for $V(I_L, V_C) < V_{lim} = \frac{C}{2}\left(\frac{\eta R P_{OUT}}{V_{IN}} - V_{IN}\right)^2$.

The stability region defined by the Lyapunov function $V(I_L, V_C) = V_{lim}$, is shown in Fig. 12.21.

Chapter 13
Examples of Optimal, Lyapunov and Direct Controls in the Field of Power Conversion

The optimal control can be used to optimise a wide range of issues in the field of power conversion. In this section, we will study some examples aimed at minimising dissipated energy during the startup phase and the time required to charge a capacitor. Before tackling this part, the reader is advised to go back to Para. 13.6 and read it carefully. In the second part of this section, we will consider interesting ways of controlling a converter directly from its averaged nonlinear system.

13.1 Minimisation of Energy in an Inrush Limitation Circuit

13.1.1 System Description

Consider the inrush current limitation circuit shown in Fig. 13.1a.

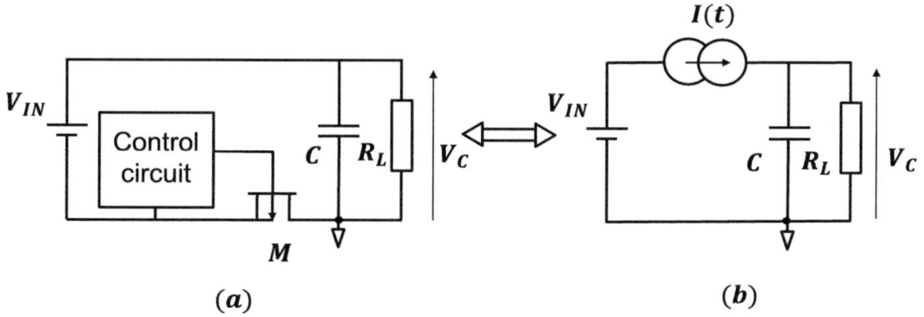

Fig. 13.1 Optimisation of a current limitation circuit

B. Rivet, *Theories of Nonlinear Systems for Power Converters*,
https://doi.org/10.1007/978-3-032-25062-9_13

The MOSFET M is used to limit the inrush current flowing through the capacitor C and the load R_L. It acts as a current generator $I(t)$ with limits given by:

$$0\,\mathrm{A} < I(t) < 1\,\mathrm{A}. \tag{13.1}$$

Our purpose is to charge the capacitor between 0 V and 40 V in a fixed duration T by minimising the energy E dissipated in the MOSFET. This system is a first-order system with the state parameter V_C and the control parameter $I(t)$. The values of the system parameters are:

$$V_{\mathrm{IN}} = 70\,\mathrm{V}, \quad R_L = 100\,\Omega, \quad C = 500\,\mu\mathrm{F}, \quad T = 100\,\mathrm{ms}. \tag{13.2}$$

The dynamics are given by:

$$\frac{\mathrm{d}V_C}{\mathrm{d}t} = -\frac{V_C}{R_L C} + \frac{I}{C}. \tag{13.3}$$

13.1.2 Optimal Control of the System

We want to minimise the energy dissipated in the MOSFET, so the cost functional to maximise is:

$$J(I(t)) = -\int_0^T (V_{\mathrm{IN}} - V_C) I \,\mathrm{d}t \tag{13.4}$$

The Lagrangian is given by:

$$L = (V_C - V_{\mathrm{IN}}) I. \tag{13.5}$$

According to the Pontryagin's theorem (Evans, 2001; Pontryagin, 1962), there exists an adjoint parameter $p(t)$ such that:

$$H(V_C, p, I) = (V_C - V_{\mathrm{IN}}) I + p\left(-\frac{V_C}{R_L C} + \frac{I}{C}\right) = \left(V_C - V_{\mathrm{IN}} + \frac{p}{C}\right) I - \frac{p V_C}{R_L C} \tag{13.6}$$

$$\frac{\mathrm{d}V_C}{\mathrm{d}t} = \frac{\partial H}{\partial p} = -\frac{V_C}{R_L C} + \frac{I}{C} \tag{13.7}$$

$$\frac{\mathrm{d}p}{\mathrm{d}t} = -\frac{\partial H}{\partial V_C} = -I + \frac{p}{R_L C} \tag{13.8}$$

H depends linearly on the bounded control I. The switching function is:

$$\varphi(t) = V_C(t) - V_{\mathrm{IN}} + \frac{p(t)}{C}. \tag{13.9}$$

As a consequence, the control is defined by:

$$\begin{aligned} &\text{if } \varphi(t) < 0 \ \text{ then } I = 0, \\ &\text{if } \text{Ø}(t) > 0 \ \text{ then } I = 1\,\text{A}, \\ &\text{if } \text{Ø}(t) = 0 \ \text{ then } I \text{ is undefined.} \end{aligned} \tag{13.10}$$

The transversality conditions give:

$$V_C(0) = 0\ \text{V} \ \text{ and } \ V_C(T) = 40\,\text{V}, \tag{13.11}$$

so $p(0)$ and $p(T)$ are free.

Thanks to the shooting method, we find that the optimal control is obtained with:

$$p^*(0) \simeq 0.0079\,\text{As}. \tag{13.12}$$

The corresponding waveforms of $\varphi^*(t)$, $V_C^*(t), p^*(t), I^*(t), E^*(t)$ are shown in Fig. 13.2. This simulation indicates that the minimum energy dissipated in M is

$$E_{\min} \simeq 1.2\,\text{J}. \tag{13.13}$$

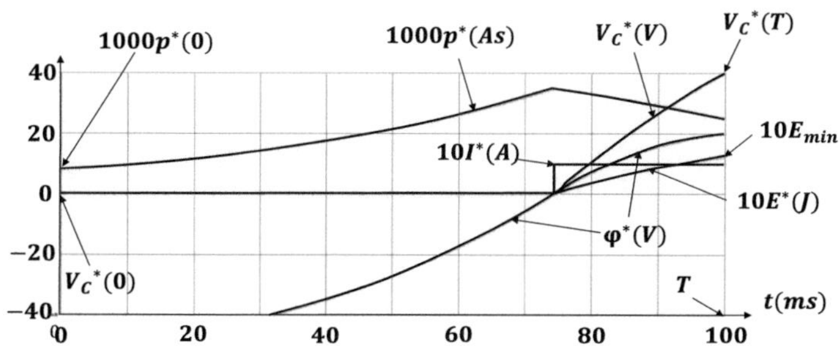

Fig. 13.2 Optimal waveforms of $\varphi(t)$, $V_C^*(t), p^*(t), I^*(t), E^*(t)$

13.2 Minimisation of the Dissipated Energy at Start up

13.2.1 System Description

Consider the buck converter depicted in Fig. 13.3.

This converter uses synchronous rectification. The two MOSFETs M_1 and M_2 are modelised in the on-state mode by resistors R_{DSON}. This converter is designed to operate only in continuous conduction mode. We will use the following averaged model to describe the dynamic behaviour of this converter. We obtain:

$$\begin{cases} \frac{\mathrm{d}V_C}{\mathrm{d}t} = -\frac{V_C}{R_L C} + \frac{I_L}{C} \\ \frac{\mathrm{d}I_L}{\mathrm{d}t} = -\frac{V_C}{L} - \frac{R_{\text{DSON}} I_L}{L} + d\,\frac{V_{\text{IN}}}{L} \end{cases}. \tag{13.14}$$

This system is second order with the state variables V_C and I_L. The duty cycle of M_1, d acts as a control parameter ($0 < \mathrm{d}(t) < 1$). The values of the system parameters are:

$$V_{\text{IN}} = 100\,\text{V}, \quad R_L = 5\,\Omega, \quad R_{\text{DSON}} = 0.1\,\Omega, \quad C = 1\,\text{mF}, \quad L = 1\,\text{mH}. \tag{13.15}$$

We want to charge the capacitor between 0 V and 130 V with a free duration T by minimising the energy dissipated in the two MOSFETs and the load R_L. At the time T, the current I_L must be equal to zero.

13.2.2 Research on the Optimal Control

The cost functional to maximise is:

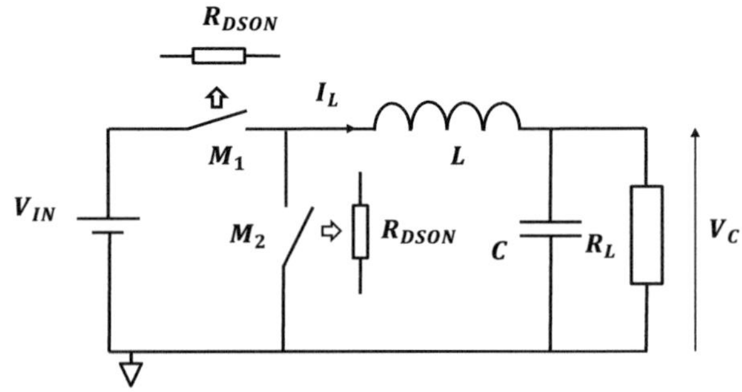

Fig. 13.3 Buck converter using synchronous rectification

$$J(I(t)) = -\int_0^T \left(R_{\mathrm{DSON}} I_L^2 + \frac{V_C^2}{R_L} \right) \mathrm{d}t. \tag{13.16}$$

The Lagrangian is given by:

$$L = -\left(R_{\mathrm{DSON}} I_L^2 + \frac{V_C^2}{R_L} \right). \tag{13.17}$$

The Pontryagin principle (Evans, 2001; Pontryagin, 1962) states:

There exists an adjoint vector $\vec{p}(t) = (p_0(t), p_1(t))$ such that:

$$\begin{aligned} H(V_C, p, I) = &-R_{\mathrm{DSON}} I_L^2 - \frac{V_C^2}{R_L} + p_0 \left(-\frac{V_c}{R_L C} + \frac{I_L}{C} \right) \\ &+ p_1 \left(-\frac{V_c}{L} - \frac{R_{\mathrm{DSON}} I_L}{L} + d\frac{V_{\mathrm{IN}}}{L} \right) \end{aligned} \tag{13.18}$$

$$\begin{cases} \frac{\mathrm{d}V_c}{\mathrm{d}t} = \frac{\partial H}{\partial p_0} = -\frac{V_c}{R_L C} + \frac{I_L}{C} \\ \frac{\mathrm{d}I_L}{\mathrm{d}t} = \frac{\partial H}{\partial p_1} = -\frac{V_c}{L} - \frac{R_{\mathrm{DSON}} I_L}{L} + d\frac{V_{\mathrm{IN}}}{L} \end{cases} \tag{13.19}$$

$$\begin{cases} \frac{\mathrm{d}p_0}{\mathrm{d}t} = -\frac{\partial H}{\partial V_C} = \frac{2V_c}{R_L} + \frac{p_0}{R_L C} + \frac{p_1}{L} \\ \frac{\mathrm{d}p_1}{\mathrm{d}t} = -\frac{\partial H}{\partial I_L} = 2R_{\mathrm{DSON}} I_L - \frac{p_0}{C} + \frac{p_1 R_{\mathrm{DSON}}}{L} \end{cases} \tag{13.20}$$

H depends linearly on the bounded control d. The switching function can be defined as:

$$\varphi(t) = p_1(t). \tag{13.21}$$

Thus, the control is given by:

$$\begin{aligned} &\text{if } \varphi(t) < 0 \text{ then } d = 0, \\ &\text{if } \varphi(t) > 0 \text{ then } d = 1, \\ &\text{if } \varphi(t) = 0 \text{ then } d \text{ is undefined.} \end{aligned} \tag{13.22}$$

The transversality conditions are given by:

$$V_C(0) = 0V,\ I_L(0) = 0\ \mathrm{A} \text{ and } V_C(T) = 130\ \mathrm{V},\ I_L(T) = 0\ \mathrm{A} \tag{13.23}$$

so, $p_0(0), p_1(0)$ and $p_0(T), p_1(T)$ are free.

Thanks to the shooting method, the optimal control is obtained by:

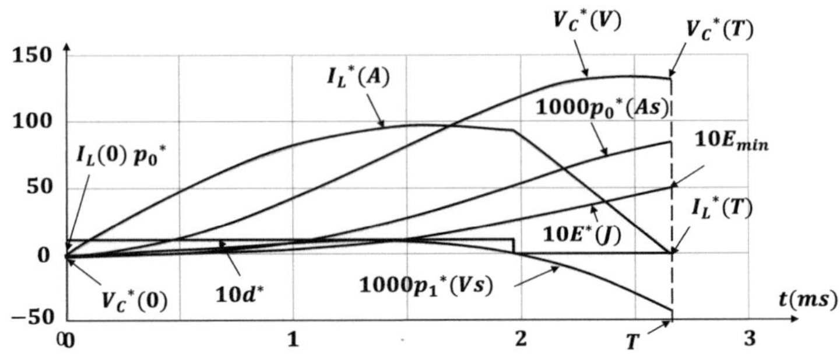

Fig. 13.4 Optimal waveforms of $V_C^*(t), I_L^*(t), p_0^*(t), p_1^*(t), d^*(t), E^*(t)$

$$p_0^*(0) = -0.001\,\text{As}, \quad p_1^*(0) = 0\,\text{AH}. \tag{13.24}$$

The corresponding waveforms of $V_C^*(t), I_L^*(t), p_0^*(t), p_1^*(t), d^*(t), E^*(t)$ are shown in Fig. 13.4. The minimum energy dissipated in the two MOSFETs and the load is:

$$E_{\min} \simeq 4.96\,\text{J}. \tag{13.25}$$

The duration is equal to:

$$T \simeq 2.66\,\text{ms}. \tag{13.26}$$

13.3 Minimisation of the Charging Time of Two Capacitors

13.3.1 System Description

In this section, we will analyse the charging circuit depicted in the Fig. 13.5.

Fig. 13.5 Charging circuit for two capacitors C_1 and C_2

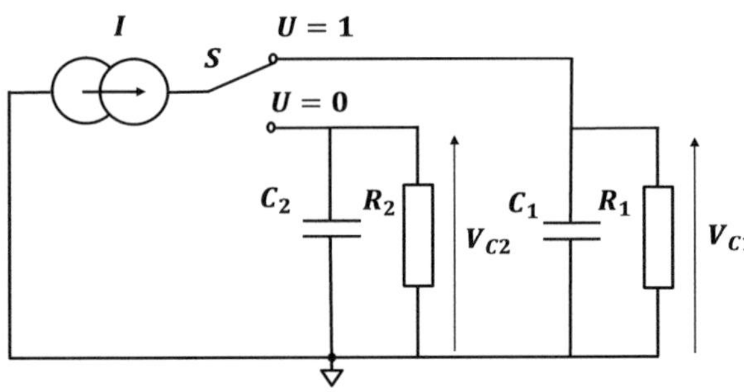

Depending on the position of the switch (S), the constant current generator I charges the capacitors C_1 or C_2. We can define the control parameter U as $U = 1$ if the current generator charges the capacitor C_1, and as $U = 0$ if the current generator charges the capacitor C_2. The dynamics of this system are given by:

$$\begin{cases} \frac{\mathrm{d}V_{C1}}{\mathrm{d}t} = -\frac{V_{C1}}{R_1 C_1} + \frac{I}{C_1} U \\ \frac{\mathrm{d}V_{C2}}{\mathrm{d}t} = -\frac{V_{C2}}{R_2 C_2} + \frac{I}{C_2}(1-U) \end{cases}. \tag{13.29}$$

This system is a second-order system with state parameters V_{C1} and V_{C2}, and the control input U. Our goal is to charge the capacitors C_1 and C_2 from 0 V to 40 V and from 0 V to 100 V, respectively in minimum time T. The analysis will be performed using the following numerical values:

$$I = 1\,\mathrm{A},\; R_1 = 1\,\mathrm{k}\Omega,\; C_1 = 100\,\mu\mathrm{F},\; R_2 = 100\,\Omega,\; C_2 = 200\,\mu\mathrm{F}. \tag{13.28}$$

13.3.2 Optimal Control of the System

The cost functional to maximise is:

$$J(I(t)) = -\int_0^T \mathrm{d}t. \tag{13.29}$$

The Lagrangian is given by:

$$L = -1. \tag{13.30}$$

The Pontryagin principle (Evans, 2001; Pontryagin, 1962) states that there exists an adjoint vector $\vec{p}(t) = (p_0(t), p_1(t))$ such that:

$$H\big(V_{C1}, V_{C2,}p_0, p_1, I\big) = -1 + p_0\left(-\frac{V_{C1}}{R_1 C_1} + \frac{I}{C_1}U\right) + p_1\left(-\frac{V_{C2}}{R_2 C_2} + \frac{I}{C_2}(1-U)\right) \tag{13.31}$$

$$\begin{cases} \frac{\mathrm{d}V_{C1}}{\mathrm{d}t} = \frac{\partial H}{\partial p_0} = -\frac{V_{C1}}{R_1 C_1} + \frac{I}{C_1} U \\ \frac{\mathrm{d}V_{C2}}{\mathrm{d}t} = \frac{\partial H}{\partial p_0} = -\frac{V_{C2}}{R_2 C_2} + \frac{I}{C_2}(1-U) \end{cases} \tag{13.32}$$

$$\begin{cases} \frac{\mathrm{d}p_0}{\mathrm{d}t} = -\frac{\partial H}{\partial V_{C1}} = \frac{p_0}{R_1 C_1} \\ \frac{\mathrm{d}p_1}{\mathrm{d}t} = -\frac{\partial H}{\partial V_{C2}} = \frac{p_1}{R_2 C_2} \end{cases} \tag{13.35}$$

H depends linearly on the bounded control U. The switching function can be defined as:

$$\varphi(t) = p_0(t) - \frac{C_1 p_1(t)}{C_2}. \tag{13.36}$$

So, the control is given by:

$$\begin{aligned} &\text{if } \varphi(t) < 0 \ \text{ then } U = 0, \\ &\text{if } \varphi(t) > 0 \ \text{ then } U = 1\,, \\ &\text{if } \varphi(t) = 0 \ \text{ then } U \text{ is undefined.} \end{aligned}$$

The transversality conditions are given by:

$$V_{C1}(0) = 0\,\text{V}, \ V_{C2}(0) = 0\,\text{V and } V_{C1}(T) = 40\,\text{V}, \ V_{C2}(T) = 100\,\text{V}. \tag{13.37}$$

so,

$$p_0(0), \ p_1(0), \ p_0(T), \ p_1(T) \text{ are free} \tag{13.36}$$

Thanks to the shooting method, the optimal control is obtained as

$$p_0^*(0) = 0.0001\,\text{s}\,\text{V}^{-1}, \ \ p_1^*(0) = 0.0001\,\text{s}\,\text{V}^{-1}. \tag{13.37}$$

leading to the minimum time:

$$T \simeq 155.4\,\text{ms} \tag{13.38}$$

The optimal waveforms of $V_{C1}^*(t)$, $V_{C2}^*(t)$, $U^*(t)$, $\varphi^*(t)$ are presented in Fig. 13.6.

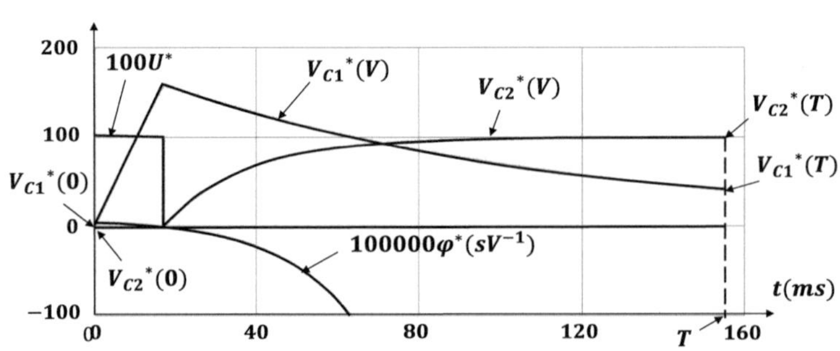

Fig. 13.6 Optimal waveforms of $V_{C1}^*(t)$, $V_{C2}^*(t)$, $U^*(t)$, $\varphi^*(t)$

13.4 Minimization of Charging Time and Junction Temperature

13.4.1 System Description

In this section, we will study the system represented in Fig. 13.7.

It consists of a current generator $I(t)$ charging a capacitor C through a MOSFET transistor modelled by its on-state resistance R_{DSON} (see Fig. 13.7a). In this system, we consider the thermal dependence of $R_{\text{DSON}}(T_J)$ on the junction temperature T_J. The thermal model of the transistor is shown in Fig. 13.7b. The dynamics of this system are given by:

$$\begin{cases} \frac{\mathrm{d}V_C}{\mathrm{d}t} = +\frac{I}{C} \\ \frac{\mathrm{d}T_J}{\mathrm{d}t} = +\frac{R_{\text{DSONref}}}{\text{C}th}\left[1+\alpha\left(T_J - T_{ref}\right)\right]I^2 + \frac{T_C - T_J}{R\text{th}C\text{th}} \end{cases}. \tag{13.39}$$

This system is a second-order system with the two state variables V_C and T_J and the control input $I(t)$. Our goal is to charge the capacitor C from 0 V to 70 V in minimum time T while limiting the final junction temperature $T_J(T)$. We will use the following numerical values:

$$\begin{aligned} &R\text{th} = 1\,^\circ\text{C}\,\text{W}^{-1}, \quad C\text{th} = 0.01\,\text{Ws}\,^\circ\text{C}^{-1}, \quad C = 1\,\text{mF}, \quad T_C = 100\,^\circ\text{C}, \\ &T_{\text{ref}} = T_C = 100\,^\circ\text{C}, \quad R_{\text{DSONref}} = 0.5\,\Omega, \quad \alpha = 0.01\,\Omega\,^\circ\text{C}^{-1}. \end{aligned} \tag{13.40}$$

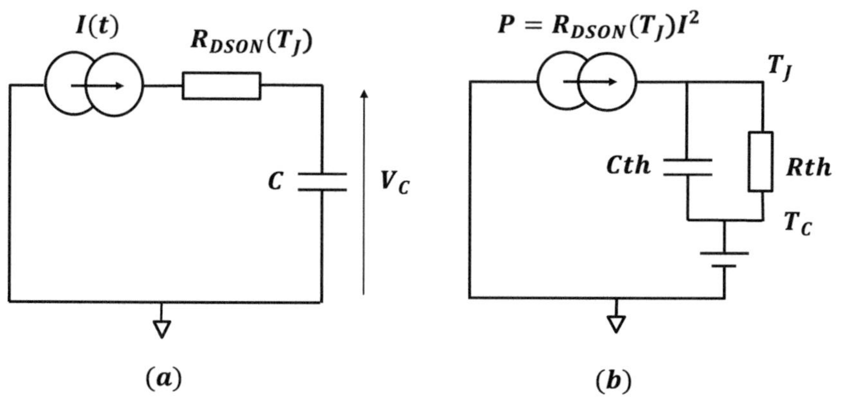

Fig. 13.7 Capacitor charging circuit with the thermal model of a MOSFET transistor

13.4.2 Optimal Control of the System

The cost functional to maximise is:

$$J(I(t)) = -\int_0^T k\,\mathrm{d}t - T_J(T). \tag{13.41}$$

The Lagrangian is given by:

$$L = -k. \tag{13.42}$$

k is a weighting coefficient. We will take:

$$k = 10000\,^\circ\mathrm{C\,s}^{-1} \tag{13.43}$$

As mentioned in the previous examples, using the Pontryaging principle (Evans, 2001; Pontryagin, 1962), we obtain the following result:

there exists an adjoint vector $\vec{p}(t) = (p_0(t), p_1(t))$, such that:

$$\begin{aligned} H(V_C, T_J, p_0, p_1, I) &= -k + p_0\left(\frac{I}{C}\right) \\ &+ p_1\left(\frac{R_{\mathrm{DSONref}}}{C\mathrm{th}}[1 + \alpha(T_J - T_{\mathrm{ref}})]I^2 + \frac{T_C - T_J}{R\mathrm{th}C\mathrm{th}}\right) \end{aligned} \tag{13.44}$$

$$\begin{cases} \frac{\mathrm{d}V_C}{\mathrm{d}t} = \frac{\partial H}{\partial p_0} = \frac{I}{C} \\ \frac{\mathrm{d}T_J}{\mathrm{d}t} = \frac{\partial H}{\partial p_1} = -\frac{T_J}{R\mathrm{th}C\mathrm{th}} + \frac{R_{\mathrm{DSONref}}}{C\mathrm{th}}[1 + \alpha(T_J - T_{\mathrm{ref}})]I^2 + \frac{T_C}{R\mathrm{th}C\mathrm{th}} \end{cases} \tag{13.45}$$

$$\begin{cases} \frac{\mathrm{d}p_0}{\mathrm{d}t} = -\frac{\partial H}{\partial V_C} = 0 \rightarrow p_0 = \text{constant} \\ \frac{\mathrm{d}p_1}{\mathrm{d}t} = -\frac{\partial H}{\partial TJ} = \frac{p_1}{R_{\mathrm{th}} C_{\mathrm{th}}} - \frac{p_1}{C_{\mathrm{th}}} R_{\mathrm{DSONref}} \alpha I^2 \end{cases} \tag{13.46}$$

H does not depend linearly on the control I, so the control is not bang-bang. The control is obtained by:

$$\frac{\partial H}{\partial I} = 0. \tag{13.47}$$

It follows that:

$$I^* = -\frac{p_0 C\mathrm{th}}{2Cp_1^*(t) R_{\mathrm{DSONref}}[1 + \alpha(T_J - T_{\mathrm{ref}})]}. \tag{13.48}$$

The transversality conditions give:

$$V_C(0) = 0,\, T_J(0) = T_{\text{ref}} = T_C, \tag{13.49}$$

so $p_0(0), p_1(0)$ are free. $V_C(T) = 70V$, so $p_0^*(T)$ is free and

$$p_1^*(T) = \frac{\partial(-T_J(T))}{\partial T_J} = -1. \tag{13.50}$$

Since T is free and the system is autonomous, we can write:

$$H(t) = 0 \quad \text{for } t \in [0, T] \tag{13.51}$$

which gives:

$$p_1^*(0) = -\frac{C\text{th}p_0^*(0)^2}{4kR_{\text{DSONref}}C^2}. \tag{13.52}$$

Using the shooting method, we obtain:

$$p_0^*(0) \simeq 1.3088\,°\text{C}\,\text{V}^{-1}, \quad p_1^*(0) \simeq 0.856\,°\text{C}\,\text{V}^{-1} \tag{13.53}$$

with

$$T \simeq 5.8\,\text{ms} \quad \text{and} \quad T_J^*(T) \simeq 138\,°\text{C}. \tag{13.54}$$

The waveforms of $V_C^*(t), T_J^*(t), p_1^*(t), I^*(t)$ corresponding to the optimal control are shown in Fig. 13.8.

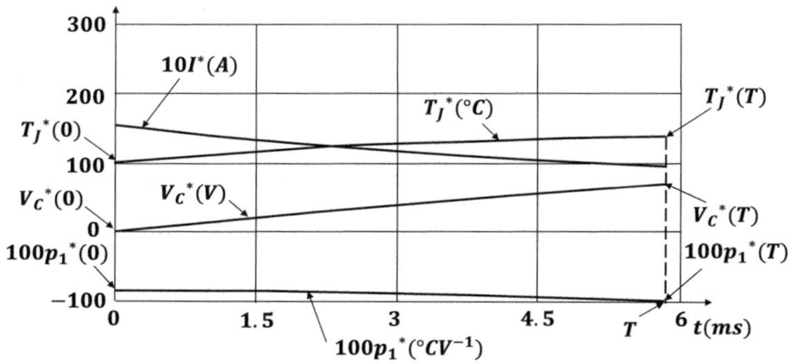

Fig. 13.8 Optimal waveforms of $V_C^*(t), T_J^*(t), p_1^*(t), I^*(t)$

13.5 Buck Converter Controlled by Lyapunov Control

Different nonlinear control strategies directly derived from the averaged model of the converter are now considered. Let us begin with the control of a buck converter using a Lyapunov function (Kokotović & Arcak, 2001; Lyapunov, 1992; Praly & Pietri, 2016).

13.5.1 Buck Converter Controlled Using a First Lyapunov Candidate Function

The averaged model of a buck converter without parasitic component is:

$$\begin{cases} \frac{\mathrm{d}V_C}{\mathrm{d}t} = -\frac{V_C}{R_L C} + \frac{I_L}{C} = F_1(V_C, I_L) \\ \frac{\mathrm{d}I_L}{\mathrm{d}t} = -\frac{V_C}{L} + d\frac{V_{\mathrm{IN}}}{L} = F_2(V_C, I_L, d) \end{cases}. \tag{13.55}$$

The equilibrium point $(V_{\mathrm{Ce}}, I_{\mathrm{Le}}, d_e)$ is given by:

$$V_{\mathrm{Ce}} = V_{\mathrm{OUTtar}}, \quad I_{\mathrm{Le}} = \frac{V_{\mathrm{Ce}}}{R_L} = \frac{V_{\mathrm{OUTtar}}}{R_L}, \quad d_e = \frac{V_{\mathrm{Ce}}}{V_{\mathrm{IN}}} = \frac{V_{\mathrm{OUTtar}}}{V_{\mathrm{IN}}}. \tag{13.56}$$

Consider the first Lyapunov function:

$$V_1(V_C, I_L) = \frac{1}{2}C(V_C - V_{\mathrm{Ce}})^2 + \frac{1}{2}L(I_L - I_{\mathrm{Le}})^2. \tag{13.57}$$

This function is a candidate Lyapunov function since:

$$V_1(V_{\mathrm{Ce}}, I_{\mathrm{Le}}) = 0 \text{ and } V_1(V_C, I_L) > 0 \quad \forall (V_C, I_L) \neq (V_{\mathrm{Ce}}, I_{\mathrm{Le}}). \tag{13.58}$$

The time derivative of $V_1(V_C, I_L)$ is given by:

$$\frac{\mathrm{d}V_1(V_C, I_L)}{\mathrm{d}t} = \frac{\partial V_1(V_C, I_L)}{\partial V_C}F_1(V_C, I_L) + \frac{\partial V_1(V_C, I_L)}{\partial I_L}F_2(V_C, I_L). \tag{13.59}$$

We obtain the following result:

$$\frac{\mathrm{d}V_1(V_C, I_L)}{\mathrm{d}t} = -\frac{V_C^2}{R_L} + 2\frac{V_C V_{\mathrm{OUTtar}}}{R_L} - V_{\mathrm{OUTtar}} I_L + d\left(I_L V_{\mathrm{IN}} - \frac{V_{\mathrm{OUTt}}}{R_L}V_{\mathrm{IN}}\right). \tag{13.60}$$

It is not possible to find a control law $d(V_C, I_L)$ in a region A around $(V_{\mathrm{Ce}}, I_{\mathrm{Le}})$ such that:

$$\frac{\mathrm{d}V_1(V_{\mathrm{Ce}}, I_{\mathrm{Le}})}{\mathrm{d}t} = 0 \ \text{ and } \ \frac{\mathrm{d}V_1(V_C, I_L)}{\mathrm{d}t} < 0$$
$$\forall (V_C, I_L) \neq (V_{\mathrm{Ce}}, I_{\mathrm{Le}}) \subset A. \tag{13.61}$$

In consequence, this Lyapunov function $V_1(V_C, I_L)$ cannot be used to stabilise a buck converter.

13.5.2 *Buck Controlled with a Second Lyapunov Candidate Function*

Now consider the second Lyapunov function:

$$V_2(V_C, I_L,) = \frac{1}{2}C(V_C - V_{\mathrm{Ce}})^2 + \frac{1}{2}L(I_L - I_{\mathrm{Le}})^2 + \frac{1}{2}k_e(d - d_e)^2. \tag{13.62}$$

This function is a candidate Lyapunov function since:

$$V_2(V_{\mathrm{Ce}}, I_{\mathrm{Le}}, d_e) = 0 \ \text{ and } \ V_1(V_C, I_L, d) > 0$$
$$\forall \ (V_C, I_L, d) \neq (V_{\mathrm{Ce}}, I_{\mathrm{Le}}, d_e). \tag{13.63}$$

The coefficient $k_e(J) > 0$ is chosen to maintain the homogeneity of the formula (13.62). Choose the control defined by:

$$\dot{d} = -\frac{1}{k_e}(I_L - I_{\mathrm{Le}})V_{\mathrm{IN}} - k(d - d_e) = F_3(V_C, I_L, d) \tag{13.64}$$

$k\left(s^{-1}\right)$ being an adjustable control parameter. The time derivative of $V_2(V_C, I_L, d)$ is:

$$\frac{\mathrm{d}V_2(V_C, I_L, d)}{\mathrm{d}t} = \frac{\partial V_2(V_C, I_L, d)}{\partial V_C}F_1(V_C, I_L, d)$$
$$+ \frac{\partial V_2(V_C, I_L)}{\partial I_L}F_2(V_C, I_L, d) + \frac{\partial V_2(V_C, I_L)}{\partial d}F_3(V_C, I_L, d) \tag{13.65}$$

After some calculations, we find:

$$\frac{\mathrm{d}V_2(V_C, I_L, d)}{\mathrm{d}t} = -kk_e(d - d_e)^2 - \frac{(V_C - V_{\mathrm{Ce}})^2}{R_L} \tag{13.66}$$

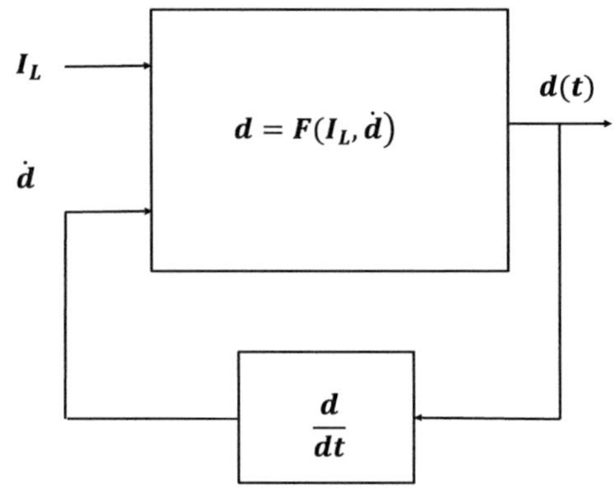

Fig. 13.9 Control block diagram

With $k > 0$, this Lyapunov function $V_2(V_C, I_L, d)$ is a good candidate function since:

$$\frac{\mathrm{d}V_2(V_{Ce}, I_{Le}, d_e)}{\mathrm{d}t} = 0 \text{ and } \frac{\mathrm{d}V_2(V_C, I_L, d)}{\mathrm{d}t} < 0$$
$$\forall\, (V_C, I_L, d) \neq (V_{Ce}, I_{Le}, d_e). \tag{13.67}$$

Thus, it is possible to stabilise a buck converter with this second Lyapunov function. From Eq. (13.64), we can deduce:

$$d = -\frac{1}{kk_e}\left(I_L - \frac{V_{\mathrm{OUTtar}}}{R_L}\right)V_{\mathrm{IN}} - \frac{1}{k}\dot{d} + \frac{V_{\mathrm{OUTtar}}}{V_{\mathrm{IN}}} = F\left(I_L, \dot{d}\right). \tag{13.68}$$

The principle of this control is illustrated in Fig. 13.9.

Note that the dynamics of this controlled system can be described by the following third-order system, where d is now considered as a state variable.

$$\begin{cases} \frac{\mathrm{d}V_C}{\mathrm{d}t} = -\frac{V_C}{R_L C} + \frac{I_L}{C} \\ \frac{\mathrm{d}I_L}{\mathrm{d}t} = -\frac{V_C}{L} + d\frac{V_{\mathrm{IN}}}{L} \\ \frac{\mathrm{d}d}{\mathrm{d}t} = -\frac{1}{k_e}\left(I_L - \frac{V_{\mathrm{OUTtar}}}{R_L}\right)V_{IN} - k\left(d - \frac{V_{\mathrm{OUTtar}}}{V_{\mathrm{IN}}}\right) \end{cases}. \tag{13.69}$$

Note that it is also possible to find a Lyapunov function using the Backstepping control method (see Paras. 7.2 and 14.2).

13.6 Sepic Controlled by a Lyapunov Function

It is a well known that the Sepic converter shown in Fig. 13.10 is difficult to control.

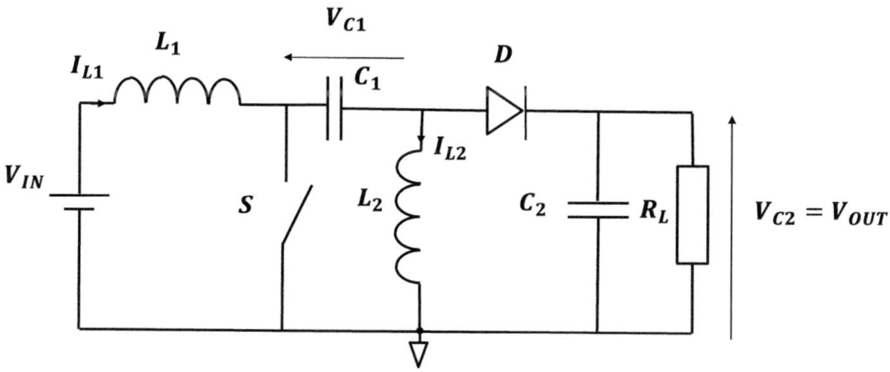

Fig. 13.10 Sepic converter

This because there is only one control parameter for the four state variables $V_{C1}, I_{l1}, V_{C2}, I_{L2}$. The averaged model of a Sepic converter without parasitic component is given by:

$$\begin{cases} \frac{dV_{C1}}{dt} = \frac{(1-d)I_{L1}}{C_1} + \frac{dI_{L2}}{C_1} \\ \frac{dI_{L1}}{dt} = -\frac{(1-d)V_{C1}}{L_1} - \frac{(1-d)V_{C2}}{L_1} + \frac{V_{IN}}{L_1} \\ \frac{dV_{C2}}{dt} = -\frac{V_{C2}}{R_L C_2} + \frac{(1-d)I_{L1}}{C_2} - \frac{(1-d)I_{L2}}{C_2} \\ \frac{dI_{L2}}{dt} = -\frac{dV_{C1}}{L_2} + \frac{(1-d)V_{C2}}{L_2} \end{cases}. \tag{13.70}$$

The equilibrium point $(V_{C1e}, I_{L1e}, V_{C2e}, I_{L2e})$ is defined by:

$$V_{C1e} = V_{\mathrm{IN}}, \quad I_{L1e} = \frac{V_{\mathrm{OUTtar}}^2}{V_{\mathrm{IN}} R_L}, \quad V_{C2e} = V_{\mathrm{OUTtar}}, \quad I_{L2e} = -\frac{V_{\mathrm{OUTtar}}}{R_L},$$
$$d_e = \frac{V_{\mathrm{OUTtar}}}{V_{\mathrm{IN}} + V_{\mathrm{OUTtar}}}. \tag{13.71}$$

Consider now the following Lyapunov function (Kokotović & Arcak, 2001; Lyapunov, 1992; Praly & Pietri, 2016):

$$V(V_{C1}, I_{L1}, V_{C2}, I_{L2}, d) = \frac{1}{2}C_1(V_{C1} - V_{C1e})^2 + \frac{1}{2}L_1(I_{L1} - I_{L1e})^2$$
$$+ \frac{1}{2}C_2(V_{C2} - V_{C2e})^2 + \frac{1}{2}L_2(I_{L2} - I_{L2e})^2 + \frac{1}{2}k_e(d - d_e)^2. \tag{13.72}$$

This function is a candidate Lyapunov function since

$$V(V_{C1e}, I_{L1e}, V_{C2e}, I_{L2e}, d_e) = 0 \tag{13.73}$$

and

$$V(V_{C1}, I_{L1}, V_{C2}, I_{L2}, d) > 0$$
$$\forall\ (V_{C1}, I_{L1}, V_{C2}, I_{L2}, d) \neq (V_{C1e}, I_{L1e}, V_{C2e}, I_{L2e}). \tag{13.74}$$

The control is defined by:

$$\dot{d} = \frac{1}{k_e}(V_{C1} + V_{C2})(I_{L1e} - I_{L2e}) + \frac{1}{k_e}(I_{L2} - I_{L1})(V_{C1e} + V_{C2e}) - k(d - d_e). \tag{13.75}$$

After some calculations, we find:

$$\frac{\mathrm{d}V(V_{C1}, I_{L1}, V_{C2}, I_{L2}, d)}{\mathrm{d}t} = -kk_e(d - d_e)^2 - \frac{(V_{C2} - V_{\mathrm{OUTtar}})^2}{R_L}. \tag{13.76}$$

The Lyapunov function $V(V_{C1}, I_{L1}, V_{C2}, I_{L2}, d)$ is a good candidate function since:

$$\frac{\mathrm{d}V(V_{C1e}, I_{L1e}, V_{C2e}, I_{L2e}, d_e)}{\mathrm{d}t} = 0 \text{ and } \frac{\mathrm{d}V(V_{C1}, I_{L1}, V_{C2}, I_{L2}, d)}{\mathrm{d}t} < 0$$
$$\forall(V_C, I_L, d) \neq (V_{\mathrm{Ce}}, I_{\mathrm{Le}}, d_e). \tag{13.77}$$

Thus, it is possible to control a Sepic converter using this Lyapunov function associated with the control $d(t)$. Figure 13.11 shows the simulation results of this control obtained with the following numerical values

$$V_{\mathrm{IN}} = 50\,\mathrm{V},\quad V_{\mathrm{OUTtar}} = 25\,\mathrm{V},\quad C_1 = C_2 = 500\,\mu\mathrm{F},\quad L_1 = L_2 = 1\,\mathrm{mH},\quad R_L = 1\,\Omega$$
$$k = 1\,\mathrm{s}^{-1},\quad k_e = 1\,\mathrm{J}. \tag{13.78}$$

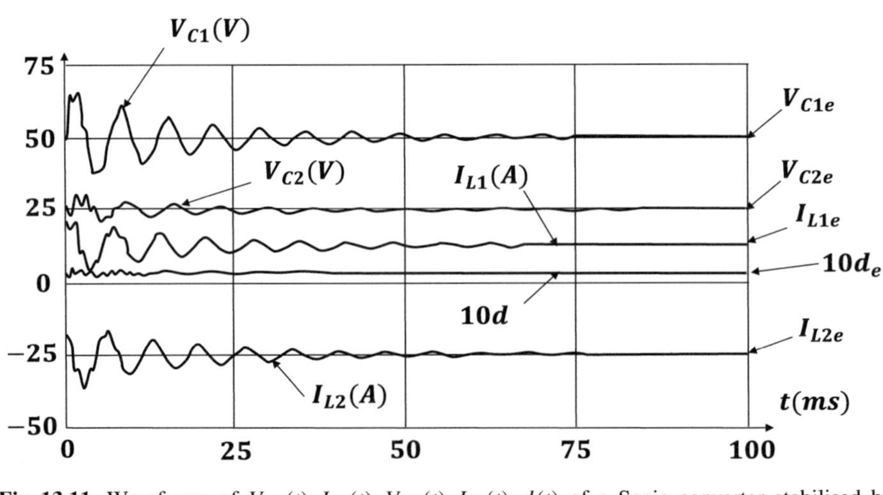

Fig. 13.11 Waveforms of $V_{C1}(t), I_{L1}(t), V_{C2}(t), I_{L2}(t), d(t)$ of a Sepic converter stabilised by Lyapunov control

13.7 Theory of the Direct Nonlinear Control

In this section, we will discuss the direct nonlinear control developed by Benoit Peron. Consider again the following averaged model of a converter

$$\begin{cases} \frac{dX_1}{dt} = F_1(X_1, \ldots, X_i, \ldots, X_n, d) \\ \vdots \\ \frac{dX_i}{dt} = F_i(X_1, \ldots, X_i, \ldots, X_n, d) \\ \vdots \\ \frac{dX_n}{dt} = F_n(X_1, \ldots, X_i, \ldots, X_n, d) \end{cases} . \quad (13.79)$$

The direct nonlinear control consists of introducing a new control parameter G_i equal to one of the nonlinear functions F_1, ..., F_i, ..., F_n :

$$G_i = F_i(X_1, \ldots, X_i \ldots, X_n, d). \quad (13.80)$$

The dynamics of this new controlled system becomes:

$$\begin{cases} \frac{dX_1}{dt} = F_1'(X_1, \ldots, X_i, \ldots, X_n, G_i) \\ \vdots \\ \frac{dX_i}{dt} = G_i \\ \vdots \\ \frac{dX_n}{dt} = F_n'(X_1, \ldots, X_i, \ldots, X_n, G_i) \end{cases} . \quad (13.81)$$

There are therefore n possibilities. We need to eliminate cases where the functions F_i do not depend on d because the former control parameter d must be expressed in term of the new control parameter G_i. In practice, we will choose the F_i function to directly control the state variable X_i of interest. This state variable is generally the voltage V_{OUT} across the output capacitor, allowing direct control the output voltage, or the courant through an inductor, allowing for current regulation.

Now consider a small variation x_i of the state variable X_i around the equilibrium point X_{ie}:

$$X_i = X_{ie} + x_i. \quad (13.82)$$

we obtain:

$$\frac{dX_i}{dt} = \frac{d(X_{ie} + x_i)}{dt} = \frac{dx_i}{dt} = G_{ie} + g_i \quad (13.83)$$

Since

$$G_{ie} = F_i(X_{1e}, \ldots, X_{ie}, \ldots, X_{ne}, d_e) = 0. \tag{13.84}$$

it results in the following interesting outcome:

$$\frac{\mathrm{d}x_i}{\mathrm{d}t} = g_i. \tag{13.85}$$

This means that the small signal control g_i acts as a first-order control on the small state variable x_i. Moreover, this control is not influenced by the variation in system parameters, such as the input voltage V_{IN} or the load. The physical control, which is always the duty cycle d will be calculated according to G_i. This will be done using Eq. (13.80).

13.8 Buck Converter Controlled with Direct Control

For a buck converter, the corresponding averaged model given by the system (13.55) shows that it is not possible to directly control the output voltage:

$$V_{\mathrm{OUT}} = V_C, \tag{13.86}$$

because d is not present in the expression of $F_1(V_C, I_L)$. Nevertheless, it is possible to perform direct control on the current I_L using the new control parameter:

$$G_2 = F_2(V_C, I_L, d) = -\frac{V_C}{L} + d\frac{V_{\mathrm{IN}}}{L}. \tag{13.37}$$

The new controlled system becomes:

$$\begin{cases} \frac{\mathrm{d}V_C}{\mathrm{d}t} = -\frac{V_C}{R_L C} + \frac{I_L}{C} = F_1(V_C, I_L) \\ \frac{\mathrm{d}I_L}{\mathrm{d}t} = G_2 \end{cases}. \tag{13.88}$$

The transfer function with the new control is:

$$\frac{I_L(s)}{G_2(s)} = \frac{1}{s}. \tag{13.89}$$

whereas with the former control, we have the more complex dependency

$$I_L(s) = -\frac{1}{Ls}V_{\mathrm{OUT}}(s) + \frac{d_e}{Ls}V_{\mathrm{IN}}(s) + \frac{V_{\mathrm{IN}}}{Ls}d(s). \tag{13.90}$$

The duty cycle will be calculated using:

$$d = \frac{G_2 L + V_C}{V_{IN}}. \tag{13.91}$$

13.9 Sepic Converter Controlled by Direct Control

In this paragraph, we focus on controlling the output voltage V_{OUT} of a Sepic converter using a direct control method. Naturally, we will choose to apply the direct control to the state variable:

$$V_{C2} = V_{OUT}. \tag{13.92}$$

It follows that:

$$\frac{dV_{C2}}{dt} = G_3 = -\frac{V_{C2}}{R_L C_2} + \frac{(1-d)I_{L1}}{C_2} - \frac{(1-d)I_{L2}}{C_2}. \tag{13.93}$$

The transfer function $\frac{V_{OUT}(s)}{G_3(s)}$ is given by

$$\frac{V_{OUT}(s)}{G_3(s)} = \frac{1}{s}. \tag{13.94}$$

This control simplifies the design of the corrector, which is usually a challenging task with this type of converter. The duty cycle d will be calculated by:

$$d = \frac{G_3 C_2 + \frac{V_{C2}}{R_L} + I_{L2} - I_{L1}}{I_{L2} - I_{L1}} \tag{13.95}$$

Figure 13.12 shows the comparison of the transfer function $\frac{V_{OUT}(s)}{G_3(s)}$ with $\frac{V_{OUT}(s)}{d(s)}$ with the numerical values given in Para. 13.6.

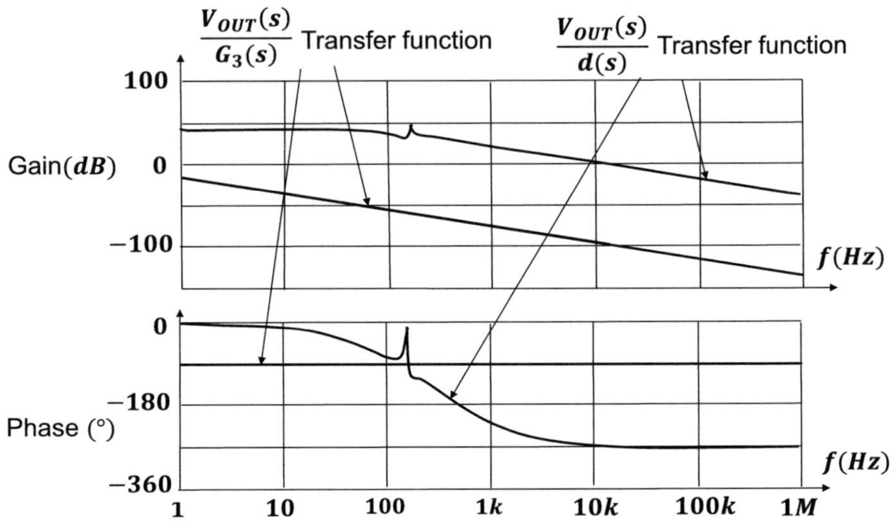

Fig. 13.12 Comparison of the transfer function $\frac{V_{\mathrm{OUT}}(s)}{G_3(s)}$ with $\frac{V_{\mathrm{OUT}}(s)}{d(s)}$

Chapter 14
Examples of Robust Controls in the Power Converter Field

In this section, we apply some of the robust control methods described in Chap. 7 to the field of power conversion. These examples were selected for their educational value.

14.1 Adaptive Lyapunov Control Applied to a Thermal System

14.1.1 System Description

Consider the thermal system presented in Fig. 14.1 along with its equivalent electrical circuit.

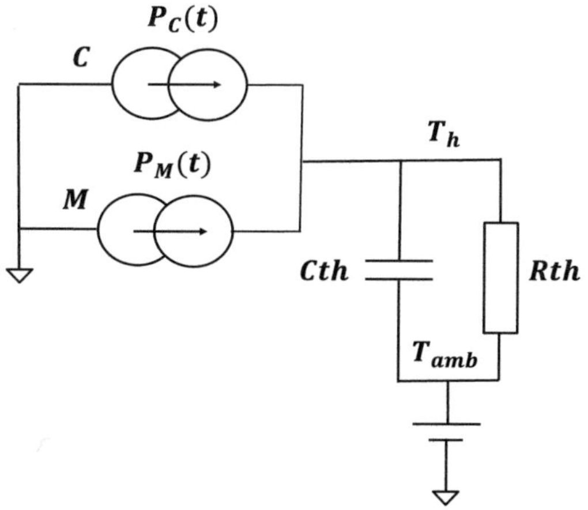

Fig. 14.1 Equivalent electrical circuit of the thermal system

B. Rivet, *Theories of Nonlinear Systems for Power Converters*,
https://doi.org/10.1007/978-3-032-25062-9_14

This thermal system consists of a MOSFET transistor M and another power component C mounted on a common heatsink. The thermal resistances of the package for M and C are negligible compared to the thermal resistance of the heatsink $R\text{th}$. So, we will assume that the junction temperature of M and C are equal to the temperature of the heatsink T_h. The power dissipated in $C(P_C(t))$ is controllable and will be our control parameter. The power dissipated in M can be expressed as:

$$P_M = R_{\text{DSON}}[1 + \alpha(T_h - T_{\text{amb}}))]I_{\text{RMS}}^2. \tag{14.1}$$

R_{DSON} is the on-state resistance of M at ambient temperature T_{amb}. R_{DSON} is a fixed but unknown parameter that belongs to $[R_{\text{DSONmin}}, R_{\text{DSONmax}}]$. This system is first-order. The state parameter is T_h. The dynamics of this system are given by:

$$\frac{\mathrm{d}T_h}{\mathrm{d}t} = \frac{R_{\text{DSON}}}{C\text{th}}[1 + \alpha(T_h - T_{\text{amb}})]I_{\text{RMS}}^2 + \frac{P_C}{C\text{th}} + \frac{T_{\text{amb}} - T_h}{R\text{th}C\text{th}} \tag{14.2}$$

Our goal is to use an adaptive Lyapunov control (Ioannou & Kokotovic, 1984; Kokotović & Arcak, 2001) to regulate the heatsink temperature at a desired value T_{he} regardless of the unknown parameter R_{DSON}.

14.1.2 Research on Adaptive Lyapunov Control

Consider the following Lyapunov function:

$$V\left(T_h, \widehat{R_{\text{DSON}}}\right) = \frac{1}{2}C\text{th}(T_h - T_{\text{he}})^2 + \frac{1}{2}k_r\left(\widehat{R_{\text{DSON}}} - R_{\text{DSON}}\right)^2. \tag{14.3}$$

$\widehat{R_{\text{DSON}}}$ is an estimated value of the unknown parameter R_{DSON}, and k_r is a coefficient used to ensure the homogeneity of the formula. According to the usual criteria, this function is a suitable candidate. We have:

$$V(T_{\text{he}}, R_{\text{DSON}}) = 0 \quad \text{and} \quad V\left(T_h, \widehat{R_{\text{DSON}}}\right) > 0 \quad \forall \; \widehat{R_{\text{DSON}}} \neq (T_{\text{he}}, R_{\text{DSON}}). \tag{14.4}$$

The time derivative of $V\left(T_h, \widehat{R_{\text{DSON}}}\right)$ is:

$$\frac{\mathrm{d}V\left(T_h, \widehat{R_{\text{DSON}}}\right)}{\mathrm{d}t} = C\text{th}(T_h - T_{\text{he}})\frac{\mathrm{d}T_h}{\mathrm{d}t} + k_r\left(\widehat{R_{\text{DSON}}} - R_{\text{DSON}}\right)\frac{\mathrm{d}\widehat{R_{\text{DSON}}}}{\mathrm{d}t}. \tag{14.5}$$

It follows that:

$$\frac{\mathrm{d}V\left(T_h, \widehat{R_{\mathrm{DSON}}}\right)}{\mathrm{d}t} = R_{\mathrm{DSON}}\left[(T_h - T_{\mathrm{he}})[1+\alpha(T_h - T_{\mathrm{amb}})]I_{\mathrm{RMS}}^2 - k_r\frac{\mathrm{d}\widehat{R_{\mathrm{DSON}}}}{\mathrm{d}t}\right]$$
$$+ (T_h - T_{\mathrm{he}})\left(P_C + \frac{T_{\mathrm{amb}} - T_h}{R_{\mathrm{th}}}\right) + k_r\widehat{R_{\mathrm{DSON}}}\frac{\mathrm{d}\widehat{R_{\mathrm{DSON}}}}{\mathrm{d}t}. \quad (14.6)$$

In order to make $\frac{\mathrm{d}V\left(T_h, \widehat{R_{\mathrm{DSON}}}\right)}{\mathrm{d}t}$ independent of the unknown parameter R_{DSON}, we take:

$$\frac{\mathrm{d}\widehat{R_{\mathrm{DSON}}}}{\mathrm{d}t} = \frac{1}{k_r}\left[(T_h - T_{\mathrm{he}})[1+\alpha(T_h - T_{\mathrm{amb}}))]I_{\mathrm{RMS}}^2\right] \quad (14.7)$$

We get:

$$\frac{\mathrm{d}V\left(T_h, \widehat{R_{\mathrm{DSO}}}\right)}{\mathrm{d}t} = (T_h - T_{\mathrm{he}})\left(P_C + \frac{T_{\mathrm{amb}} - T_h}{R_{\mathrm{th}}} + \widehat{R_{\mathrm{DSON}}}[1+\alpha(T_h - T_{\mathrm{amb}})]I_{\mathrm{RMS}}^2\right). \quad (14.8)$$

the control $P_C(t)$ is chosen to achieve:

$$\frac{\mathrm{d}V\left(T_h, \widehat{R_{\mathrm{DSON}}}\right)}{\mathrm{d}t} = -k(T_h - T_{\mathrm{he}})^2 \quad (14.9)$$

k is an adjustable control parameter. This condition gives:

$$P_C(t) = -k(T_h - T_{\mathrm{he}}) + \frac{T_h - T_{\mathrm{amb}}}{R_{\mathrm{th}}} - \widehat{R_{\mathrm{DSON}}}[1+\alpha(T_h - T_{\mathrm{amb}})]I_{\mathrm{RMS}}^2. \quad (14.10)$$

The control is well defined with formulas (14.7) and (14.10). Figure 14.2 shows the simulations of $T_h(t)$ and $\widehat{R_{\mathrm{DSON}}}$ performed with $R_{\mathrm{DSON}} = R_{\mathrm{DSONmin}}$ and $R_{\mathrm{DSON}} = R_{\mathrm{DSONmax}}$. This simulation confirms that $T_h(t)$ tends to T_{he} and $\widehat{R_{\mathrm{DSON}}}(t)$ tends to R_{DSON} regardless of the values of the unknown parameter R_{DSON}.
This simulation was performed with the following numerical values:

$$R_{\mathrm{DSON}} \in [0.05\Omega, 0.2\Omega],\ \alpha = 5\,\mathrm{m}\Omega\,{}^{\circ}\mathrm{C}^{-1},\ T_{\mathrm{he}} = 100\,{}^{\circ}\mathrm{C},\ T_{\mathrm{amb}} = 25\,{}^{\circ}\mathrm{C},\ I_{\mathrm{RMS}} = 3\,\mathrm{A},$$
$$Rth = 10\,{}^{\circ}\mathrm{C\,W}^{-1},\quad Cth = 10{}^{\circ}\mathrm{sW}\,{}^{\circ}\mathrm{C}^{-1},\quad k = 100\,\mathrm{W}\,{}^{\circ}\mathrm{C}^{-1},\ k_r = 3\,\mathrm{J}\,{}^{\circ}\mathrm{CO}^{-2}. \quad (14.11)$$

We note that at start up, $\widehat{R_{\mathrm{DSON}}}$ is negative, which is possible because it is not a physical parameter. However, $P_C(t)$ must be positive, which is indeed the case.

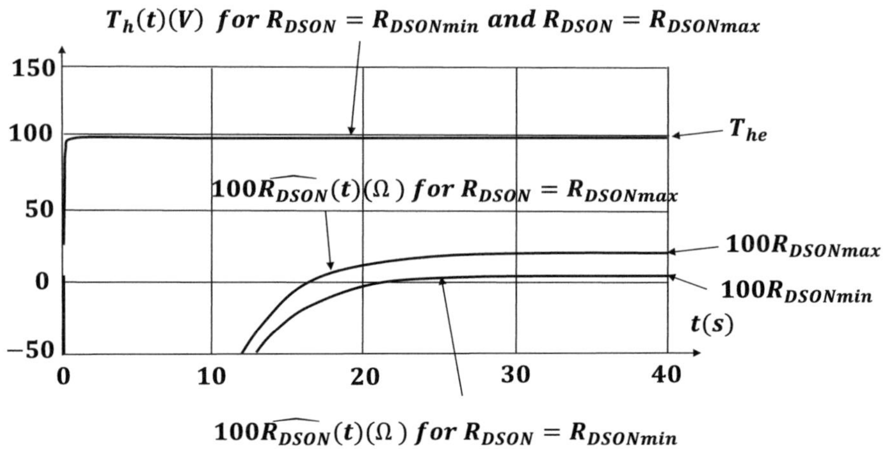

Fig. 14.2 Waveforms of $T_h(t)$ and $\widehat{R_{\mathrm{DSON}}}(t)$ for $R_{\mathrm{DSON}} = R_{\mathrm{DSONmin}}$ and $R_{\mathrm{DSON}} = R_{\mathrm{DSONmax}}$

14.2 Adaptive Backstepping Control Applied to a Buck Converter

In this paragraph, we use adaptive backstepping control (Chen, 1996; Krstić et al., 1995) to regulate the output voltage of a buck converter operating in continuous conduction mode at V_{OUTtar}. In this example, the load R_L is treated as an unknown parameter. The dynamics of this converter are given by:

$$\begin{cases} \frac{\mathrm{d}V_C}{\mathrm{d}t} = -\theta\frac{V_C}{C} + \frac{I_L}{C} \\ \frac{\mathrm{d}I_L}{\mathrm{d}t} = -\frac{V_C}{L} + d\frac{V_{\mathrm{IN}}}{L} \end{cases} \tag{14.12}$$

with

$$\theta = \frac{1}{R_L} \tag{14.13}$$

as an unknown constant parameter. Our goal is to regulate V_C at the targeted level V_{OUTtar}. The first step consists of stabilising the subsystem:

$$\frac{\mathrm{d}V_C}{\mathrm{d}t} = -\theta\frac{V_C}{C} + \frac{I_L}{C} \tag{14.14}$$

with a virtual control I_L. Consider the following Lyapunov function:

$$V_1\left(V_C, \tilde{\theta}\right) = \frac{C}{2}(V_C - V_{\mathrm{OUTtar}})^2 + \frac{k_\theta}{2}\tilde{\theta}^2, \tag{14.15}$$

which depends on V_C and the estimation error:

$$\tilde{\theta} = \theta - \hat{\theta}. \tag{14.16}$$

$\hat{\theta}$ is the estimated value of θ. Its time derivative is:

$$\frac{\mathrm{d}V_1}{\mathrm{d}t} = (V_C - V_{\mathrm{OUTtar}})(-\theta V_C + I_L) + k_\theta \tilde{\theta}\dot{\tilde{\theta}} \tag{14.17}$$

with the virtual control

$$I_L = -k_1(V_C - V_{\mathrm{OUTtar}}) + \hat{\theta} V_C = \alpha_1\big(V_C, \hat{\theta}\big) \tag{14.18}$$

and with

$$\dot{\tilde{\theta}} = \frac{\mathrm{d}\tilde{\theta}}{\mathrm{d}t} = \frac{V_C(V_C - V_{\mathrm{OUTtar}})}{k_\theta}. \tag{14.19}$$

$\dot{V}_1$ becomes independent of both θ and $\tilde{\theta}$. We obtain:

$$\frac{\mathrm{d}V_1}{\mathrm{d}t} = -k_1(V_C - V_{\mathrm{OUTtar}})^2. \tag{14.20}$$

If the control parameter k_1 is positive, then $\frac{\mathrm{d}V_1}{\mathrm{d}t}$ is a negative definite function. Since I_L is not a real control, we need to define the deviation of I_L from $\alpha_1\big(V_C, \hat{\theta}\big)$:

$$z = I_L - \alpha_1\big(V_C, \hat{\theta}\big). \tag{14.21}$$

With this new parameter, the dynamics of the system can be written as:

$$\begin{cases} \frac{\mathrm{d}V_C}{\mathrm{d}t} = -\frac{k_1}{C}(V_C - V_{\mathrm{OUTtar}}) - \tilde{\theta}\frac{V_C}{C} + \frac{z}{C} \\ \frac{\mathrm{d}z}{\mathrm{d}t} = -\frac{V_C}{L} + d\frac{V_{\mathrm{IN}}}{L} - \dot{\alpha}_1 \end{cases}. \tag{14.22}$$

Let us add the error variable to the Lyapunov function

$$V_2\big(V_C, z, \tilde{\theta}\big) = V_1\big(V_C, \tilde{\theta}\big) + \frac{k_Z}{2} z^2. \tag{14.23}$$

The new dynamics become:

$$\frac{\mathrm{d}V_1}{\mathrm{d}t} = -k_1(V_C - V_{\mathrm{OUTtar}})^2 + z(V_C - V_{\mathrm{OUTtar}}) - \tilde{\theta}(V_C(V_C - V_{\mathrm{OUTtar}}) + k_\theta \dot{\hat{\theta}}). \tag{14.24}$$

The time derivative of $V_2\big(V_C, z, \tilde{\theta}\big)$ is given by

$$\frac{\mathrm{d}V_2}{\mathrm{d}t} = \frac{\mathrm{d}V_1}{\mathrm{d}t} + k_Z z\dot{z}. \tag{14.25}$$

The time derivative of z is:

$$\frac{\mathrm{d}z}{\mathrm{d}t} = -\frac{V_C}{L} + d\frac{V_{\mathrm{IN}}}{L} + \left[-\frac{k_1}{C}(V_C - V_{\mathrm{OUTtar}}) - \tilde{\theta}\frac{V_C}{C} + \frac{z}{C}\right](k_1 - \hat{\theta}) - V_C\dot{\hat{\theta}}. \tag{14.26}$$

After some calculations, we obtain the following result:

$$\begin{aligned}\frac{\mathrm{d}V_2}{\mathrm{d}t} &= -k_1(V_C - V_{\mathrm{OUTtar}})^2 + \tilde{\theta}\left(-V_C(V_C - V_{\mathrm{OUTtar}}) - k_\theta\dot{\hat{\theta}} + \frac{k_Z V_C z}{C}(\hat{\theta} - k_1)\right)\\ &+ z\bigg(V_C - V_{\mathrm{OUTtar}} - \frac{k_Z V_C}{L} + d\frac{k_Z V_{\mathrm{IN}}}{L} - \frac{k_Z k_1^2}{C}(V_C - V_{\mathrm{OUTtar}})\\ &+\frac{k_1 k_z z}{C} - k_z\dot{\hat{\theta}}V_C + \frac{k_1 k_Z\hat{\theta}}{C}(V_C - V_{\mathrm{OUTtar}}) - \frac{k_z z}{C}\hat{\theta}\bigg).\end{aligned} \tag{14.27}$$

To make $\dot{V}_2$ negative and independent of both θ and $\tilde{\theta}$, we can choose the control law defined by:

$$\begin{aligned}d = \frac{L}{k_Z V_{\mathrm{IN}}}\bigg(&-k_2 z + V_{\mathrm{OUTtar}} - V_C + \frac{k_Z V_C}{L} + \frac{k_Z k_1^2}{C}(V_C - V_{\mathrm{OUTtar}})\\ &-\frac{k_1 k_z z}{C} + k_z\dot{\hat{\theta}}V_C - \frac{k_1 k_Z\hat{\theta}}{C}(V_C - V_{\mathrm{OUTtar}}) + \frac{k_z z}{C}\hat{\theta}\bigg)\end{aligned} \tag{14.28}$$

and

$$\dot{\hat{\theta}} = \frac{1}{k_\theta}\left(\frac{k_Z V_C z}{C}(\hat{\theta} - k_1) - V_C(V_C - V_{\mathrm{OUTtar}})\right). \tag{14.29}$$

The time derivate of $\dot{V}_2$ becomes:

$$\frac{\mathrm{d}V_2}{\mathrm{d}t} = -k_1(V_C - V_{\mathrm{OUTtar}})^2 - k_2 z^2 \le 0. \tag{14.30}$$

Given that $k_2 > 0$, it results that:

$$V_C(t) \to V_{\mathrm{OUTtar}} \text{ and } \hat{\theta}(t) \to \theta = \frac{1}{R}. \tag{14.31}$$

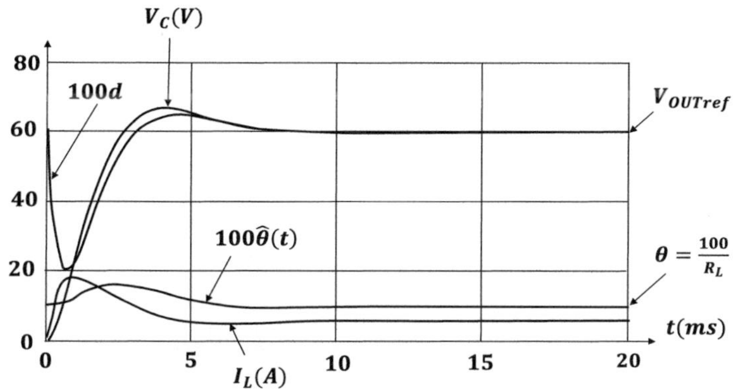

Fig. 14.3 Waveforms of $d(t)$, $V_C(t)$, $I_L(t)$ and $\hat{\theta}(t)$

Figure 14.3 shows the simulation results for the following numerical values:

$$V_{\mathrm{IN}} = 100\,\mathrm{V},\, V_{\mathrm{OUTtar}} = 60\,\mathrm{V},\;\; R_L = 10\,\Omega,\;\; C = 500\,\mu\mathrm{F},\;\; L = 1\,\mathrm{mH},\;\; k_1 = 1\,\Omega^{-1},$$
$$k_2 = 3\,\Omega,\quad k_z = 4\,\mathrm{mH},\quad k_\theta = 100\,\mathrm{J}\Omega^2. \tag{14.32}$$

14.3 Robust Backstepping Control Applied to a Buck Converter

Once again, we consider a Buck converter where the load consists of a known resistor R_L in parallel with an unknown current source $I_p(t)$ (see Fig. 14.4). This uncertainty $I_p(t)$ is bounded by Δ, so:

$$\left|I_p(t)\right| < \Delta. \tag{14.33}$$

The dynamics of this system are given by:

$$\begin{cases} \frac{\mathrm{d}V_C}{\mathrm{d}t} = -\frac{V_C}{R_L C} + \frac{I_L}{C} - \frac{I_p(t)}{C} \\ \frac{\mathrm{d}I_L}{\mathrm{d}t} = -\frac{V_C}{L} + d\frac{V_{\mathrm{IN}}}{L} \end{cases}. \tag{14.34}$$

Our goal is to stabilise the output voltage V_C around V_{OUTtar}. We choose the robust Control Lyapunov Function (CLF) (Chen, 1996; Kokotović & Arcak, 2001) as:

$$V_1(V_C) = \frac{C}{2}(V_C - V_{\mathrm{OUTtar}})^2. \tag{14.35}$$

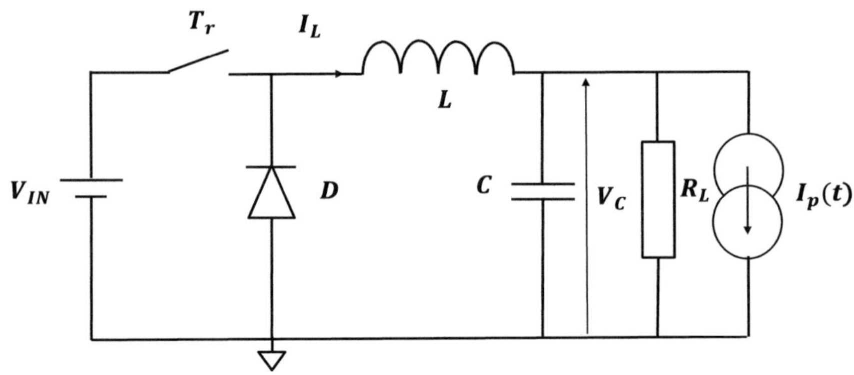

Fig. 14.4 Buck converter with unknown load

The virtual control:

$$I_L = \alpha(V_C) \tag{14.36}$$

will have to consider worst-case scenario of $I_p(t)$. Then, to make $\frac{dV_1}{dt}$ negative, we have to look for the value of $\alpha(V_C)$ which (with $V_C \neq V_{\text{OUTref}}$ and all uncertainties $I_p(t)$ bounded by Δ), satisfies:

$$\dot{V}_1(V_C) = (V_C - V_{\text{OUTtar}})\left(-\frac{V_C}{R_L} + \alpha(V_C) - I_p\right)$$

$$< -\frac{V_C(V_C - V_{\text{OUTtar}})}{R_L} + \alpha(V_C)(V_C - V_{\text{OUTtar}}) + |(V_C - V_{\text{OUTtar}}|\Delta < 0. \tag{14.37}$$

A possible choice for the virtual control is:

$$\alpha(V_C) = \frac{V_C}{R_L} - k_1(V_C - V_{\text{OUTtar}}) - \Delta \text{sgn}(V_C - V_{\text{OUTtar}}) \quad \text{with}\, k_1 > 0 \tag{14.38}$$

Giving:

$$\dot{V}_1(V_C) < -k_1(V_C - V_{\text{OUTtar}})^2. \tag{14.39}$$

In order to regulate $I_L = \alpha(V_C)$ to zero, we suggest considering

$$V_2(V_C, I_L) = V_1(V_C) + \frac{k_r}{2}(I_L - \alpha(V_C))^2 \text{ with } k_r > 0 \tag{14.40}$$

as a robust Control Lyapunov Function (CLF). Our objective now is to find a control d such that:

$$\dot{V}_2 = (V_C - V_{\text{OUTtar}})\left(-\frac{V_C}{R_L} + I_L - I_p\right) + k_r(I_L - \alpha(V_C)) \\ \times\left(-\frac{V_C}{L} + d\frac{V_{\text{IN}}}{L} - \left(\frac{1}{R_L} - k_1\right)\left(-\frac{V_C}{R_L \cdot C} + \frac{I_L}{C} - \frac{I_p(t)}{C}\right)\right) < 0. \quad (14.41)$$

For $V_C \neq V_{\text{OUTtar}}$ and all uncertainties $I_p(t)$, the choice of $\alpha(V_C)$ implying

$$(V_C - V_{\text{OUTtar}})\left(-\frac{V_C}{R_L} + \alpha(V_C) - I_p\right) < -k_1(V_C - V_{\text{OUTtar}})^2 \quad (14.42)$$

leads to

$$\dot{V}_2(X) < -k_1(V_C - V_{\text{OUTtarf}})^2 + (I_L - \alpha(V_C)) \\ \times\left[V_C - V_{\text{OUTtar}} - \frac{k_r V_C}{L} + d\frac{k_r V_{\text{IN}}}{L} - \left(\frac{1}{R_L} - k_1\right)\left(-\frac{k_r V_C}{R_L C} + \frac{k_r I_L}{C} - \frac{k_r I_p(t)}{C}\right)\right] \quad (14.43)$$

with

$$d = \frac{L}{V_{\text{IN}}}\left[\frac{V_{\text{OUTtar}} - V_C}{k_r} + \frac{V_C}{L} + \left(\frac{1}{R_L} - k_1\right)\left(-\frac{V_C}{R_L C} + \frac{I_L}{C}\right) \\ - \left|\frac{1}{R_L} - k_1\right|\frac{\Delta}{C}\text{sgn}(I_L - \alpha(V_C)) - \frac{k_2}{k_r}(I_L - \alpha(V_C))\right] \quad (14.44)$$

We obtain:

$$\dot{V}_2(X) < -k_1(V_C - V_{\text{OUTar}})^2 - k_2(I_L - \alpha(V_C))^2. \quad (14.45)$$

Then, with $k_2 > 0$, this control allows to stabilising the converter at V_{OUTtar} for all uncertainties $I_p(t)$ bounded by Δ. Figure 14.5 shows the simulation results of $d(t)$, $V_C(t)$, $I_L(t)$ and $\alpha(t)$ performed with the following numerical values:

$$V_{\text{IN}} = 100\,\text{V}, \quad V_{\text{OUTtar}} = 60\,\text{V}, \quad R_L = 10\,\Omega, \quad C = 500\,\mu\text{F}, \quad L = 1\,\text{mH}, \\ k_1 = 0.05\,\Omega^{-1}, \; k_2 = 3\,\Omega, \; k_r = 4\,\text{mH}, \; \Delta = 1\,\text{A}, \; I_P(t) = 0.5\sin(1000t)\,\text{A}. \quad (14.46)$$

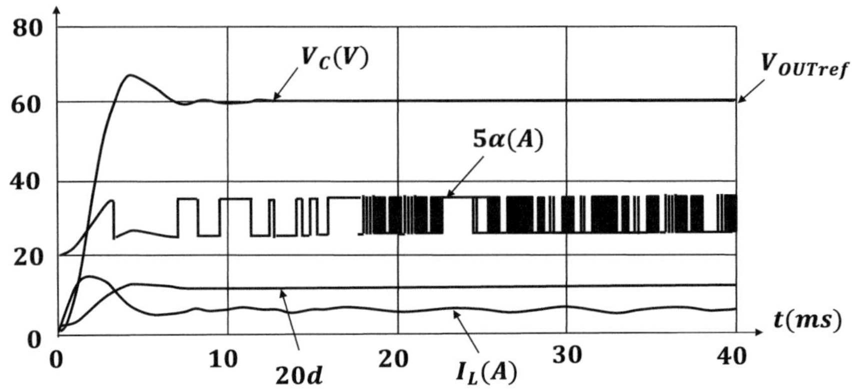

Fig. 14.5 Waveforms of $d(t)$, $V_C(t)$, $I_L(t)$ and $\alpha(t)$

14.4 Inverse Optimal Control Applied to an *L,R,C,I,P* Circuit

In this paragraph, we use inverse optimal control (Kalman, 1964; Primbs, 1999; Primbs et al., 1999) to drive the circuit presented in Fig. 14.6.

This is a second-order system with two state parameters (V_C, I_L) and one control parameter (I_C). The uncertainty

$$W(t) = \frac{1}{r(t)} \tag{14.47}$$

is bounded by:

$$|W(t)| < W_B = \frac{1}{R}. \tag{14.48}$$

Fig. 14.6 L,R,C,I,P circuit with uncertainty $W(t) = \frac{1}{r(t)}$

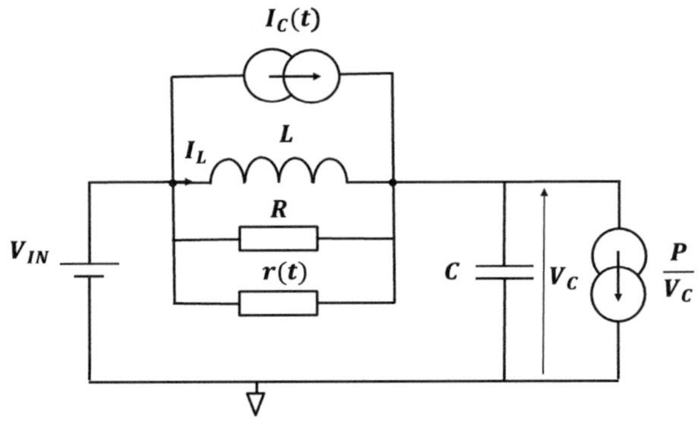

The dynamics are given by:

$$\begin{cases} \frac{dV_C}{dt} = \frac{V_{IN}-V_C}{RC} + \frac{V_{IN}-V_C}{r(t)C} - \frac{P}{CV_C} + \frac{I_L}{C} + \frac{I_C}{C} \\ \frac{dI_L}{dt} = -\frac{V_C}{L} + \frac{V_{IN}}{L} \end{cases}. \tag{14.49}$$

The equilibrium point is:

$$(V_{Ce}, I_{Le}) = \left(V_{IN}, \frac{P}{V_{IN}}\right). \tag{14.50}$$

Our purpose is to define a state feedback control $I_C(V_C, I_L)$ such that the point (V_{Ce}, I_{Le}) is a globally asymptotically stable equilibrium point of the resulting closed-loop system. Moreover, this control law has to minimise the following cost functional:

$$J = \int_0^\infty (q(V_C, I_L) + k_r I_C^2)\mathrm{d}t. \tag{14.51}$$

These two objectives must be achieved for all uncertainties $W(t)$. The difference with direct optimal control is that the cost functional is not given a priori but will be deduced. Only the general form of the cost functional is proposed. Here, $q(V_C, I_L)$ is unknown. In this approach, the control law is obtained with a known Control Lyapunov Function ($V(V_C, I_L)$. $V(V_C, I_L)$), which is in fact the value function of the considered optimal control problem. The considered Lyapunov function is:

$$V(V_C, I_L) = \frac{C}{2}(V_C - V_{IN})^2 + \frac{L}{2}\left(I_L - \frac{P}{V_{IN}}\right)^2. \tag{14.52}$$

$V(V_C, I_L)$ must be the solution of the following HJB equation, taking into account uncertainty $W(t)$. It can be formulated using the notation of Para. 7.42 as follows:

$$\inf_{I_C} \sup_{W} \left(q(V_C, I_L) + k_r I_C^2 + \frac{\partial V}{\partial \vec{X}} F\left(\vec{X}\right) + \frac{\partial V}{\partial \vec{X}} G\left(\vec{X}\right) I_C + \frac{\partial V}{\partial \vec{X}} H\left(\vec{X}\right) W \right) = 0. \tag{14.53}$$

The minimum is obtained by:

$$I_C(V_C, I_L) = -\frac{1}{2k_r}\frac{\partial V}{\partial \vec{X}} G\left(\vec{X}\right) = \frac{V_{IN} - V_C}{2k_r}. \tag{14.54}$$

The worst-case uncertainty is given by:

$$W = \operatorname{sgn}\left(\frac{\partial V}{\partial \vec{X}} H(\vec{X})\right) W_B = \operatorname{sgn}\left(-(V_C - V_{\text{IN}})^2\right) W_B = -W_B. \tag{14.55}$$

It follows that:

$$q(V_C, I_L) = (V_C - V_{\text{IN}})^2 \left(\frac{1}{4k_r} - \frac{P}{V_C V_{\text{IN}}} + \frac{1}{R} - W_B\right). \tag{14.56}$$

With this function, the control given by Eq. (14.54) is optimal. Now, we need to check that this optimal control stabilises the system around the equilibrium point $(V_{\text{Ce}}, I_{\text{Le}}) \cdot V(V_C, I_L)$ is a C^1 positive definite function with:

$$V(V_{\text{Ce}}, I_{\text{Le}}) = 0. \tag{14.57}$$

Its time derivative is given by:

$$\dot{V} = \frac{\partial V}{\partial \vec{X}} F\left(\vec{X}\right) + \frac{\partial V}{\partial \vec{X}} G\left(\vec{X}\right) I_C + \frac{\partial V}{\partial \vec{X}} H\left(\vec{X}\right) W. \tag{14.58}$$

It follows that:

$$\dot{V}(V_C, I_L) < -(V_C - V_{\text{IN}})^2 \left(\frac{1}{R} - W_B + \frac{1}{2k_r} - \frac{P}{V_C V_{\text{IN}}}\right). \tag{14.59}$$

The system will be globally asymptotically stable for all uncertainties $W(t)$ bounded by W_B if

$$\dot{V}(V_C, I_L) < 0 \quad \forall (V_C, I_L) \neq \left(V_{\text{IN}}, \frac{P}{V_{\text{IN}}}\right) \tag{14.60}$$

Leading to the condition:

$$V_C > V_{\text{Clim}} = \frac{P}{V_{\text{IN}}\left(\frac{1}{R} - W_B + \frac{1}{2k_r}\right)}. \tag{14.61}$$

With this condition in mind, note that $q(V_C, I_L) > 0$. Figure 14.7 shows the simulated waveforms of $V_C(t)$ and $I_L(t)$ performed with the following numerical values:

$$\begin{gathered} V_{\text{IN}} = 400\,\text{V}, \quad P = 4\,\text{kW}, \quad R = 100\,\Omega, \quad C = 1\,\text{mF}, \quad L = 1\,\text{mH}, \quad k_r = 10\,\Omega \\ r(t) = r_0(1.5 + \sin(1000\text{t}))\Omega, \quad r_0 = 300\,\Omega, \quad W_B = \frac{1}{0.5 r_0} S. \end{gathered} \tag{14.62}$$

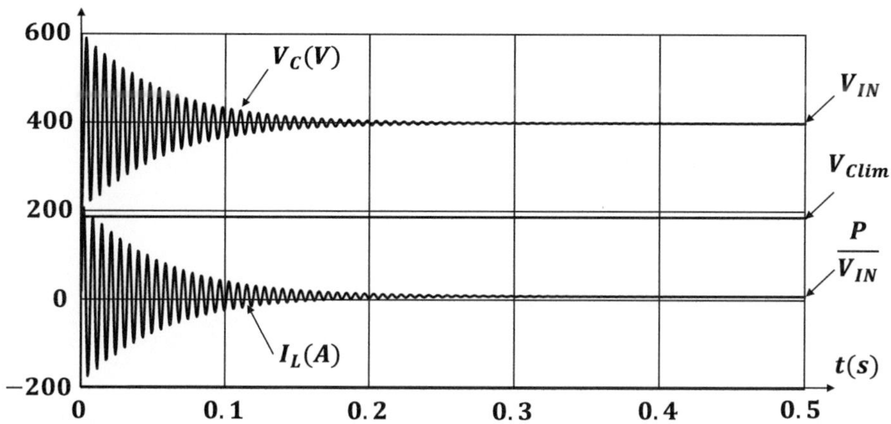

Fig. 14.7 Waveforms of $V_C(t)$ and $I_L(t)$

14.5 Robust Control of a Buck Converter Using Sliding Mode Control

In this paragraph, we will use sliding mode control to achieve robust control of a Buck converter (Edwards & Spurgeon, 1998; Ferrara et al., 2008; Utkin, 1977). Our goal is to regulate V_C at V_{OUTtar} regardless of the uncertainty v_{IN}.

14.5.1 Representation of a Buck Converter Using a Hybrid Model

We consider the Buck converter shown in Fig. 14.9, with the uncertainty v_{IN} bounded by

$$|v_{\text{IN}}(t)| < W_{\text{IN}}. \tag{14.63}$$

Here, the dynamics of this converter are given by its hybrid representation:

$$\begin{cases} \frac{\mathrm{d}V_C}{\mathrm{d}t} = -\frac{V_C}{R_L C} + \frac{I_L}{C} \\ \frac{\mathrm{d}I_L}{\mathrm{d}t} = -\frac{V_C}{L} + u\frac{V_{\text{IN}}+v_{\text{IN}}(t)}{L} \end{cases} . \tag{14.64}$$

u is the discontinuous control, which can assume the value 0 when the switch S is off or 1 when the switch is on.

14.5.2 Choice of the Sliding Line

Referring to Para. 7.5, let us first choose a sliding line so that the system response reaches the targeted equilibrium point:

$$V_{\mathrm{Ce}} = V_{\mathrm{OUTtar}}, \quad I_{\mathrm{Le}} = \frac{V_{\mathrm{Ce}}}{R_L} = \frac{V_{\mathrm{OUTtar}}}{R_L}. \tag{14.65}$$

Consider the sliding line defined by:

$$\sigma(I_L, V_C) = (I_L - I_{\mathrm{Le}}) + k_c(V_C - V_{\mathrm{Ce}}) = 0. \tag{14.66}$$

$k_c > 0$ is a control parameter. Assume that a control law $u((I_L, V_C))$ has been defined to obtain a sliding mode on the line $\sigma(I_L, V_C) = 0$. For each initial condition belonging to the region of attraction R_A, there exists a finite time t_i for which the trajectory of the closed-loop system intersects the sliding line $\sigma(I_L, V_C) = 0$. Assuming the sliding mode exists for all instants $t > t_i$, we have:

$$\sigma(I_L, V_C) = 0 \text{ and } \frac{\mathrm{d}\sigma(I_L, V_C)}{\mathrm{d}t} = 0. \tag{14.67}$$

These conditions yield the equivalent control u_{eq}:

$$u_{\mathrm{eq}} = \frac{V_C}{V_{\mathrm{IN}} + v_{\mathrm{IN}}(t)} + \frac{k_c}{V_{\mathrm{IN}} + v_{\mathrm{IN}}(t)}\left(\frac{LV_C}{R_L C} - \frac{LI_L}{C}\right). \tag{14.68}$$

Using Eqs. (14.66) and (14.68) we can thus infer the dynamics of the system on the sliding line

$$\begin{cases} \frac{\mathrm{d}V_C}{\mathrm{d}t} = (V_{\mathrm{OUTtar}} - V_C)\left(\frac{1}{R_L C} + \frac{k_c}{C}\right) \\ \frac{\mathrm{d}I_L}{\mathrm{d}t} = \frac{k_c V_C}{R_L C} - \frac{k_c I_L}{C} \end{cases}. \tag{14.69}$$

It is easy to see that, for each initial condition on the sliding curve and regardless of the value of the uncertainty v_{IN}, the system will slide towards the equilibrium point $(I_{\mathrm{Le}}, V_{\mathrm{Ce}})$, with the control at the equilibrium given by:

$$u_{\mathrm{eqe}} = \frac{V_{\mathrm{OUTtar}}}{V_{\mathrm{IN}} + v_{\mathrm{IN}}(t)}. \tag{14.70}$$

14.5.3 Determination of the Control Law

Now we have to find the control u that drives the trajectory onto the sliding line $\sigma = 0$ and maintains the system on it. The control must ensure:

When $\sigma > 0$, $((I_L - I_{\text{Le}}) + k_c(V_C - V_{\text{Ce}}) > 0)$, we require:

$$\dot{\sigma} = \dot{I}_L + k_c\dot{V}_C = -\frac{V_C}{L} + u\frac{V_{\text{IN}}}{L} + k_c\left(\frac{I_L}{C} - \frac{V_C}{R_L C}\right) < 0, \tag{14.71}$$

thus, $u = 0$ is a possibility, provided that

$$V_C > \frac{k_c L R_L I_L}{R_L C + k_c L}. \tag{14.72}$$

When $\sigma < 0$, $((I_L - L_{\text{Le}}) + k_c(V_C - V_{\text{Ce}}) < 0)$, we require:

$$\dot{\sigma} = \dot{I}_L + k_c\dot{V}_C = -\frac{V_C}{L} + u\frac{V_{\text{IN}}}{L} + k_c\left(\frac{I_L}{C} - \frac{V_C}{R_L C}\right) > 0 \tag{14.73}$$

thus, $u = 1$ is a possibility, provided that

$$V_C < \frac{R_L C V_{\text{IN}} + L R_L I_L k_c}{R_L C + k_c L}. \tag{14.74}$$

14.5.4 Domain of Attraction R_A

A domain of attraction R_A can be defined by as the region where:

$$\sigma\dot{\sigma} < 0. \tag{14.75}$$

Taking into account the condition (14.63) on the uncertainty v_{IN} and based on the analysis in the previous paragraph, R_A can be defined by:

$$(I_L - I_{\text{Le}}) + k_c(V_C - V_{\text{Ce}}) > 0 \;\text{ and }\; V_C > \frac{k_c L R_L I_L}{R_L C + k_c L} \tag{14.76}$$

or

$$(I_L - I_{\text{Le}}) + k_c(V_C - V_{\text{Ce}}) < 0 \;\text{ and }\; V_C < \frac{R_L C(V_{\text{IN}} - W_{\text{IN}}) + L R_L I_L k_c}{R_L C + k_c L} \tag{14.77}$$

The region that is not part of R_A is represented with hatch lines in Fig. 14.8. We will study this Buck converter using the following numerical values:

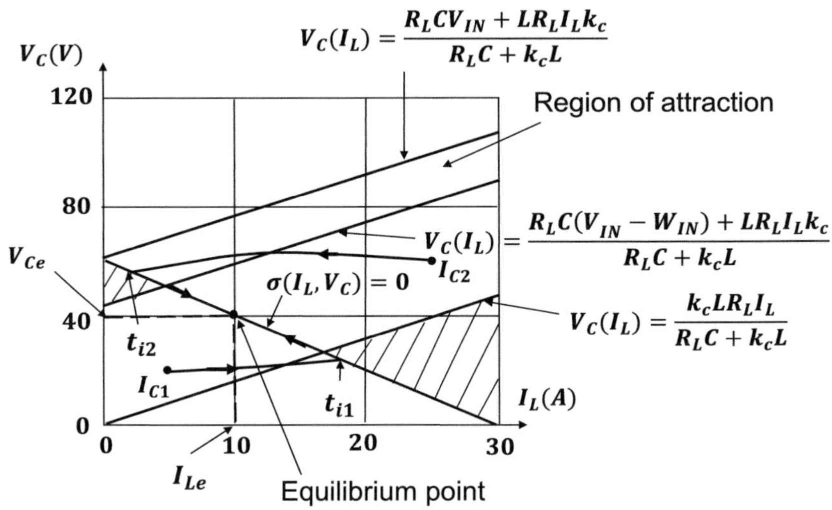

Fig. 14.8 Domain of attraction R_A

$$V_{IN} = 100\,\text{V},\; V_{OUTtar} = 40\,\text{V},\; R_L = 4\,\Omega,\; C = 200\,\mu\text{F},\; L = 1\,\text{mH},\; k_c = 0.5\,\Omega^{-1}$$
$$W_{IN} = 30\,\text{V},\quad v_{IN}(t) = 30\sin(10000t)\,\text{V}. \tag{14.78}$$

14.5.5 Reduction of the Gathering Effect

In practice, to reduce the switching frequency, we will consider two sliding lines: σ^+ and σ^- defined by:

$$\sigma^+ = (I_L - I_{Le}) + k_c(V_C - V_{Ce} + 0.2) = 0 \tag{14.79}$$

$$\sigma^- = (I_L - I_{Le}) + k_c(V_C - V_{Ce} - 0.2) = 0. \tag{14.80}$$

The switch S turns off when $\sigma^+ > 0$ and turns on when $\sigma^- < 0$. Figure 14.9 shows one way to implement this control using a latch circuit.

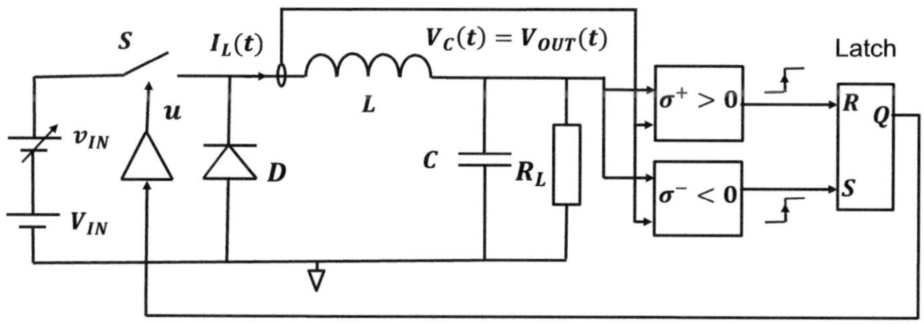

Fig. 14.9 Control circuit for reducing the gathering effect

14.5.6 Simulation Results

In Fig. 14.8, two trajectories with initial conditions: I_{C1} and I_{C2} inside the domain of attraction R_A are represented in the phase plan. These two trajectories reach the sliding line $\sigma(I_L, V_C) = 0$ at times t_{i1} and t_{i2}, respectively, and then slide along this line to reach the equilibrium point (V_{Ce}, I_{Ce}). Figure 14.10 shows the time-domain simulation of the trajectory obtained for the initial conditions $V_{C0} = 20\,\text{V}$ and $I_{L0} = 5\,\text{A}$, illustrating how the system evolves over time from these starting points towards the equilibrium state.

The switching frequency is approximately 100 kHz. It is interesting to note that, in this example, robust control was achieved simply by accounting for the uncertainty and reducing the domain of attraction R_A.

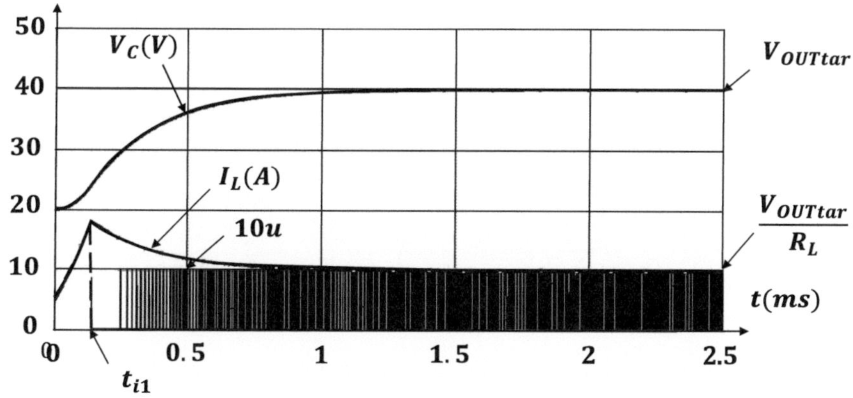

Fig. 14.10 Time-domain simulation of a sliding mode controlled system during the transient phase

14.6 Robust Control of a Buck Converter Using Predictive Control

Here, we use predictive control to achieve the same objective as in the previous paragraph (Fernández Camacho & Bordons Alba, 1999; Pekař, 2010; Qin & Badgwell, 1998). We will keep the same numerical values for the parameters of the Buck converter and the uncertainty v_{IN}. The dynamics of the Buck converter can be described by the following system of difference equations:

$$\begin{cases} \vec{X}_{k+1} = \vec{F}\left(\vec{X}_k, u_k\right) \\ \vec{Y}_k = C\left(\vec{X}_k\right) \end{cases}. \tag{14.81}$$

The state vector $\vec{X}_k$ is given by:

$$\vec{X}_k = \begin{pmatrix} V_{\mathrm{Ck}} \\ I_{\mathrm{Lk}} \end{pmatrix} = \begin{pmatrix} V_C(kT_S) \\ I_L(kT_S) \end{pmatrix}. \tag{14.82}$$

The sampling time T_S is equal to 2 μs. F is defined based on the system (14.64). The C matrix is:

$$C = \begin{pmatrix} 1 & 0 \\ 0 & 1 \end{pmatrix} \tag{14.83}$$

thus, the output vector is:

$$\vec{Y}_k = \vec{X}_k = \begin{pmatrix} V_{\mathrm{Ck}} \\ I_{\mathrm{Lk}} \end{pmatrix}. \tag{14.84}$$

The current control vector u_k is obtained by solving an optimal control problem that minimise the following cost functional:

$$J(k) = \sum_{i=0}^{N_P} \left(\hat{\vec{Y}}_{k+i} - \vec{r}_{k+i}\right)^T Q \left(\hat{\vec{Y}}_{k+i} - \vec{r}_{k+i}\right). \tag{14.85}$$

We will set the predictive horizon $N_P = 2$ and the control horizon $N_C = 2$. The predicted process $\hat{\vec{Y}}_k$ output is obtained from the system model (14.81) without considering the uncertainty v_{INk}. The reference signals $\vec{r}_k$ and the matrix Q are given by:

$$\vec{r}_k = \begin{pmatrix} V_{\mathrm{OUTtar}} \\ \frac{V_{\mathrm{OUTar}}}{R} \end{pmatrix}, \quad Q = \begin{pmatrix} 1F & 0 \\ 0 & 1H \end{pmatrix}. \tag{14.86}$$

The cost functional becomes:

$$J(k) = \sum_{i=0}^{N_P} \left(\widehat{V_{Ck+i}} - V_{\text{OUTtar}} \right)^2 + \left(\widehat{I_{Lk+i}} - \frac{V_{\text{OUTtar}}}{R_L} \right)^2. \tag{14.87}$$

Using the algorithm described in Para. 7.7, with the initial conditions

$$V_{C0} = 39\,\text{V} \quad \text{and} \quad I_{L0} = 8\,\text{A}, \tag{14.88}$$

we obtain the simulation results shown in Fig. 14.11. These waveforms confirm the robustness of the predictive control against the uncertainty v_{IN}.

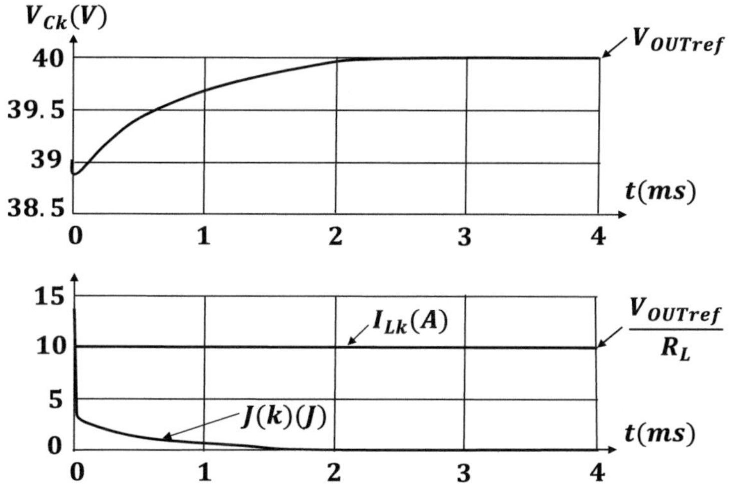

Fig. 14.11 Time evolution of V_{Ck}, I_{Lk} and $J(k)$

Chapter 15
Examples of Chaotic Controls in the Power Converter Field

In this chapter examples of chaotic controls in the power converter field will be studied.

15.1 Example 1: Chaotic Behaviour in a Thyristor Rectifier Circuit

15.1.1 System Description

Consider the thyristor rectifier circuit shown in Fig. 15.1.

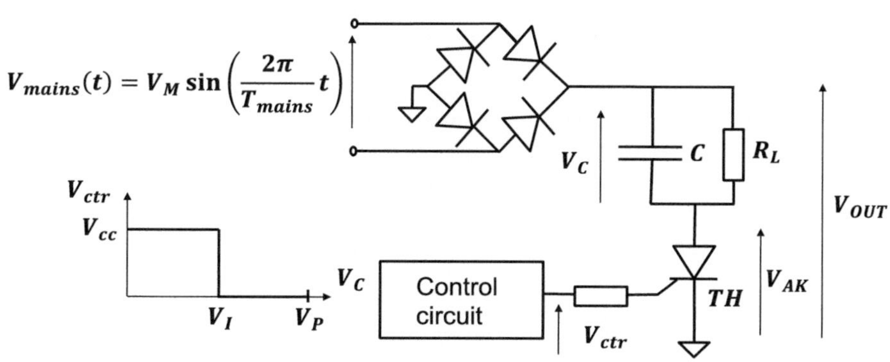

Fig. 15.1 Thyristor rectifier circuit

B. Rivet, *Theories of Nonlinear Systems for Power Converters*,
https://doi.org/10.1007/978-3-032-25062-9_15

This circuit improves the power factor of an AC/DC converter compared to the basic configuration of a bridge rectifier and a bulk capacitor. Today the solution shown in Fig. 15.1 is rarely used. The most common solution is the PFC circuit (described in Para. 11.4), which provides a power factor close to unity. However, from a pedagogical perspective, this circuit is interesting. The mains voltage $V_{\text{mains}}(t)$ is given by:

$$V_{\text{mains}}(t) = V_M \sin\left(\frac{2\pi}{T_{\text{mains}}} t\right) \tag{15.1}$$

with $V_M = 310\,\text{V}$ and $T_{\text{mains}} = 20\,\text{ms}$.

The control law is simple and defined as follows:

- For $0 < V_C < V_I = 160\,\text{V}$ then $V_{\text{ctr}} = V_{\text{cc}}$
- For $V_I < V_C < V_M$ then $V_{\text{ctr}} = 0$.

The thyristor turns on only if:

$$V_{\text{ctr}} = V_{\text{cc}} \text{ and } V_{\text{AK}} > 0(|V_{\text{mains}}(t)| > V_C). \tag{15.2}$$

When it turns on, it remains on even if $V_{\text{ctr}} = 0$ and will turn off only when its current decreases to zero.

15.1.2 Representation as an Autonomous Hybrid System

This circuit can be represented as an autonomous hybrid system. Following the generic representation given in Para. 3.1.1, we obtain:

- $Q = \{q_1, q_2\}$ is a set of two discrete states ($n_Q = 2$).
- $V_C : \mathbb{R} \to \mathbb{R}$ is a one-dimensional state parameter ($n = 1$).
- Init $\in Q \times \mathbb{R}$ is a set of $1 + n = 2$ initial conditions.
- Dom $= \{D_{q_1}, D_{q_2}\}$ is a set of continuous domains with $D_{q_1} = D_{q_2} = [0, V_P]$.
- U = empty set (no control vector).
- $F_{q_1} : D_{q_1} \to \mathbb{R}$ and $F_{q_2} : D_{q_2} \to \mathbb{R}$ are functions describing the dynamics of the system:
  $q = q_1 : V_C(t) = |V_{\text{mains}}(t)|$ and $q = q_2 : \frac{dV_C}{dt} = -\frac{V_C}{R_L C}$.
- $E = \{e_1 = (q_1, q_2), e_2 = (q_2, q_1)\}$ is a set of two discrete transitions.
- $G = \{G_{e_1}, G_{e_2}\}$ is a set of guards prescribing when a discrete state transition occurs. We have:

  $G_{e_1}(q_1, q_2) = \{V_C : V_C > |V_{\text{mains}}(t)|\}$ and

$$G_{e_2}(q_2, q_1) = \{V_C : V_C = V_I \text{ and } V_C < V_{\text{mains}}(t)\}.$$

- R_{e_2} is a reset state function that allows the state parameter to switch discontinuously from V_I to $V_{\text{mains}}(t)$ when the system transitions from q_2 state to q_1 state.

15.1.3 Bifurcation Diagram

From this representation, a one-dimensional mapping results that gives the sequence of V_{Ck}. V_{Ck} is the value of V_C just after the thyristor turns on (see Fig. 15.3). It is therefore possible to develop an algorithm that provides $V_{C(k+1)}$ according to V_{Ck} with the following system parameters V_M, T_S, C and R_L as input parameters. With this algorithm, we can draw the bifurcation diagram shown in Fig. 15.2 (with $C = 100\,\mu\text{F}$) considering the load R_L as the bifurcation parameter. R_L varies between 300 Ω and 400 Ω. For R_L values close to 300 Ω, the system is stable with one equilibrium point. A bifurcation phenomenon appears around $R_L = 310\,\Omega$. Near $R_L = 340\,\Omega$, there is a small window where the system becomes chaotic. For $R_L > 350\,\Omega$, the circuit becomes stable again with one equilibrium point.

Figure 15.3 shows the different dynamics of the circuit described above. We observe that the time between each sample V_{Ck} is different from one configuration to another: 20 ms for $R_L = 305\,\Omega$, 30 ms for $R_L = 320\,\Omega$, 10, 20, 30 ms for $R_L = 339\,\Omega$ and 30 ms for $R_L = 360\,\Omega$.

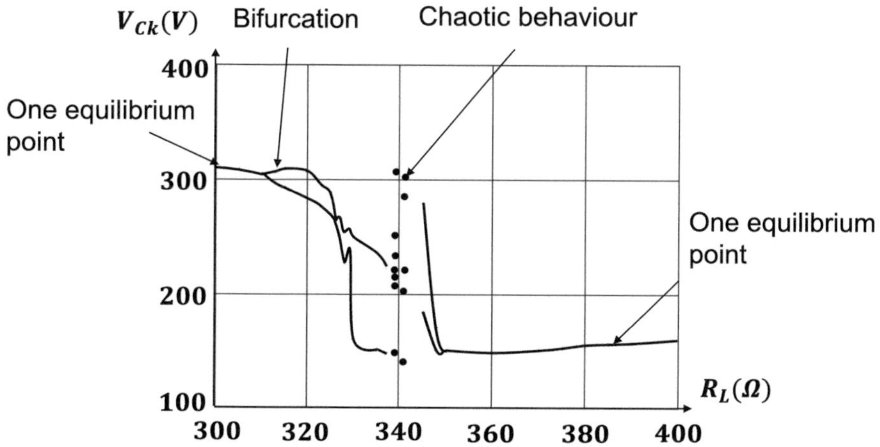

Fig. 15.2 Bifurcation diagram with R_L as the bifurcation parameter

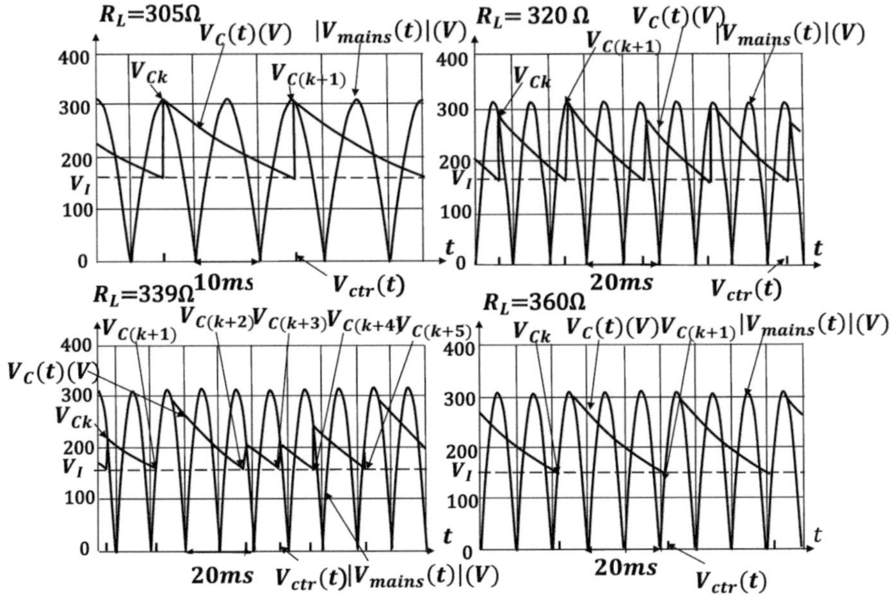

Fig. 15.3 $|V_{\text{mains}}(t)|$, $V_C(t)$, $V_{\text{ctr}}(t)$ for $R_L = 305\,\Omega$, $R_L = 320\,\Omega$, $R_L = 339\,\Omega$, $R_L = 360\,\Omega$

15.2 Chaotic Behaviour in a CCM Buck Converter

Consider the Buck converter represented in Fig. 15.4, along with its control circuit (Fossas & Olivar, 1996; Mehrizi-Sani et al., 2007; Angulo et al., 2007).

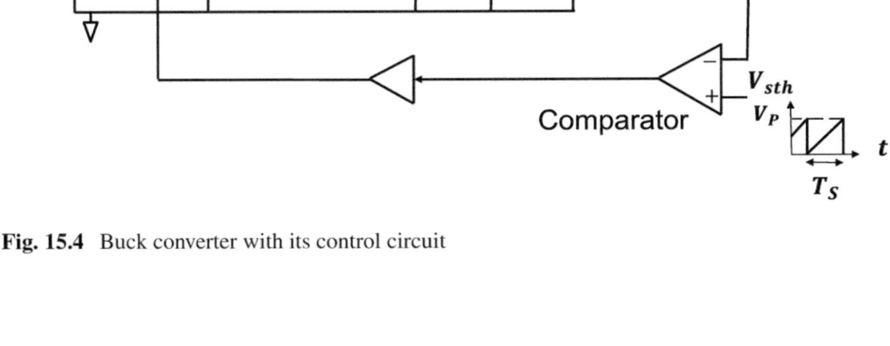

Fig. 15.4 Buck converter with its control circuit

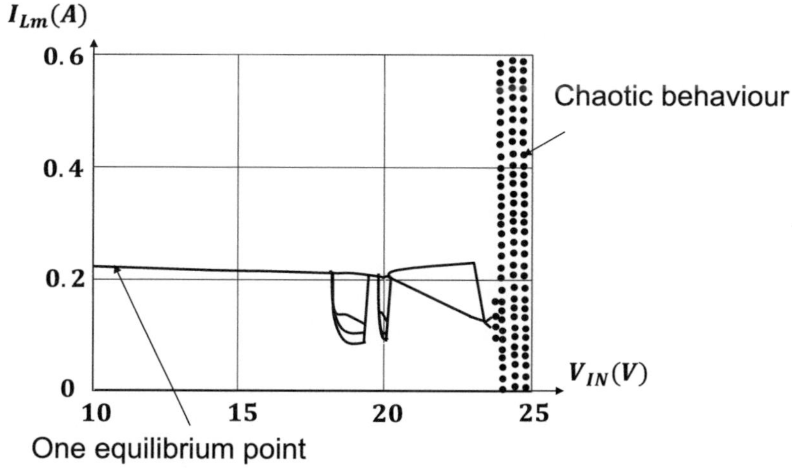

Fig. 15.5 Bifurcation diagram with V_{IN} as the bifurcation parameter

The numerical values of the various parameters are:

$$V_{\mathrm{IN}} \in [10\,\mathrm{V}, 25\,\mathrm{V}],\ V_{\mathrm{ref}} = 5\,\mathrm{V},\ V_P = 10\,\mathrm{V},\ C = 50\,\mu\mathrm{F},$$
$$L = 10\,\mathrm{mH},\ R_L = 20\,\Omega,\ T_S = 400\,\mu\mathrm{s},\ A = 10. \tag{15.3}$$

This converter operates in continuous conduction mode. Bifurcation and chaos phenomena appear in this converter as the input voltage V_{IN} increases. So, V_{IN} is naturally chosen as the bifurcation parameter. Figure 15.5 illustrates the corresponding bifurcation diagram where I_{Lm} is the minimum value of I_L.

As shown in Fig. 15.5, for V_{IN} approximately lower than 18.5 V, the converter is stable with one equilibrium point. For V_{IN} higher than 18.5 V, the converter progressively moves towards chaotic behaviour, displaying a few period domains.

Figure 15.6b is an example of period $6T_S$ behaviour and Fig. 15.6c an example of chaotic behaviour.

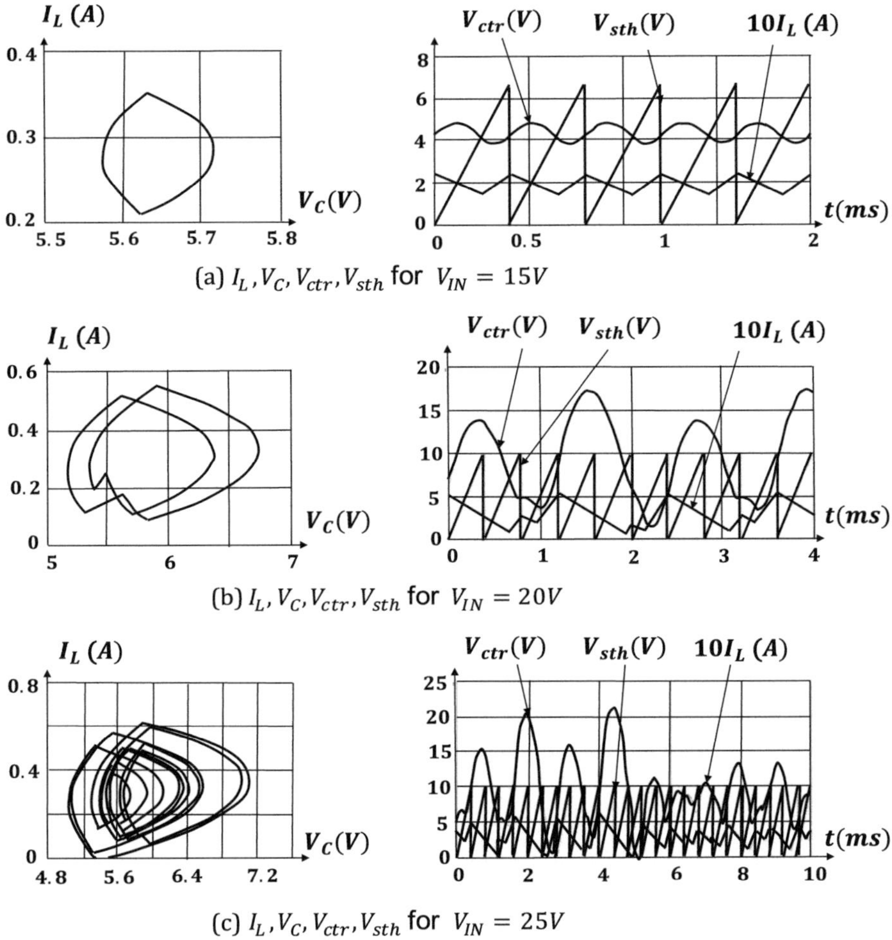

Fig. 15.6 Behaviour of a CCM buck converter

15.3 Chaotic Behaviour of a CCM Boost Converter Using Current Mode Control

Consider the boost converter shown in Fig. 15.7. It is operating in continuous conduction mode with current mode control. Parameters values are:

$$V_{\text{IN}} = 10\,\text{V},\, V_{\text{OUTtar}} = 30\,\text{V},\, V_{\text{ref}} = 10\,\text{V},\, C = 100\,\mu\text{F},\, C_1 = 3\,\mu\text{F},\, L = 100\,\mu\text{H},$$
$$R_L = 10\,\Omega,\, R_S = 0.01\,\Omega,\, R_1 = 20\,\text{k}\Omega,\, R_2 = 10\,\text{k}\Omega,\, R_3 = 1\,\text{M}\Omega,\, T_S = 20\,\mu\text{s}. \tag{15.4}$$

We reach the equilibrium:

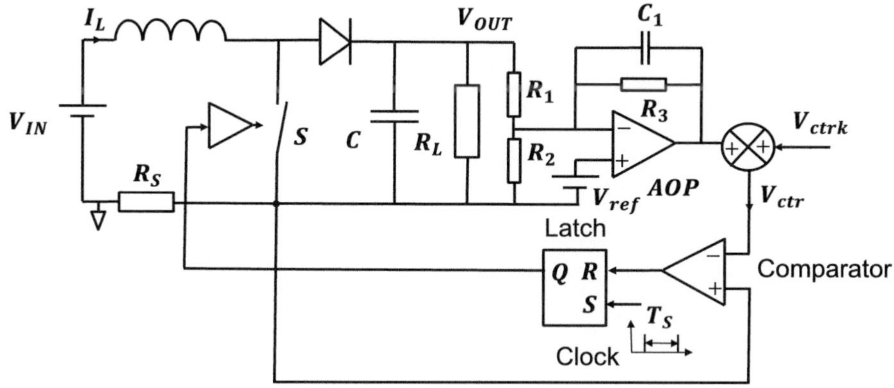

Fig. 15.7 Boost converter using current-mode control

$$d_e = \frac{V_{\text{OUTtar}} - V_{\text{IN}}}{V_{\text{OUTtar}}}, I_{\text{LMe}} = \frac{V^2_{\text{OUTtar}}}{R_L V_{\text{IN}}} + \frac{V_{\text{IN}}}{2L} \frac{V_{\text{OUTtar}} - V_{\text{IN}}}{V_{\text{OUTtar}}} T_S \tag{15.5}$$

$$V_{\text{Ctre}} = R_s I_{\text{LMe}}, I_{\text{Lme}} = \frac{V^2_{\text{OUTtar}}}{R_L V_{\text{IN}}} - \frac{V_{\text{IN}}}{2L} \frac{V_{\text{OUTtar}} - V_{\text{IN}}}{V_{\text{OUTtar}}} T_S \tag{15.6}$$

Using the same assumptions as those in Chap. 9, we derive the recurrence equation around the equilibrium point:

$$I_{\text{Lm}(k+1)} = \frac{V_{\text{Ctre}} + V_{\text{Ctrk}}}{R_s} + \frac{V_{\text{IN}} - V_{\text{OUTtar}}}{L} \left[1 - \left(\frac{V_{\text{Ctre}} + V_{\text{Ctrk}}}{R_s} - I_{\text{Lmk}} \right) \frac{L}{T_S V_{\text{IN}}} \right] T_S. \tag{15.7}$$

V_{Ctrk} is a discreet auxiliary control. From the last equation, taking $V_{\text{Ctrk}} = 0$, It is possible to draw the bifurcation diagram shown in Fig. 15.8, with V_{IN} as the bifurcation parameter. This bifurcation diagram shows that the Boost converter becomes unstable for V_{IN} lower than 15.2 V. In the next three paragraphs, we will use some of the control techniques for chaotic systems presented in Chap. 8 to eliminate chaotic behaviour.

These controls will be performed using the auxiliary control V_{Ctrk} and the numerical values provided by (15.4).

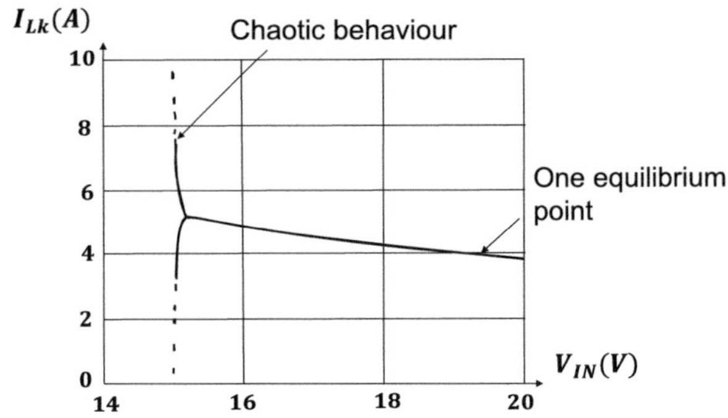

Fig. 15.8 Bifurcation diagram with V_{IN} as the bifurcation parameter

15.4 Control of the Chaotic Behaviour Using Predictive Mode Control

Here, we will follow the method described in Para. 8.3 (Ott et al., 1990; Boukabou & Mansouri, n.d.). Our objective is to stabilise the system around the unstable equilibrium point I_{Lme} defined by Eq. (15.6). We will use the following predictive model when $V_{\mathrm{Ctrk}} = 0$:

$$I_{\mathrm{LPm}(k+1)} = \frac{V_{\mathrm{Ctre}}}{R_s} + \frac{V_{\mathrm{IN}} - V_{\mathrm{OUTtar}}}{L}\left[1 - \left(\frac{V_{\mathrm{Ctre}}}{R_s} - I_{\mathrm{Lmk}}\right)\frac{L}{T_S V_{\mathrm{IN}}}\right]T_S. \tag{15.8}$$

The control is defined as follows:

$$\begin{aligned} &V_{\mathrm{Ctrk}} = k\left(I_{\mathrm{LPm}(k+1)} - I_{\mathrm{Lmk}}\right) && \text{if } |I_{\mathrm{Lmk}} - I_{\mathrm{Lme}}| < \varepsilon = 0.5\,\mathrm{A} \\ &V_{\mathrm{Ctrk}} = 0 && \text{otherwise.} \end{aligned} \tag{15.9}$$

Gain k will be determined by the linearisation technique to stabilise the system around the equilibrium point. When the system approaches I_{Lme}, the dynamics becomes:

$$I_{\mathrm{Lm}(k+1)} = \frac{V_{\mathrm{Ctre}} + V_{\mathrm{Ctrk}}}{R_s} + \frac{V_{\mathrm{IN}} - V_{\mathrm{OUTar}}}{L}\left[1 - \left(\frac{V_{\mathrm{Ctre}} + V_{\mathrm{Ctrk}}}{R_s} - I_{\mathrm{Lmk}}\right)\frac{L}{T_S V_{\mathrm{IN}}}\right]T_S \tag{15.10}$$

with

$$V_{\mathrm{Ctrk}} = k\left[\frac{V_{\mathrm{Ctre}}}{R_s} + \frac{V_{\mathrm{IN}} - V_{\mathrm{OUTtar}}}{L}\left[1 - \left(\frac{V_{\mathrm{Ctre}}}{R_s} - I_{\mathrm{Lmk}}\right)\frac{L}{T_S V_{\mathrm{IN}}}\right]T_S - I_{\mathrm{Lmk}}\right]. \tag{15.11}$$

The next step is to determine $\frac{\mathrm{d}\left(I_{\mathrm{Lm}(k+1)}\right)}{\mathrm{d}(I_{\mathrm{Lmk}})}$. The calculation yields:

$$\frac{\mathrm{d}\left(I_{\mathrm{Lm}(k+1)}\right)}{\mathrm{d}(I_{\mathrm{Lmk}})} = \frac{-kV_{\mathrm{OUTtar}}^2 + (V_{\mathrm{IN}} - V_{\mathrm{OUTtar}})R_s V_{\mathrm{IN}}}{R_s V_{\mathrm{IN}}^2}. \tag{15.12}$$

In order to stabilise the system, the following condition must be met:

$$-1 < \frac{\mathrm{d}I_{\mathrm{Lm}(k+1)}}{\mathrm{d}I_{\mathrm{Lmk}}} < 1. \tag{15.13}$$

in order to accelerate the stabilisation, k is calculated to obtain:

$$\frac{\mathrm{d}I_{\mathrm{Lm}(k+1)}}{\mathrm{d}I_{\mathrm{Lmk}}} = 0. \tag{15.14}$$

It follows that:

$$k = \frac{(V_{\mathrm{IN}} - V_{\mathrm{OUTtar}})R_s V_{\mathrm{IN}}}{V_{\mathrm{OUTtar}}^2}. \tag{15.15}$$

It is necessary to add the following two conditions for activating the control signal:

$$|V_{\mathrm{OUT}} - V_{\mathrm{OUTtar}}| < \varepsilon_1 = 0.5\,\mathrm{V} \tag{15.16}$$

and

$$|V_{\mathrm{ctr}} - V_{\mathrm{ctre}}| < \varepsilon_2 = 0.05\,\mathrm{V}. \tag{15.17}$$

Figure 15.9 shows the simulation results obtained with the predictive control V_{Ctrk}.

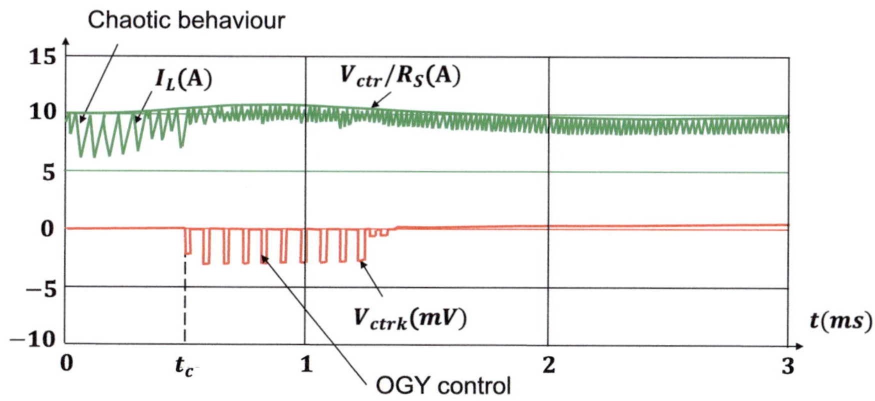

Fig. 15.9 Stability obtained with model predictive control

The predictive control is activated at time t_c when the steady state is reached. In steady-state mode, these waveforms show that the chaotic behaviour has been eliminated.

15.5 Control of Chaotic Behaviour with Linear Feedback Control

As in the previous example, we will try to stabilise the Boost converter around the unstable equilibrium point I_{Lme} by using linear feedback control (Chen & Dong, 1992; Auteur(s), 1993; Ogorzalek, 1993). As described in the Para. 8.5, the control signal is given by:

$$\begin{aligned} V_{\text{Ctrk}} &= k(I_{\text{Lme}} - I_{\text{Lmk}}) \text{ if } |I_{\text{Lmk}} - I_{\text{Lme}}| < \varepsilon = 2\,\text{A} \\ V_{\text{Ctrk}} &= 0 \qquad\qquad\qquad\quad \text{otherwise.} \end{aligned} \tag{15.18}$$

The gain k will also be determined by the linearisation technique in order to stabilise the system around the equilibrium point. When the system gets close to I_{Lme}, the dynamics become:

$$\begin{aligned} I_{\text{Lm}(k+1)} = {} & \frac{V_{\text{Ctre}} + k(I_{\text{Lme}} - I_{\text{Lmk}})}{R_s} \\ & + \frac{V_{\text{IN}} - V_{\text{OUTtar}}}{L}\left[1 - \left(\frac{V_{\text{Ctre}} + k(I_{\text{Lme}} - I_{\text{Lmk}})}{R_s} - I_{\text{Lmk}}\right)\frac{L}{T_S\, V_{\text{IN}}}\right]T_S. \end{aligned} \tag{15.19}$$

The calculation of $\frac{\mathrm{d}(I_{\text{Lm}(k+1)})}{\mathrm{d}(I_{\text{Lmk}})}$ gives:

$$\frac{\mathrm{d}(I_{\text{Lm}(k+1)})}{\mathrm{d}(I_{\text{Lmk}})} = \frac{-kV_{\text{OUTtar}} + (V_{\text{IN}} - V_{\text{OUTtar}})R_s}{R_s V_{\text{IN}}}. \tag{15.20}$$

In order to stabilise, the system we need to have:

$$-1 < \frac{\mathrm{d}I_{\text{Lm}(k+1)}}{\mathrm{d}I_{\text{Lmk}}} < 1. \tag{15.21}$$

k is calculated to obtain:

$$\frac{\mathrm{d}(I_{\text{Lm}(k+1)})}{\mathrm{d}(I_{\text{Lmk}})} = 0 \tag{15.22}$$

in order to accelerate the stabilisation. The calculation yields:

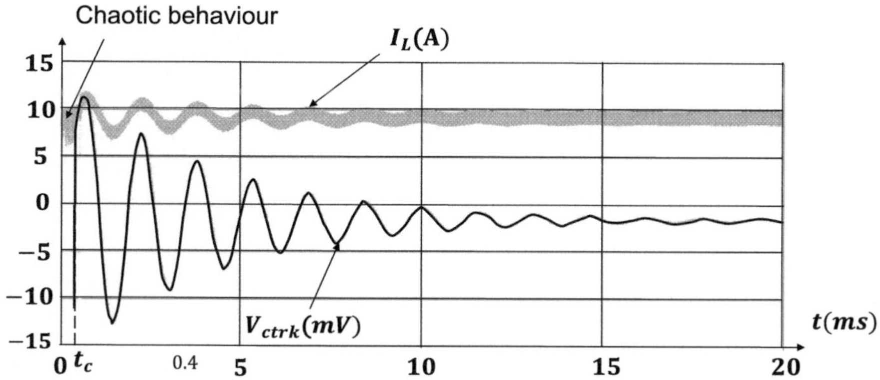

Fig. 15.10 Stabilty obtained with linear feedback control

$$k = \frac{(V_{\text{IN}} - V_{\text{OUTtar}})R_s}{V_{\text{OUTtar}}}. \tag{15.23}$$

Thus, the control law is:

$$V_{\text{Ctrk}} = \frac{(V_{\text{IN}} - V_{\text{OUTtar}})R_s}{V_{\text{OUtar}}}(I_{\text{Lme}} - I_{\text{Lmk}}). \tag{15.24}$$

The remarks about the control activation region made in the previous paragraph are also applicable to the linear feedback control. In Fig. 15.10, the simulation results show that the converter has been stabilised.

15.6 Control of Chaotic Behaviour Using OGY Control

In this paragraph, we apply the OGY control theory presented in Para. 8.7 (Ott et al., 1990) to suppress chaotic behaviour in the Boost converter. The dynamics of this system can be written as:

$$I_{\text{Lm}(k+1)} = F(I_{\text{Lmk}}, U) = \frac{U_k}{R_s} + \frac{V_{\text{IN}} - V_{\text{OUTtar}}}{L}\left[1 - \left(\frac{U_k}{R_s} - I_{\text{Lmk}}\right)\frac{L}{T_S V_{\text{IN}}}\right]T_S \tag{15.25}$$

with

$$U_k = V_{\text{Ctre}} + V_{\text{Ctrk}}. \tag{15.26}$$

The first step consists of linearising this equation around the equilibrium point $(I_{\mathrm{Lme}}, V_{\mathrm{Ctre}})$. We obtain:

$$i_{\mathrm{Lm}(k+1)} = \left(\frac{\partial F(I_{\mathrm{Lmk}}, U_k)}{\partial I_{\mathrm{Lmk}}}\right)_{(I_{\mathrm{Lme}}, V_{\mathrm{Ctre}})} i_{\mathrm{Lmk}} + \left(\frac{\partial F(I_{\mathrm{Lmk}}, U_k)}{\partial U_k}\right)_{(I_{\mathrm{Lme}}, V_{\mathrm{Ctre}})} V_{\mathrm{Ctrk}} \tag{15.27}$$

with

$$i_{\mathrm{Lmk}} = I_{\mathrm{Lmk}} - I_{\mathrm{Lme}} \text{ and } V_{\mathrm{Ctrk}} = U_k - V_{\mathrm{Ctre}}. \tag{15.28}$$

The calculations yield:

$$\left(\frac{\partial F\left(I_{\mathrm{Lm}(k+1)}, U_k\right)}{\partial I_{\mathrm{Lmk}}}\right)_{(I_{\mathrm{Lme}} V_{\mathrm{Ctre}})} = \frac{V_{\mathrm{IN}} - V_{\mathrm{OUTtar}}}{V_{\mathrm{IN}}} \tag{15.29}$$

and

$$\left(\frac{\partial F\left(I_{\mathrm{Lm}(n+1)}, U_k\right)}{\partial U_k}\right)_{(I_{\mathrm{Lme}} V_{\mathrm{Ctre}})} = \frac{V_{\mathrm{OUTtar}}}{R_S V_{\mathrm{IN}}}. \tag{15.30}$$

Thus, it follows that:

$$i_{\mathrm{Lm}(k+1)} = \frac{V_{\mathrm{IN}} - V_{\mathrm{OUTtar}}}{V_{\mathrm{IN}}} i_{\mathrm{Lmk}} + \frac{V_{\mathrm{OUTar}}}{R_S V_{\mathrm{IN}}} V_{\mathrm{Ctrk}}. \tag{15.31}$$

In this example, there is only one eigenvalue:

$$\lambda_u = \left(\frac{\partial F(I_{\mathrm{Lmk}}, U_k)}{\partial I_{\mathrm{Lmk}}}\right)_{(I_{\mathrm{Lme}} V_{\mathrm{Ctre}})} = \frac{V_{\mathrm{IN}} - V_{\mathrm{OUTtar}}}{V_{\mathrm{IN}}} = -2. \tag{15.32}$$

We obtain:

$$|\lambda_u| > 1. \tag{15.33}$$

Thus, this eigenvalue is unstable. Since the system under study is first-order, it is not useful to employ a transition matrix. The control is directly obtained by setting:

$$i_{\mathrm{Lm}(k+1)} = 0 \tag{15.34}$$

in the Eq. (13.31). It follows that:

$$V_{\text{Ctrk}} = \frac{(V_{\text{IN}} - V_{\text{OUref}})R_s}{V_{\text{OUTref}}}(I_{\text{Lme}} - I_{\text{Lmk}}). \tag{15.35}$$

We find the same control law as in the linear feedback control. The magnitudes of the control signals V_{Ctrk} from the predictive, linear feedback, and OGY methods are very small compared to the nominal control signal V_{Ctre}. It allows achieving a pure current-mode control. With the usual technique of using an artificial ramp, current mode control tends to behave like voltage-mode control as the ramp magnitude increases. In this way, the benefit of current-mode control is lost.

15.7 Elimination of Chaotic Behaviour Using Sliding-Mode Control

In this paragraph, we show that the chaotic behaviour of the previously discussed Boost converter can be eliminated by using sliding-mode control (Edwards & Spurgeon, 1998; Ferrara et al., 2008; Utkin, 1977).

15.7.1 Representation of the Boost Converter as a Hybrid System

The dynamics of this Boost converter can be expressed as follows:

$$\begin{pmatrix} \dot{I}_L \\ \dot{V}_C \end{pmatrix} = A\begin{pmatrix} I_L \\ V_C \end{pmatrix} + B(1-u) + D \tag{15.36}$$

with

$$A = \begin{pmatrix} 0 & 0 \\ 0 & -\frac{1}{R_L C} \end{pmatrix}, B = \begin{pmatrix} -\frac{V_C}{L} \\ \frac{I_L}{C} \end{pmatrix}, D = \begin{pmatrix} \frac{V_{\text{IN}}}{L} \\ 0 \end{pmatrix}. \tag{15.37}$$

u is the discontinuous control, which can take the value 0 when the switch S is off and 1 when the switch is on. This model is valid for $I_L > 0$.

15.7.2 Choice of a Sliding Surface $\sigma(I_L, V_C) = 0$

Referring to Para. 7.5, we begin by choosing a sliding surface of dimension $n-1 = 1$ such that the system response reaches the desired equilibrium point:

$$V_{\mathrm{Ce}} = V_{\mathrm{OUTtar}} = 30\,\mathrm{V}, I_{\mathrm{Le}} = \frac{V_{\mathrm{Ce}}^2}{V_{\mathrm{IN}} R_L} = 9\,\mathrm{A}. \tag{15.38}$$

Consider the sliding line defined by:

$$\sigma(I_L, V_C) = (I_L - I_{\mathrm{Le}}) + k_c(V_C - V_{\mathrm{Ce}}) = 0 \tag{15.39}$$

with k_c as the control parameter. Assume that a control law $u((I_L, V_C))$ has been defined to establish a sliding mode on the line $\sigma(I_L, V_C) = 0$. For each initial condition belonging to R_A, there exists a finite time t_i at which the trajectory of the closed-loop system intercepts the line $\sigma(I_L, V_C) = 0$. Since the sliding mode exists for all instants $t > t_i$, we have:

$$\sigma(I_L, V_C) = 0, \text{ and } \frac{\mathrm{d}\sigma(I_L, V_C)}{\mathrm{d}t} = 0 \tag{15.40}$$

These conditions yield the equivalent control u_{eq}:

$$u_{\mathrm{eq}} = 1 - \frac{V_{\mathrm{IN}} R_L C - k_c V_C L}{R_L(V_C C - k_c L I_L)}. \tag{15.41}$$

By integrating this expression over Eq. (15.36) we obtain the system dynamics on the sliding line:

$$\begin{cases} \frac{\mathrm{d}I_L}{\mathrm{d}t} = -\frac{V_C}{L}\frac{V_{\mathrm{IN}} R_L C - k_c V_C L}{R_L(V_C C - k_c L I_L)} + \frac{V_{\mathrm{IN}}}{L} \\ \frac{\mathrm{d}V_C}{\mathrm{d}t} = -\frac{V_C}{R_L C} + \frac{I_L}{C}\frac{V_{\mathrm{IN}} R_L C - k_c V_C L}{R_L(V_C C - k_c L I_L)} \end{cases}. \tag{15.42}$$

It can be readily shown that, for each initial condition on the sliding curve, the system will slide towards the equilibrium point $(I_{\mathrm{Le}}, V_{\mathrm{Ce}})$, with the control at equilibrium equal to:

$$u_{\mathrm{eqe}} = \frac{V_{\mathrm{Ce}} - V_{\mathrm{IN}}}{V_{\mathrm{Ce}}}. \tag{15.43}$$

15.7.3 Determination of the Control Law: u

Now we need to determine u that brings the trajectory to the line $\sigma = 0$ and keeps it there. The control must yield the following results:

When $\sigma > 0$, that is:

$$((I_L - I_{\mathrm{Ce}}) + k_c(V_C - V_{\mathrm{Ce}}) > 0). \tag{15.44}$$

We want to have:

$$\dot{\sigma} = \dot{I}_L + k_c\dot{V}_C = -\frac{V_C}{L}(1-u) + \frac{V_{\mathrm{IN}}}{L} + k_c\left(-\frac{V_C}{R_L C} + \frac{I_L}{C}(1-u)\right) < 0 \tag{15.45}$$

so,

$$u = 0 \tag{15.46}$$

is a possible solution, provided that the following condition is satisfied:

$$V_C > \frac{R_L(CV_{\mathrm{IN}} + k_c L I_L)}{R_L C + k_c I_L} \tag{15.47}$$

When $\boldsymbol{\sigma} < 0$, that is:

$$(I_L - I_{\mathrm{Ce}}) + k_c(V_C - V_{\mathrm{Ce}}) < 0 \tag{15.48}$$

we want to have

$$\dot{\sigma} = \dot{I}_L + k_c\dot{V}_C = -\frac{V_C}{L}(1-u) + \frac{V_{\mathrm{IN}}}{L} + k_c\left(-\frac{V_C}{R_L \cdot C} + \frac{I_L}{C}(1-u)\right) > 0 \tag{15.49}$$

so,

$$u = 1 \tag{15.50}$$

is a possible solution, provided that the following condition is satisfied:

$$V_C < \frac{R_L \mathrm{CV}_{\mathrm{IN}}}{\mathrm{Lk}_c}. \tag{15.51}$$

15.7.4 Region of Attraction R_A

A region of attraction is defined as the area where:

$$\sigma\dot{\sigma} < 0. \tag{15.52}$$

According to the analysis of the previous paragraph, this condition can be expressed by:

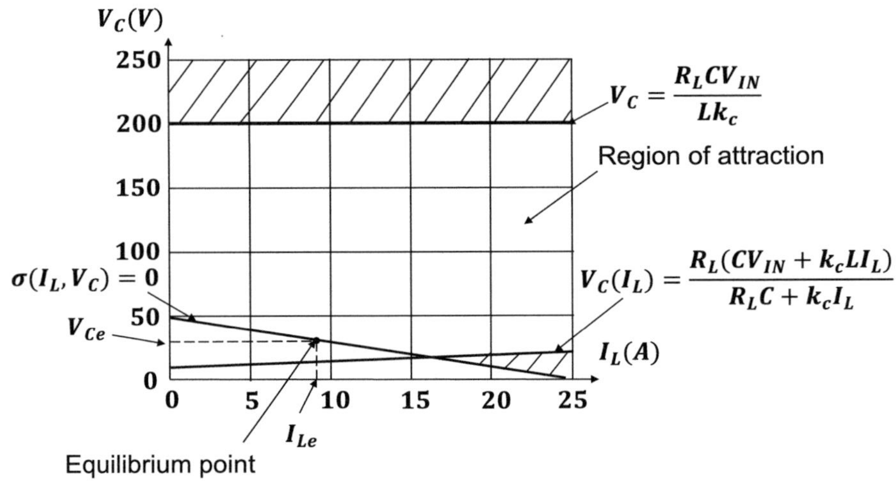

Fig. 15.11 Region of attraction with $k_c = 0.5$ S

$$(I_L - I_{\text{Ce}}) + k_c(V_C - V_{\text{Ce}}) > 0 \text{ and } V_C > \frac{R_L(CV_{\text{IN}} + k_c L I_L)}{R_L C + k_c L} \tag{15.53}$$

or

$$(I_L - I_{\text{Ce}}) + k_c(V_C - V_{\text{Ce}}) < 0 \text{ and } V_C < \frac{R_L \text{CV}_{\text{IN}}}{\text{Lk}_c}. \tag{15.54}$$

The region that does not belong to R_A is shown with hatching in Fig. 15.11.

15.7.5 Reduction of the Gathering Effect

In practical designs, to reduce the switching frequency, we consider two sliding lines: σ^+ and σ^- defined by

$$\sigma^+ = (I_L - I_{\text{Ce}}) + k_c(V_C - V_{\text{Ce}} + 0.6) = 0 \tag{15.55}$$

$$\sigma^- = (I_L - I_{\text{Ce}}) + k_c(V_C - V_{\text{Ce}} - 0.6) = 0. \tag{15.56}$$

The switch S turns off when $\sigma^+ < 0$, and will turn on when $\sigma^- > 0$. Figure 15.12 illustrates a method of implementing this control using a latch circuit.

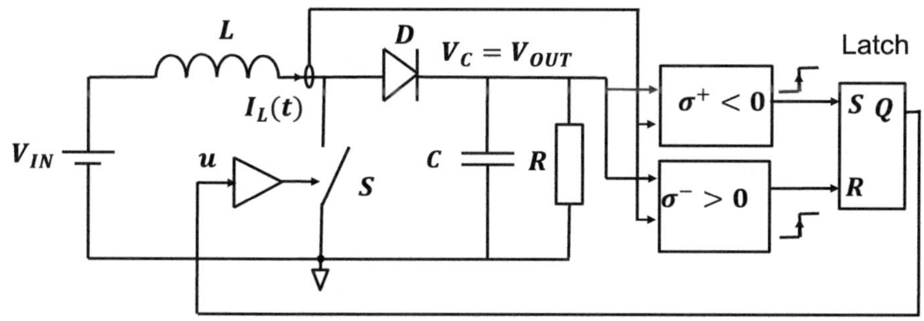

Fig. 15.12 Control circuit to reduce the gathering effect

15.7.6 Temporal Simulation

Figure 15.13 shows a time-domain simulation of the studied system during its transient phase. The simulation starts with the initial conditions:

$$I_{C0} = 30\,\text{A},\ V_{C0} = 100\,\text{V}. \tag{15.57}$$

Thus, $\sigma^+ > 0, \dot{\sigma} < 0$ and $u = 0$ (S off). At time $t_1 = 0.04$ ms, the diode turns off and remains off until t_I. At time $t_I = 0.8$ ms, the trajectory of the closed-loop system intercepts the line $\sigma^+(I_L, V_C) = 0$, the system begins to slide between σ^+ and σ^- towards the desired equilibrium point: $I_{\text{Le}} = 9$ A, $V_{\text{Ce}} = 30$ V.

Figure 15.14 shows the steady state-mode of the sliding control. The switching frequency is equal to 93 kHz. As predicted, no chaotic behaviour is observed.

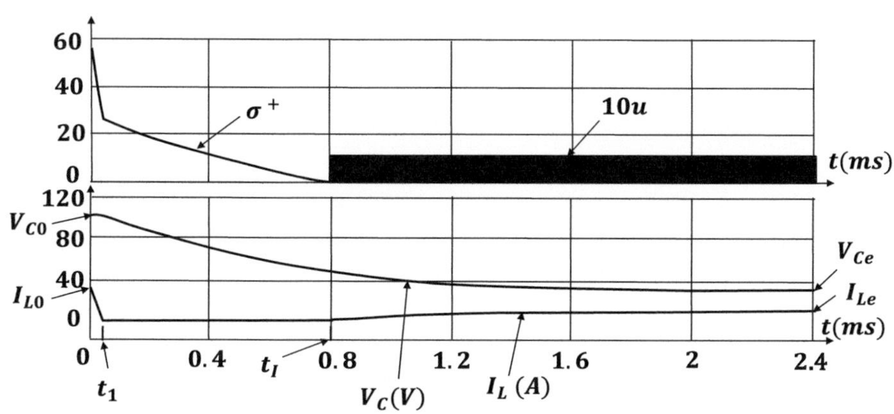

Fig. 15.13 Time-domain simulation of the sliding-mode controlled system

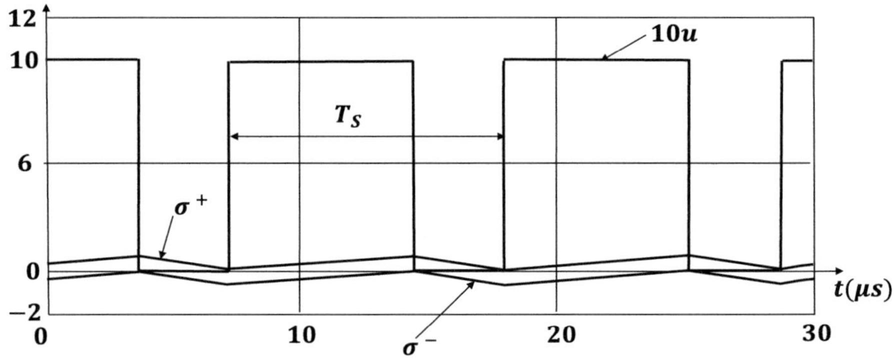

Fig. 15.14 Time-domain simulation of the sliding-mode controlled system in steady state mode

References

Alur, R., Courcoubetis, C., Henzinger, T. A., & Ho, P.-H. (1993). Hybrid automata: An algorithmic approach to the specification and verification of hybrid systems. *Theoretical Computer Science, 138*(1), 3–34. https://doi.org/10.1016/0304-3975(94)90010-2

Angulo, F., Fossas, E., & Olivar, G. (2007). Transition phenomena in the boundary of the continuous conduction mode in DC–DC converters. *Nonlinear Dynamics, 47*(1–3), 125–140.

Anonyme. (s.d.). *Contribution à la modélisation et à la commande d'un convertisseur CC-CC* (Thèse de doctorat). Université d'Oran. http://www.univusto.dz/theses_en_ligne/doc_num.php?explnum_id=386

Arzelier, D. (n.d). *Commande optimale des systèmes dynamiques* (Cours). LASS-CNRS.

Asahara, H., & Kousaka, T. (2021). Stability analysis based on monodromy matrix for switched dynamical systems. *Nonlinear Theory and Its Applications, IEICE, 12*(3), 237–252.

Auteur(s). (1989). Remerging bifurcation sequences in an autonomous electronic circuit. In *Proceedings of the 1989 European conference on circuit theory and design* (pp. 142–146). Brighton, UK.

Auteur(s). (1989). Order and chaos in RC-ladder network with nonlinear feedback. *IEEE Transactions on Circuits and Systems, 36*, 1221–1230.

Auteur(s). (1993). Controlling Chua's circuit. *Circuits, Systems, and Computers, 3*, 139–148.

Auteur(s). (1993). Ordering chaos in Chua's circuit—A feedback control approach. In *Proceedings of the IEEE international symposium on circuits and systems* (Vol. 4, pp. 2604–2607). Chicago, IL, USA.

Aybar, O. O., Kusbeyzi Aybar, I., & Hacinliyan, A. S. (2013). Stability and bifurcation in the Hénon map and its generalizations. *Chaotic Modeling and Simulation (CMSIM), 4*, 529–538.

Barrade, P. (1997). *Comportement dynamique des ensembles convertisseur—filtre* (Thèse de doctorat non publiée). Institut National Polytechnique de Toulouse.

Bellman, R. E. (1957). *Dynamic programming*. Princeton University Press.

Boukabou, A., & Mansouri, N. (s.d.). *Contrôle prédictif des systèmes discrets à comportement chaotique*. Département d'électronique, Faculté des Sciences de l'Ingénieur, Université Mentouri Constantine.

Camacho, E. F., & Bordons, C. (1999). *Model Predictive Control*. Springer.

Chen, C. (1996). *Backstepping control design and its applications to vehicle lateral control in automated highway systems* (Doctoral dissertation). University of California at Berkeley.

Chen, G., & Dong, X. (1992). On feedback control of chaotic dynamic systems. *International Journal of Bifurcation and Chaos, 2*(2), 407–411. https://doi.org/10.1142/S0218127492000281

Chen, G., & Dong, X. (1994). Time-delayed feedback control of chaos in the Rössler system. *Physical Review E, 50*(1), 120–125.

B. Rivet, *Theories of Nonlinear Systems for Power Converters*,
https://doi.org/10.1007/978-3-032-25062-9

D'Azzo, J. J., & Houpis, C. H. (2003). *Linear control system analysis and design with MATLAB* (5th ed.). Marcel Dekker, Inc.

De Schutter, B (2005). *Models for hybrid systems.* 1st HYCON Summer School on Hybrid Systems Siena, Italy.

Edwards, C., & Spurgeon, S. (1998). *Sliding mode control: Theory and applications.* Taylor & Francis.

El-Farra, & N. H. (2006). Robust inverse optimal control laws for nonlinear systems. *International Journal of Robust and Nonlinear Control, 16*(6), 333–346.

El-Farra, N. H., & Christofides, P. D. (2003). Robust inverse optimal control laws for nonlinear systems. *International Journal of Robust and Nonlinear Control, 13*(3–4), 251–275.

Erickson, R. W., & Maksimovic, D. (2001). *Fundamentals of power electronics* (2nd ed.). Springer.

Erickson, R. W., & Maksimović, D. (2020). *Fundamentals of power electronics* (3rd ed.). Springer.

Erickson, R. W. (1996). *Fundamentals of power electronics.* Kluwer Academic Publishers.

Evans, L. C. (2001). *An introduction to mathematical optimal control theory.* University of California, Berkeley.

Ferrara, A., Giacomini, L., & Vecchio, C. (2008). Modern sliding mode control theory: New perspectives and applications. In G. Bartolini, A. Ferrara, L. Giacomini, & E. Usai (Eds.), *Modern sliding mode control theory: New perspectives and applications* (Lecture Notes in Control and Information Sciences, Vol. 375, pp. 223–245). Springer.

Ferrieux, J.-P., & Forest, F. (1976). *Alimentations à découpage: Modélisation discrète et applications aux convertisseurs DC–DC (Thèse d'université).* Université de Toulouse.

Fossas, E., & Olivar, G. (1996). Study of chaos in the buck converter. *IEEE Transactions on Circuits and Systems I: Fundamental Theory and Applications, 43*(1), 13–25.

Freeman, R. A., & Kokotović, P. V. (1996). *Robust nonlinear control design: State-space and Lyapunov techniques.* Birkhäuser.

Furman, B. (2011). *A two-tank liquid level control system case study.* SJSU

Goebel, R., Sanfelice, R. G., & Teel, A. R. (2012). *Hybrid dynamical systems: Modeling, stability, and robustness.* Princeton University Press.

Goudarzian, A., Khosravi, A., & Raeisi Heidar, A. (2022). Modeling, design and control of a modified flyback converter with ability of right-half-plane zero alleviation in continuous conduction mode. *Renewable Energy, 191*, 141–152. https://doi.org/10.1016/j.renene.2020.05.088

Hénon, M. (1976). A two-dimensional mapping with a strange attractor. *Communications in Mathematical Physics, 50*(1), 69–77.

Hiskens, I. A. (2001, January). Stability of limit cycles in hybrid systems. In *Proceedings of the 34th annual Hawaii international conference on system sciences* (pp. 1–8). IEEE.

Htibler, A., & Ltischer, E. (1989). Resonant stimulation and control of nonlinear oscillators. *Naturwissenschaften, 76*, 76.

Htischer, E., & Htibler, A. (1989). Resonant stimulation of complex systems. *Helvetica Physica Acta, 62*, 543.

Infineon Technologies (2016). *Design guide PFC CCM boost converter* (Application note DN 2013-01, Rev. 1.1). Infineon Technologies AG. https://www.infineon.com/dgdl/InfineonApplicationNote_PFCCCMBoostConverterDesignGuide-ApplicationNotes-v02_00-EN.pdf

Ioannou, P. A., & Kokotovic, P. V. (1984). Adaptive systems with reduced models. *IEEE Transactions on Automatic Control, 29*(3), 234–242.

Jackson, E. A. (1991). On the control of complex dynamic systems. *Physica D: Nonlinear Phenomena, 50*(3), 341–366. https://doi.org/10.1016/0167-2789(91)90004-S

Johns, A. T., & Sudhoff, K. R. (1985). Discrete-time modeling of switching converters for control analysis. *IEEE Transactions on Power Electronics, 1*(3), 157–163. https://doi.org/10.1109/TPEL.1985.7335362

Jury, E. I. (1961). A simplified stability criterion for linear discrete systems. *Proceedings of the IRE, 50*(6), 1493–1500.

Kalman, R. E. (1964). When is a linear control system optimal? *Journal of Basic Engineering, 86*(1), 51–60.

Kapitaniak, T., Kocarev, L., & Chua, L. O. (1993). Controlling chaos without feedback and control signals. *International Journal of Bifurcation and Chaos, 3*(2), 459–468. https://doi.org/10.1142/S0218127493000362

Khalil, H. K. (2002). *Nonlinear systems* (3rd ed.). Prentice Hall.

Kokotović, P. V., Khalil, & H. K., O'Reilly, J. (1996). *Singular perturbation methods in control:* Analysis and design. SIAM.

Kokotović, P., & Arcak, M. (2001). Nonlinear and adaptive control: An abbreviated status report. In *The 9th mediterranean conference on control and automation, Dubrovnik, Croatia.*

Kolar, J. W., Schweizer, M., & Trattner, A. (2012). Design and implementation of a highly efficient three-level T-type converter for low-voltage applications. *IEEE Transactions on Power Electronics, 28*(2), 899–907. https://doi.org/10.1109/TPEL.2012.2188580

Krein, P. T. (1998). *Elements of power electronics.* Oxford University Press.

Krstić, M., Kanellakopoulos, I., & Kokotović, P. V. (1995). *Nonlinear and adaptive control design.* Wiley.

LAAS-CNRS (2004). Commande optimale des systèmes dynamiques. In *Cours 5: Applications du principe de Pontryagin (Support de cours).* LAAS-CNRS.

Lavigne, É. (2019). *Analyse et commande avancée des convertisseurs résonants LLC* (Thèse de doctorat). École de technologie supérieure, Montréal.

Llibre, J., & Zhang, X. (2019). Non–existence, existence and uniqueness of limit cycles for quadratic systems. *Annals of the University of Bucharest (Mathematics Series), 149*(1), 3–20.

Lorenz, E. N. (1963). Deterministic nonperiodic flow. *Journal of the Atmospheric Sciences, 20*(2), 130–140.

Lyapunov, A. M. (1992). *The General problem of the stability of motion.* Taylor & Francis.

Mammano, R. (1994). *Design note: Voltage mode vs. current mode.* Texas Instruments. https://www.ti.com/lit/pdf/slua119

Manneville, P. (2005). Dynamique non lineaire et chaos. In *Seminaire E2PHY.*

Martinon, P. (2005). *Résolution numérique de problèmes de contrôle optimal par une méthode homotopique simpliciale* (Doctoral dissertation). École Polytechnique, HAL.

Mehrizi-Sani, A., Kinsner, W., & Filizadeh, S. (2007). On the chaotic behaviour of buck converters. In *Proceedings of the IEEE Canadian conference on electrical and computer engineering* (pp. 1564–1567). IEEE.

Middlebrook, R. D., & C Ćuk, S. (1976). A general unified approach to modeling switching power converter stages. In *IEEE power electronics specialists conference record* (pp. 18–34).

Middlebrook, R. D. (1976). *Modeling and analysis of switching regulators* (Report No. 76-4). California Institute of Technology.

Naslin, P. (1962). *Les régimes variables dans les systèmes linéaires et non linéaires.* Dunod.

Nyquist, H. (1932). Regeneration theory. *Bell System Technical Journal, 11*(1), 126–147.

Obaro, A. Z., Munda, J. L., & Siti, M. W. (2018). Optimal energy management of an autonomous hybrid energy system. *International Journal of Renewable Energy Development.*

Ogata, K. (2010). *Modern control engineering* (5th ed.). Pearson.

Ogorzalek, M. J. (1993). Taming chaos: Part II—Control. *IEEE Transactions on Circuits and Systems I: Fundamental Theory and Applications, 40*(11), 700–706. https://doi.org/10.1109/81.246146

Ott, E., Grebogi, C., & Yorke, J. A. (1990). Controlling chaos. *Physical Review Letters, 64*(11), 1196–1199. https://doi.org/10.1103/PhysRevLett.64.1196

Panardo, I. (2022). *Stability of periodic systems and Floquet Theory* (Master's thesis). Università degli Studi di Padova.

Patre, P. (2009). *Lyapunov-based robust and adaptive control of nonlinear systems using a novel feedback structure* (Doctoral dissertation). University of Florida.

Pekař, J. (2010, May). *Short introduction to model based predictive control.* Brno University of Technology.

Pelegrin, M., Gille, J. C., & Decaulne, P. (1960). *Système asservi non linéaire.* Dunod.

Poincaré, H. (1892–1899). *Les méthodes nouvelles de la mécanique céleste* (Vol. 1–3). Gauthier-Villars, Paris.
Pontryagin, L. S., Boltyanskii, V. G., Gamkrelidze, R. V., & Mishchenko, E. F. (1962). *The mathematical theory of optimal processes.* Interscience Publishers.
Praly, L., & Bresch-Pietri, D. (2016). *Fonctions de Lyapunov, stabilité, stabilisation et atténuation de perturbations: Partie 1, Stabilité.* EDP Sciences.
Primbs, J. A., & Freeman, R. A. (1996, December). Control Lyapunov functions: New ideas from an old source. In *Proceedings of the 35th IEEE conference on decision and control* (Vol. 4, pp. 3926–3931). IEEE.
Primbs, J. A. (1999). *Non-linear optimal control: A receding horizon approach* (Doctoral dissertation). California Institute of Technology.
Primbs, J. A., Nevistić, V., & Doyle, J. C. (1999). Nonlinear optimal control: A control Lyapunov function and receding horizon perspective. *Asian Journal of Control, 1*(1), 14–24.
Priyadarshi, N., Sanjeevikumar, P., Farooque, A., Bharatiraja, S., & Singh, R. (2024). *Advanced power electronics converters for future renewable energy systems.* Routledge.
Pyragas, K. (1992). Continuous control of chaos by self-controlling feedback. *Physics Letters A, 170*(6), 421–428. https://doi.org/10.1016/0375-9601(92)90745-8
Qin, S. J., & Badgwell, T. A. (1998). A survey of industrial model predictive control technology. *Control Engineering Practice, 11*(7), 733–764.
Ramana, B. V. (2006). *Calculus of Variations.*
Rickenbach, R., Lahr, A., & Zeilinger, M. N. (2025). *Inverse optimal control with constraint relaxation.* arXiv preprint arXiv:2507.11392
Rossler, O. E. (1976). An equation for continuous chaos. *Physics Letters A, 57*(5), 397–398.
Roussel, M. R. (2005). *Maps and differential equations.*
Routh, E. J. (1877). *A treatise on the stability of a given state of motion.* Macmillan.
Sanfelice, R. G. (2011). *Hybrid systems: Stability and control.* UCSC preprint.
Sethi, S. P., & Thompson, G. L. (2000). *Optimal control theory: Applications to management science and economics* (2nd ed.). Springer.
Shin, S.-K., Kong, B. S., Lee, C.-G., Jun, Y.-H., & Kim, J.-W. (2005). A high current driving charge pump with current regulation method. In *Proceedings of the IEEE custom integrated circuits conference* (pp. 207–210). https://doi.org/10.1109/CICC.2005.1506848
Skogestad, S., & Postlethwaite, I. (2005). *Multivariable feedback control: Analysis and design* (2nd ed.). Wiley.
Souleres, T (2022). *Le système de Lorenz.* Bureau d'étude, Instabilités hydrodynamiques.
Strogatz, S. H. (2015). *Nonlinear dynamics and chaos: With applications to physics, biology, chemistry, and engineering* (2nd ed.). Westview Press.
Texas Instruments. (2010). *Feedback loop design of an LLC resonant power converter (Application note SLUA582A).* Texas Instruments Incorporated. https://www.ti.com/lit/pdf/slua582
Utkin, V. I. (1977). Variable structure systems with sliding modes. *IEEE Transaction on Automatic Control, 22*(2), 212–222. https://doi.org/10.1109/TAC.1977.1101446
Van der Pol, B. (1926). On relaxation-oscillations. *The London, Edinburg and Dublin Philosophical Magazine and Journal of Science, 2*(11), 978–992.
van der Schaft, A., & Schumacher, J. M. (2000). *An introduction to hybrid dynamical systems.* Springer.
Vecchio, C. (2008). *Sliding mode control: theoretical developments and applications to uncertain mechanical systems.*
Vossen, G. (2010). Switching time optimization for bang-bang and singular controls. *Journal of Optimization Theory and Applications, 144*(2), 409–429.
Williams, R. J. (n.d.). *Dynamic programming: The optimality equation.* Statistical Laboratory, University of Cambridge.
Witvoet, G. (2005). *Control of chaotic dynamical systems using OGY* (DCT rapporten; Vol. 2005.036). Technische Universiteit Eindhoven.

Wolf, A., Swift, J. B., Swinney, H. L., & Vastano, J. A. (1985). Determining Lyapunov exponents from a time series. *Physica D: Nonlinear Phenomena, 16*(3), 285–317. https://doi.org/10.1016/0167-2789(85)90011-9

Yan, X., Shu, Z., Sharkh, S. M., Wu, Z.-G., & Chen, M. Z. Q. (n.d.). Sampled-data control with adjustable switching frequency for DC–DC Converters. *IEEE Transactions on Industrial Electronics*. Retrieved from https://eprints.soton.ac.uk/425377/1/ALL_17_TIE_2109.pdf

MIX
Papier aus verantwortungsvollen Quellen
Paper from responsible sources
FSC® C105338

If you have any concerns about our products,
you can contact us on
ProductSafety@springernature.com

In case Publisher is established outside the EU,
the EU authorized representative is:
Springer Nature Customer Service Center GmbH
Europaplatz 3, 69115 Heidelberg, Germany

Printed by Libri Plureos GmbH
in Hamburg, Germany